T0222658

Magnetoresistive Oxides
and Related Materials

MATERIALS RESEARCH SOCIETY
SYMPOSIUM PROCEEDINGS VOLUME 602

Magnetoresistive Oxides and Related Materials

Symposium held November 29–December 2, 1999, Boston, Massachusetts, U.S.A.

EDITORS:

M.S. Rzchowski
University of Wisconsin-Madison
Madison, Wisconsin, U.S.A.

M. Kawasaki
Tokyo Institute of Technology
Yokohama, Japan

A.J. Millis
Rutgers University
Piscataway, New Jersey, U.S.A.

S. von Molnár
Florida State University
Tallahassee, Florida, U.S.A.

M. Rajeswari
University of Maryland
College Park, Maryland, U.S.A.

Materials Research Society
Warrendale, Pennsylvania

CAMBRIDGE UNIVERSITY PRESS
Cambridge, New York, Melbourne, Madrid, Cape Town,
Singapore, São Paulo, Delhi, Mexico City

Cambridge University Press
32 Avenue of the Americas, New York NY 10013-2473, USA

Published in the United States of America by Cambridge University Press, New York

www.cambridge.org
Information on this title: www.cambridge.org/9781107413245

Materials Research Society
506 Keystone Drive, Warrendale, PA 15086
http://www.mrs.org

First published 2001
First paperback edition 2013

Single article reprints from this publication are available through
University Microfilms Inc., 300 North Zeeb Road, Ann Arbor, MI 48106

CODEN: MRSPDH

ISBN 978-1-107-41324-5 Paperback

CONTENTS

SPIN POLARIZATION AND
TUNNELING IN MAGNETIC OXIDES

NOVEL MAGNETIC OXIDES

*Invited Paper

v

TRANSPORT AND OPTICAL
PROPERTIES

CHARGE AND ORBITAL
ORDERING EFFECTS

TWO-PHASE COEXISTENCE IN
THE MANGANITES

*Invited Paper

STRAIN EFFECTS IN
MANGANITE THIN FILMS

*Invited Paper

POSTER SESSION II

MAGNETIC OXIDE THIN FILMS
AND HETEROSTRUCTURES

MAGNETIC OXIDE
HETEROSTRUCTURES AND DEVICES

*Invited Paper

PREFACE

Advances in epitaxial and crystal growth techniques, improved understanding of strongly-correlated systems, and prospects for spin-based electronic device structures have all created new opportunities for magnetic oxide materials. The proceedings from Symposium JJ, "Magnetoresistive Oxides and Related Materials," held November 29–December 2 at the 1999 MRS Fall Meeting in Boston, Massachusetts, range from basic research investigating microscopic electronic and magnetotransport mechanisms to applications such as bolometric-based detectors and magnetoresistive sensors.

Materials exhibiting colossal magnetoresistance (CMR) comprise a significant portion of the systems explored in Symposium JJ. As in several other magnetic oxide systems, substitutional doping on the A-site of the perovskite lattice leads to a rich magnetic phase diagram. In addition, strong coupling to the lattice and to electronic charge results in an even more varied array of magnetotransport phenomena, including phase-transition or crossover behavior reminiscent of metal-insulator type phase transitions. The lattice and structure match between CMR materials and common oxide substrates has opened avenues to study strain-dependence of the phase diagram. By varying dimensionality, the Ruddlesden-Popper series of layered analogs to these materials has provided additional insight into basic mechanisms, as well as introducing new phenomena.

Spin-based electronic devices depend critically on electronic structure, spin polarization at the Fermi energy, and interfacial structure. Near half-metallic behavior of CMR materials, lattice match with common oxide substrates and interlayers, and high spin polarization makes these good candidates for such structures. Other exciting possibilities, such as CrO_2, are also being explored in epitaxial thin films.

Magnetic oxides have a long history and an exciting future. The Materials Research Society meetings and proceedings provide an essential forum for mapping the route.

M.S. Rzchowski
M. Kawasaki
A.J. Millis
S. von Molnár
M. Rajeswari

March 2001

MATERIALS RESEARCH SOCIETY SYMPOSIUM PROCEEDINGS

MATERIALS RESEARCH SOCIETY SYMPOSIUM PROCEEDINGS

Prior Materials Research Society Symposium Proceedings available by contacting Materials Research Society

Spin Polarization and
Tunneling in Magnetic Oxides

Magnetotransport and interface magnetism in manganite heterostructures : implications for spin polarized tunneling

MOON-HO JO*, N.D. MATHUR, J.E. EVETTS AND M.G. BLAMIRE
Department of Materials Science, University of Cambridge, Pembroke Street, Cambridge CB2 3QZ, United Kingdom

ABSTRACT

The electronic and magnetic structure at surfaces and interfaces in mixed valence manganites is an active area of research. To study localized properties of interface magnetism in heterostructures we have conducted both current in-plane and current perpendicular-plane measurements in multilayers and trilayer tunnel junctions, respectively. Initial study of the transport in multilayers, consisting of alternating $La_{0.7}Ca_{0.3}MnO_3$ and $SrTiO_3$ with different layer thickness provided information on vertical inhomogeniety in the manganite layers and disordered interfaces. Issues of lattice mismatched strain, structural inhomogeniety were addressed. Trilayer tunnel junctions have been fabricated with the aim of correlating tunneling magnetoresistance to the interface magnetism deduced from multilayer measurements. We describe devices incorporating a novel barrier material which have yielded single domain switchings and very high TMR values above 77 K.

INTRODUCTION

The mixed valence manganites have been recognised for several years as being good candidates for tunnelling magnetoresistance (TMR) devices because of their high spin polarisation and their good low temperature metallic properties[1,2,3] However, the magnetic properties of these materials are highly sensitive to local crystal structure and the extrinsic strain fields induced by the lattice mismatch with the substrates or tunnel barrier can be sufficient to severely degrade the ferromagnetic order in the surface layers which are critical for tunneling.[4] Previous studies of ultra-thin manganite films have shown strong suppression of magnetic properties as the thickness is reduced.[5,6] Heteroepitaxial multilayer structures can be utilized not only for artificially tailoring electronic properties but are useful tools to investigate interface magnetism. In this article we report the study of the electrical transport and magnetic structure of manganite heterostructures. Results from series of $La_{0.7}Ca_{0.3}MnO_3(LCMO)/SrTiO_3(STO)$ indicate the existence of a spin disordered region at lattice matched LCMO/STO interfaces. We have therefore incorporated a novel tunnel barrier which has the minimum lattice mismatch into the integration of manganite tunnel junctions and have observed much enhanced tunnel junction performance such as single domain switching and a very TMR.

EXPERIMENT

LCMO/STO multilayers were grown *in-situ* by pulsed laser deposition (KrF laser, 248 nm) using stoichiometric targets on (100) STO substrates at 600°C in a 15Pa flowing oxygen atmosphere. Subsequently they were annealed at 600°C in 50kPa oxygen for an hour and cooled down to room temperature. A series of multilayers were deposited with the individual LCMO

3

layer thickness of 2.5, 5, 8.3, 12.5, 25 nm with different numbers of layers ranging from 10 to 1 (keeping the total LCMO thickness at 25 nm) with the individual STO layer thickness 7 nm in each case. Electrical transport and magnetoresistance were measured using a four point probe and magnetic measurements were performed with a commercial SQUID magnetometer. In both measurements the magnetic field was applied parallel to the plane of the substrate. For trilayer tunnel junctions, LCMO/NdGaO$_3$/LCMO trilayers were grown on NdGaO$_3$ (001) substrates with layer thicknesses scale of 80nm(bottom)/2.5-3nm/60nm(top) by the same deposition process. Devices were patterned from these trilayers using optical lithography and Ar ion milling to produce devices with a range of areas between 36 and 600 μm^2. The tunneling conductance was measured as a function of magnetic field and temperature using a four terminal measurement with the magnetic field was applied parallel to the plane of the substrate.

RESULTS

1. La$_{0.7}$Ca$_{0.3}$MnO$_3$(LCMO)/SrTiO$_3$(STO) multilayers

Sharp interfaces and uniformly continuous layers in our multilayers were confirmed by low magnification transmission electron microscopy as in Fig. 1. X-ray θ-2θ diffraction scans around the LCMO (002) peaks of several multilayers showed satellite peaks adjacent to the main peaks arising from chemical modulation of multilayer structure indicate coherent heterostructure growth.[4]

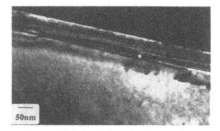

Fig. 1. Cross-sectional TEM photograph of LCMO/STO multilayers of (2.5nm/7nm)$_{10}$ on STO (001) substrate.

Figure 2 shows the zero-field temperature dependence of the resistance: with reducing individual LCMO layer thickness the peak resistance temperature (T$_p$) is monotically lowered and the breaths of the metal-insulator transitions broaden. Despite the substantial changes in low temperature properties, the resistivity above T$_p$ is almost unaffected by the change in individual. layer thickness. The inset shows that the maximum temperature coefficient of resistance, a measure of the sharpness of the metallic transition,[7] decreases with lower T$_p$ for thinner multilayers. In mixed valence manganites, it is accepted that external hydrostatic pressure or internal pressure induced by different cation ionic size alters the transfer interaction between Mn ions.[5,6] As a consequence of the superlattice structure can be reasonably postulated that a larger strain is accommodated in thinner multilayers.[5,6,8] Thus the observed R(T) behavior of heteroepitaxial multilayers is likely to be related to pronounced tensile strain in the LCMO layers from the lattice mismatched STO layers

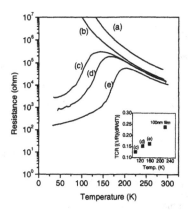

FIG. 2. Temperature dependent resistance of multilayers for (a) $(LCMO/STO)_n$: $(2.5nm/7nm)_{10}$, (b) $(5nm/7nm)_5$, (c) $(8.3nm/7nm)_3$, (d) $(12.5nm/7nm)_2$, and (e) 25nm thick LCMO film. It should be noted that the total LCMO thickness in (f) is twice that of the other samples. In the inset, the maximum temperature coefficients of resistance were presented for several multilayers along with 100nm thick plain LCMO film. The resistance of sample (a) and (b) at low temperatures are beyond our experimental set-up.

However, the high field magnetization measurements in Fig. 3 showed that the saturation magnetization of the multilayer of $(2.5/7)_3$ reduced to only ~150emu/cm³ whilst the single 25nm LCMO layer is ~400emu/cm³ which is comparable to the value of the typical epitaxial film.[9] The loss of the total magnetic moment in the thinnest multilayer implies appreciable magnetic disorder at the interfaces. Then this is clearly related to the loss in low temperature conductivity observed in Fig. 2, while at high temperatures where the conductivity is dominated by thermally-induced magnetic disorder and rather independent of layer thickness. Figure 4 shows the temperature dependence of the normalized magnetization for three different samples. Whilst T_p is monotically lowered as the LCMO layer thickness is decreased (Fig.2), the Curie temperature (T_c) in (a) and (b) is reduced from the bulk value $(T_c = \text{~}250 K)$ by a much smaller amount. Even in the thinnest multilayer, although T_c is strongly suppressed, ferromagnetic behavior is still observed at low temperatures despite a monotonic increase in transport resistance with decreasing temperature. It should be noted that the M(T)/M(5K) of the multilayers decays much faster with increasing temperature and in particular the thinnest multilayer showed similar behavior to the magnetization of the surface boundary in thick $La_{0.7}Sr_{0.3}MnO_3$ films measured by spin resolved photoemission.[10] The decoupling of T_p from T_c, which is evident in the comparison of Fig. 2 and Fig. 4, indicates a break-up of the one-to-one correlation between ferromagnetic and metallic transition implicit in the double exchange interaction.[2] The localized interface magnetism observed in LCMO/STO layers implies a strong correlation between inhomogeneous transport and magnetically disordered interfaces. The fact that the transport properties in multilayers are primarily dependent upon the LCMO layer thickness even under the different strain field indicates the existence of vertical inhomogeniety in the electrical and magnetic structure at interfaces.

5

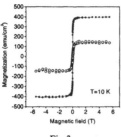

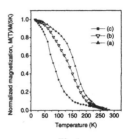

Fig.3.   Fig.4.

Fig. 3. Magnetic hysteresis loops measured at 10 K for (a) 25nm thick plain LCMO film and (b) $(2.5nm/7nm)_{10}$ multilayers.

Fig. 4. Temperature dependence of magnetization of (a) 25nm thick plain LCMO film,
(b) $(LCMO/STO)_n$: $(8.3nm/7nm)_3$, and (c) $(2.5nm/7nm)_{10}$, measured at 0.5 T after zero-field cooling

2. $La_{0.7}Ca_{0.3}MnO_3/NdGaO_3(NGO)/La_{0.7}Ca_{0.3}MnO_3$ tunnel junctions

The interfacial magnetic disorder we have observed in lattice mismatched manganite multilayers has serious implications for the optimisation of spin tunneling junctions based upon heterostructures with half-metallic oxides. In fact, materials problems have so far limited to the TMR obtained from half-metallic devices to significant values only at the lowest temperatures and even these are well below those predicted on the basis of the independently measured polarisation. Here we selected $NdGaO_3$ for both the substrate and tunnel barrier because of its low lattice mismatch with LCMO of less than 0.08%. The LCMO film growth on $NdGaO_3$ substrate shows a typical step flow growth mode for substantial thicknesses which then transforms to two dimensional island growth beyond the critical thickness of about 90 nm. The Curie temperature (T_c) of the base electrode was 265 K which is identical to the T_c of unpatterned single films. Our measurement of the base electrode resistance showed that the sheet resistance of the bottom electrode was several orders of magnitude lower than the intrinsic junction resistance, implying a uniform current distribution across the barrier. Devices were patterned from these trilayers using optical lithography and Ar ion milling to produce devices with a range of areas between 36 and 600 μm^2; the structure of the complete device is shown schematically in the Fig. 5. The tunneling conductance was measured as a function of magnetic field and temperature using a four terminal measurement with the magnetic field applied parallel to the plane of the substrate. All the devices measured showed a large magnetoresistance at low temperatures with extremely sharp switching between the high and low resistance states which we assume to correspond to parallel and anti-parallel alignment of the moments of the two LCMO electrodes. Figure 5(a) shows the resistance at 77 K as a function of magnetic field of the highest impedance junction measured to date. Two striking characteristics are evident: firstly, the measured TMR is 86% with the definition of $\Delta R/R_{AP} = (R_{AP}-R_P)/R_{AP}$; (i.e. a factor of 7.3 between the parallel resistance (R_P) and antiparallel resistance (R_{AP}) states); secondly, there is extremely sharp and coherent switching at both coercive fields (field sensitivity defined by $1/R \cdot (dR/dH)$ is 400 %/Oe). The TMR value is close to the highest ever measured and has been achieved at a much higher temperature than any other device. The sensitivity, which is a measure of the coherence of the switching of the electrodes, is higher than has been reported in any magnetic tunnel junction system.

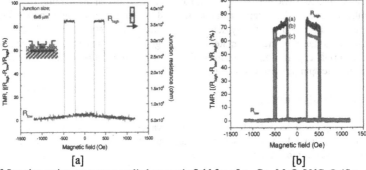

[a]   [b]

Fig.5. [a] Junction resistance versus applied magnetic field for a $La_{0.7}Ca_{0.3}MnO_3/NdGaO_3/La_{0.7}Ca_{0.3}MnO_3$ junction at 77 K, showing the resistance change of a factor of 7.3 between parallel and anti-parallel alignment and the coherently sharp switching at H_{C1} and H_{C2}. [b] The TMR values for several junctions with different junction size and aspect ratio in the same chip; (a) $8 \times 8 \mu m$, (b) $8 \times 16 \mu m$ and (c) $20 \times 30 \mu m$.

Sharp switching and distinct resistance states (R_P and R_{AP}) strongly suggest single domain switching in both top and bottom electrode. Devices on the same chip which have different junction size and the aspect ratio show similar TMR values and identical values of H_{c1}, and this value agrees well with the coercive field of a plain LCMO film measured using a SQUID magnetometer as in Figure 5(b). This suggests that H_{C1} is associated with the switching of the common base electrode, whereas devices whose top electrodes are of different sizes and aspect ratios show different values of H_{C2}. Indeed the R(H) curves were quite stable for both magnetic history and thermal cycles and also reproducible in the laboratory time scale. A simple elastic model for spin tunnelling was first formulated by Julliere.[11] This model is based only on the effective spin polarisation at the Fermi energy (E_F) so that for identical ferromagnetic electrodes the zero bias conductance is given by

$$R_P^{-1} = M\{D\uparrow^2(E_F) + D\downarrow^2(E_F)\}, \; R_{AP}^{-1} = M\{2D\uparrow(E_F)D\downarrow(E_F)\} \quad (1)$$

where R_P and R_{AP} are the resistances in parallel and anti-parallel orientations and $D\uparrow(E)$, $D\downarrow(E)$ are the spin up and spin down density of states and M is the tunnelling probability. Thus the TMR is given by

$$\Delta R / R_{AP} = (R_{AP} - R_P)/R_{AP} = 2PP' \; /1 + PP' \quad (2)$$

where P is the spin polarisation given by $P = [D\uparrow(E_F) - D\downarrow(E_F)]/[D\uparrow(E_F) + D\downarrow(E_F)]$.
The maximum polarisation deduced via (2) for the LCMO electrodes from the data in Fig. 1(b) was 0.84. This value is the much higher than has been previously measured in the manganites[12,13,14] at this temperature.

CONCLUSIONS

The magnetically disordered interfaces which we have observed in the lattice mismatched LCMO/STO multilayers has serious implications for the optimisation in spin tunneling junctions based upon heterostructures with half-metallic oxides. The tunnel junctions based upon LCMO/NdGaO$_3$/LCMO trilayers structure show much enhanced junction performance ever reported ; the measured TMR was 86%; (i.e. a factor of 7.3 between the parallel resistance (R_P) and antiparallel resistance (R_{AP}) states) and they showed extremely sharp and coherent switching at both coercive fields (field sensitivity defined by $1/R \cdot (dR/dH)$ is 400 %/Oe). Although we are still to integrate another half-metallic materials of a higher T_c, our results demonstrated that by appropriate materials engineering the intrinsic high polarisation of half-metallic materials can be fully utilized for the spin polarised tunnelling.

ACKNOWLEDGEMENTS

This work was supported by EPSRC through the Advanced Magnetics Programme, the UK - Korea Science & Technology Collaboration Fund, and the Royal Society. The authors are indebed to Dr. S. Newcomb for the TEM micrograph.

*Corresponding author. E-mail : mhj24@cam.ac.uk

REFERENCES

[1] J.Z. Sun, *Philos.Trans. R. Soc. London, Ser. A* **356,** 1663 (1998).

[2] M. Viret, M. Drouet, J. Nassar, J.P. Coutour, C. Fermon, and A. Fert, *Europhys. Lett.* **39,** 545 (1997).

[3] T. Obata, T. Manako, Y. Shimakawa, and Y. Kubo, *Appl. Phys. Lett.* **74,** 290 (1999).

[4] M.-H. Jo, N.D. Mathur, J.E. Evetts, M.G. Blamire, M. Bibes, J. Fontcuberta, *Appl. Phys. Lett.* **75,** (1999) 3689

[5] J. Aarts, S. Freisem, R. Hendrix and H.W. Zandbergen, *Appl. Phys. Lett.* **72,** 2975 (1998).

[6] M.G. Blamire, B.-S. Teo, J.H. Durrell, N.D. Mathur, Z.H. Barber, J.L. McManus Driscoll, L.F. Cohen and J.E. Evetts, *J. Magn. Magn. Mater.* **191,** 359 (1998).

[7] T. Venkatesan, M. Rajeswari, Z.W. Dong, S.B. Ogale and R. Ramesh, *Philos.Trans. R. Soc. London, Ser. A* **356,** 1661 (1998).

[8] J.Z. Sun, D.W. Abraham, R.A. Rao and C.B. Eom, *Appl. Phys. Lett.* **74,** 3017 (1999).

[9] A. Gupta, G. Q. Gong, G. Xiao, P. R. Duncombe, P. Lecoeur, P. Trouilloud, Y. Y. Wang, V. P. Dravid and J. Z. Sun, *Phys. Rev. B* **54,** 15629 (1996).

[10] J.-H. Park, E. Vescovo, H.-J. Kim, C. Kwon, R. Ramesh and T. Venkatesan, *Phys. Rev. Lett.* **81,** 1953 (1998).

[11] M. Julliere, *Phys. Lett.* **54A,** 225 (1975).

[12] R.J. Soulen Jr, J.M. Byers, M.S. Osofsky, B. Nadgorny, T. Ambrose, S.F. Cheong, P.R. Broussard, C.T. Tanaka, J. Nowak, J.S. Moodera, A. Barry, J.M.D. Coey, *Science* **282,** 85 (1998)

[13] Y. Okimoto, K. Katsufuji, T. Ishikawa, A. Urushibara, T. Arima and Y. Tokura, *Phys. Rev. Lett.* **75,** 109 (1995).

[14] J.Y.T. Wei, N.-C.Yeh, and R.P. Vasquez, *Phys. Rev. Lett.* **79,** 5150 (1997).

Growth of "colossal" magnetoresistance heterostructures by molecular beam epitaxy

J. O'Donnell, A. E. Andrus, S. Oh, E. Colla, M. Warusawithana, B. A. Davidson, and J. N. Eckstein

University of Illinois - Department of Physics, 1110 West Green Street, Urbana, IL 61801

We discuss the heteroepitaxial growth of $La_{1-x}Sr_xMnO_3$ films and $CaTiO_3$ insulating barriers by molecular beam epitaxy. We find that the surface morphology and residual resistivity of the manganite electrodes is critically dependent on the film stoichiometry. The most important parameter is the concentration of La+Sr (cubic perovskite A-site cations) to that of Mn (B-site cation). If La+Sr is supplied in slight excess, the films grow with atomically flat surfaces, but the residual resistivity at 4.2K is high (as high as 6500 $\mu\Omega$-cm), and Curie temperature (T_c) low (<300 K). If Mn is supplied in slight excess, the films have high T_c (370 K) and residual resistivity (35 $\mu\Omega$-cm) better than bulk single crystal values, but the surface is no longer atomically flat. There appears to be a very narrow region of phase space where it is possible to have low resistivity, high T_c films with atomically flat surfaces. This is precisely where one would like to place heterostructure devices.

Introduction

The ground state of the manganite oxides ($La_{0.67}A_{0.33}MnO_3$ where A=divalent dopant) is ferromagnetic and very nearly half-metallic. That is, the conduction electrons are nearly 100% spin-polarized. Experiments indicate that the spin-polarization, $P=(n_\uparrow - n_\downarrow)/(n_\uparrow + n_\downarrow)$, ($n_\uparrow$, $n_\downarrow$ are the density of states for spin +1/2 and spin −1/2 electrons respectively) is at least 0.83.[1,2,3], compared with 0.42 for the canonical ferromagnet, iron.[3] This is a result of the unusual ferromagnetic interaction known as double exchange[4] in which the core Mn 3d-electron spins on adjacent sites are coupled by the hopping of the conduction 3d-electrons between Mn^{3+} and Mn^{4+} ions via the intermediate O^{2-} ion. Details of the nature of the ferromagnetic state, metal-insulator transition, and colossal magnetoresistance exhibited by these compounds remains a subject of controversy.[5] However, it is clear that these materials offer an interesting tool for scientists – a solid state source of highly (possibly 100%) spin-polarized electrons in a system which is structurally, chemically, and epitaxially compatible with many other interesting perovskite oxide materials.

What can be done with such a tool? Applications include spin-valve type magnetic tunnel junctions for disk drive read heads[1,2,6,7,8], spin-injection devices[9,10], and spin-polarized electron tunneling spectroscopy of High T_c superconductors. All of these applications involve heteroepitaxy between the manganites and other perovskite oxide compounds. In this paper we address fundamental questions regarding manganite growth as it relates to heterostructure applications. We approach this topic from the context of the development of "colossal magnetoresistance" spin-valve sensors. However, our conclusions apply to manganite heteroepitaxy problems in general.

We have found that it is possible to grow $La_{1-x}Sr_xMnO_3$ (LSMO) films with atomically flat surfaces, ideal for well-ordered interfaces. We have also found that LSMO can be grown with extremely low residual resistivity (35 $\mu\Omega$-cm) and high Curie temperature (380K). However, there is only a very narrow region of stoichiometric phase space where these two regions (atomically flat and low resistivity/high T_c) coexist. Outside the narrow range, either the interfaces are so rough that disorder dominates the properties, or the material is so highly resistive that transport measurements are nearly impossible. This is probably the explanation for the wide variety of results seen in LSMO magnetic tunnel junction

9

experiments, and can also account for the rapid decrease in magnetoresistance in LSMO magnetic tunnel junctions with increasing temperature.

Growth of La$_{1-x}$Sr$_x$MnO$_3$ electrodes

We have used an oxide molecular beam epitaxy (MBE) technique to fabricate trilayer magnetic tunnel junctions with La$_{1-x}$Sr$_x$MnO$_3$ electrodes (x=0.25-0.32) and insulating barriers of CaTiO$_3$ (CTO). Details of the MBE system have been described elsewhere.[11] MBE allows in-situ monitoring of the growth via reflection high-energy electron diffraction (RHEED), as well as atomic (monolayer) level control of film stoichiometry.

For the growth of the manganite films, the cations are codeposited one unit cell at a time on single crystal [100] oriented STO substrates in a beam of flowing ozone at a chamber pressure of 4 x 10^{-6} Torr (base pressure ~ 3 x 10^{-9} Torr) with a short (5-15 second) anneal between every 5 unit cell group. The substrate is heated to 690 – 710C. A quartz crystal oscillator and/or atomic absorption spectroscopy[11] measures the flux from each Knudsen cell thermal source. Atomic absorption spectroscopy can be done continuously and non-invasively during the growth. This allows us to correct source flux drift in real-time and is particularly useful for Sr and Ca sources which tend to drift on the ± 5% level and the 60 minute time scale despite accurate source temperature stabilization of |ΔT| < 0.2 C. Rutherford backscattering spectroscopy performed on samples recently grown on MgO substrates calibrates the quartz crystal oscillator and atomic absorption measurements. Careful analysis of the Rutherford backscattering spectrum gives numerical factors to convert quartz oscillator and/or atomic absorption spectroscopy measurements into atomic flux in atoms/(cm^2·s) with an absolute accuracy of ~ 3%. The sources are carefully flux matched to satisfy F$_{Mn}$ = F$_{La}$/0.67 = F$_{Sr}$/0.33 to within ± 5% accuracy. Each monolayer takes roughly 25 seconds to deposit.

We do not rotate the sample continuously during deposition. Consequently, there is a composition gradient of each constituent of roughly ± 2.5% across the surface of the substrate. Our approach is to sacrifice the stoichiometry of the perimeter of the sample in order to gain information about the growth by scanning the film surface with RHEED, and to expand the tolerances on flux calibration (see below). We parameterize this composition gradient by two terms, x and y as (La$_{1-x}$,Sr$_x$)$_y$ MnO$_3$. Thus,

$$x = \frac{[Sr]}{[La] + [Sr]} \qquad (1)$$

is the doping level, typically targeted for x≈0.3, and

$$y = \frac{[La] + [Sr]}{[Mn]} \qquad (2)$$

is the perovskite A-site occupation, typically targeted for y=1.0. The notation [α] represents the concentration of species α.

The positions of the La, Sr, and Mn cells on the source flange of the vacuum system, determine the gradient in x across the surface of the 1.4 cm x 1.4 cm substrate of ± 1.3% from the nominal value of 0.33, and the gradient in y of ± 3.8% from the nominal value of 1.0. Where y<1.0 the film will most likely have vacancies on the A-site which will be compensated for by Mn^{3+}→Mn^{4+} and possibly oxygen vacancies.[12] For y>1.0, the situation is not as clear. Defect chemistry studies of the manganites differ in whether or not the system can tolerate Mn vacancies.[12,13,14] Neither is there an obvious site in the lattice where La/Sr interstitials could be accommodated. We see no evidence for second phase precipitation of La/Sr oxides in atomic force micrographs or x-ray diffraction.

The gradient in y has pronounced effects on important film properties. The surface morphology undergoes a dramatic change from atomically flat with unit cell terraces spaced at the substrate miscut

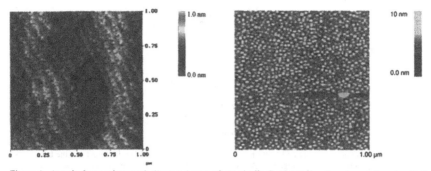

Figure 1 Atomic force micrograph (1μm x 1 μm) of atomically flat LSMO surface in y<1.0 region (left) and "bumpy" surface in y>1.0 region (right).

angle for y>1.0, to rough with oriented, faceted bumps having a characteristic lateral dimension of 300 Å and a height of 30 Å for y<1.0. The two regions are usually present on the same sample, and are easily distinguishable even with the naked eye. The bumpy region has a dark purple hue, which can be seen in reflection, whereas the smooth region simply looks black. Atomic force microscope images of the two regions are shown in Figure 1. The transition region between these two distinct surface morphologies spans only 100 μm of the sample surface. Given our known flux gradients, this distance translates to a miniscule change in stoichiometry of $\Delta y=0.0005$. Thus the surface morphology does not evolve smoothly and continuously with y. Rather, there are two distinct surface morphologies separated by an extremely narrow transition region. We do not know for certain whether the critical value for y is exactly the stoichiometric value 1.0. We estimate $y_{crit}=1.00\pm0.03$. We observe this behavior on virtually all LSMO samples and can shift the border up and down reproducibly even within the same growth by changing the stoichiometry slightly.

The bumpy region is smooth enough to produce two-dimensional streaked RHEED images. The bumpy region is easily identified in RHEED by the appearance of faint elevated ½-order streaks as shown in Figure 2. Close inspection reveals that these 1/2-order streaks are really part of a complete 1/2 order Laue circle such as that which would arise from a $\sqrt{2} \times \sqrt{2}$ surface reconstruction. We do not yet have a real-space model for this reconstruction. Crystallographically, the two regions are indistinguishable in x-ray reciprocal lattice maps of both [002] and [103] peaks. Both regions are coherently strained through

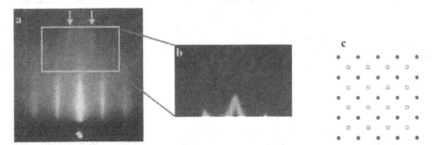

Figure 2 (a) RHEED image of bumpy region along [100]. Arrows mark elevated 1/2-order streaks. Intensities are scaled logarithmically to bring out dim features. (b) False color zoomed view of boxed region of (a). (c) Reciprocal lattice corresponding to RHEED image. Filled circles represent reciprocal lattice rods of bulk-terminated Mn-O_2 surface. Open circles represent additional rods due to surface reconstruction.

11

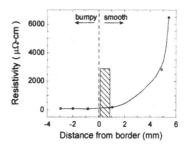

Figure 3 Residual resistivity of LSMO film at 4.2 K as a function of the distance from the border between smooth and bumpy regions. All data points are from the same sample. Hatched area represents region of film where low resistivity coexists with atomically flat surface.

the entire film thickness with a=b=3.90Å and c=3.85Å. Rocking curves on 1000 Å films give $\Delta\omega$ <0.05°, comparable to the substrate.

The most remarkable effect of the gradient in y is in transport properties. The resistivity vs. temperature ($\rho(T)$) from the bumpy region looks almost identical to bulk $\rho(T)$ data the only difference being that the resistivity of the films is slightly lower over the entire temperature range. In the bumpy region we have measured residual resistivity ($\rho_0 \equiv \rho(T=0)$) as low as 35 $\mu\Omega$-cm at 4.2 K with critical temperatures (T_c, estimated as resistivity peak temperatures) of 370K. The $\rho(T)$ of the smooth region is characterized by a decrease in T_c (to as low as 280 K), an increase in ρ_0 (to as much as 6500 μ Ω-cm), and a slight upturn in resistivity as the temperature is lowered below 30 K. In these respects, the change in transport properties is similar to the effect of reduced doping, x. Our $\rho(T)$ for y<1.0, however lack of a sharp drop in ρ at T_c such as that seen in the data of Urushibara et al.[15]. Instead, the resistivity obeys the following form for $T<T_c$

$$\ln\left[\frac{\rho(T)}{\rho(T_c)}\right] = B(T - T_c) \qquad (3)$$

where $B \approx +0.01$ K^{-1}.

We have measured the residual resistivity of a single film taken at several locations on the surface at varying distances from the smooth/bumpy border. Figure 3 shows a narrow region (hatched) where low resistivity coexists with atomically flat surfaces. This is where one would like to place a heterostructure device. The region is narrow in real space (~1mm), and very narrow in terms of composition (although not nearly as sharp a transition as the critical behavior of the surface morphology). One mm of the film surface corresponds to $\Delta y=0.005$. Given our calibration uncertainty, we cannot purposely target so narrow a stoichiometry range. Instead we rely on the flux gradients to give us a continuous stoichiometry range on the same film. We can reproducibly place the critical region in y somewhere on the sample. Furthermore, we are relieved of the necessity to know the exact value of y_{crit} (although this is an important question).

Growth of the barrier

The growth of the insulating barrier, especially the interfacial layers, is just as critical as the growth of the manganite electrodes. Indeed, most speculation regarding the rapid loss of magnetoresistance in colossal magnetoresistance magnetic tunnel junctions with increasing temperature has focused on barrier and interfacial disorder (structural, magnetic, and/or compositional).[2,6,8] By comparing RHEED oscillations before and after the growth of the insulating barrier we have shown that even starting from an atomically flat base electrode, disorder can nucleate in the barrier heteroepitaxy and propagate into the top electrode. The result is a decrease in both the magnitude of the tunneling magnetoresistance, and the temperature at which the magnetoresistance vanishes.[1] We also observe a zero bias anomaly in the tunneling conductance at low temperatures of the form predicted by Altshuler et al. for electron-electron interaction induced renormalization of the density of states in a disordered metal.[1,16] We have investigated $SrTiO_3$, $CaTiO_3$, and $LaAlO_3$ barriers. Of the three, $CaTiO_3$ grows the best as judged by the RHEED image of the barrier surface, and RHEED oscillations during barrier growth. Our highest magnetoresistance devices with $\Delta R/R = 4.5$ have also had $CaTiO_3$ barriers.[1]

For the growth of the insulating barrier, the substrate temperature is decreased by 90 C in constant ozone pressure for the first two interfacial layers to prevent interdiffusion, then increased 90 C for the remainder of the barrier growth, then decreased again by 90 C for the initial layers of the manganite counter electrode. We typically grow 4,5, or 6 unit cell insulating barriers. Five unit cell barriers give a typical tunneling conductance of 5 x 10^{-7} $\Omega^{-1}/\mu m^2$ in tunnel junctions with an area of 200 μm^2. The tunneling conductance varies exponentially with barrier thickness with each unit cell decreasing the conductance by roughly a factor of 12.

The interfacial $La_{1-x}Sr_xO$ / CaO layer between the terminating MnO_2 plane of the manganite and the TiO_2 plane of the $CaTiO_3$ barrier is deposited in a layer-by-layer mode with composition 1/2 ($La_{1-x}Sr_xO$) + 1/2 (CaO). The interfacial plane donates electrons to the adjacent MnO_2 plane. In a simple ionic picture growing this mixed composition at the interface maintains a constant Mn^{4+}/Mn^{3+} ratio of x:(1-x). If the interfacial layer is not doped in this way at the atomic layer level, the Mn^{4+}/Mn^{3+} ratio of the interfacial MnO_2 plane will be increased, effectively increasing the local doping level. It is not clear at this time whether the effect is significant. A simple ionic picture neglects the fact that the conduction electrons are partially delocalized and short length scale doping modulations will tend to be averaged out. We can say however, that to date our highest magnetoresistance devices are those with this atomic layer engineering of the interface. A recent experiment on a buried $LSMO/YBa_2Cu_3O_{7-\delta}$ interface lends some credence to this notion of local doping effects. Stadler et al. observed a decrease in the x-ray magnetic circular dichroism signal with increasing $YBa_2Cu_3O_{7-\delta}$ thickness and interpreted this in terms of cation displacement/exchange changing the effective doping at the interface.[17]

Summary

It should be clear from the preceding discussion that growth of manganite heterostructures with the desired atomically abrupt interfaces, high T_c, and low resistivity is an experimental challenge. It is not obvious that the stoichiometric compound (i.e. y=1.00) is the optimal LSMO composition for heterostructure applications. Whatever the optimal y value is, the tolerance $|\Delta y|=0.005$ for achieving the desired LSMO properties is so small that it is not plausible to control the stoichiometry with this precision. Thus stoichiometry gradients, which normally are removed by rotating the sample during growth, become essential.

We have previously shown that barrier and interfacial disorder seen in RHEED correlates with decreased overall magnetoresistance, premature loss of magnetoresistance at $T \ll T_c$, and the appearance of a zero-bias anomaly in the tunneling conductance at low temperatures.[1] The difficulties in barrier heteroepitaxy combined with the criticality of the electrode stoichiometry, may very well explain the discrepancies in the data among various groups investigating colossal magnetoresistance spin-valve type heterostructures.

There is a degree of freedom that we have not yet explored - substrate temperature during growth. We expect that the substrate growth temperature may affect the surface morphology through the kinetics of surface adatom mobility. Perhaps it will be possible to expand the region of atomically flat surfaces and low resistivity by optimizing the substrate growth temperature. This will be investigated in future work.

References

[1] J. O'Donnell, A. E. Andrus, S. Oh, E. Colla, and J. N. Eckstein (submitted to Appl.Phys. Lett.)

[2] M. Viret, M.Drouet, J. Nassar, J.P. Contour, C. Fermon, and A. Fert, Europhys. Lett. **39**, 545 (1997).

[3] R. J. Soulen, Jr., M. S. Osofsky, B. Nadgorny, T. Ambrose, P. Broussard, S. F. Cheng, J. Byers, C. T. Tanaka, J. Nowack, J. S. Moodera, G. Laprade, A. Barry and M. D. Coey, Journ. Appl. Phys. **85**, 4589 (1999).

[4] C. Zener, Phys. Rev. **82**, 403 (1951).

[5] For a review of the physics of the manganites see M. Imada, A. Fujimori, Y. Tokura, Rev. Mod. Phys. **70**, 1309 (1998).

[6] J. Z. Sun, L. Krusin-Elbaum, P. R. Duncombe, A. Gupta, and R. B. Laibowitz, Appl. Phys. Lett. **70** ,1769 (1997); J. Z. Sun, D. W. Abraham K. Roche, and S. S. P. Parkin, Appl. Phys. Lett. **73**, 1008 (1998).

[7] T. Obata, T. Manako, Y. Shimakawa, and Y. Kubo, Appl. Phys. Lett. **74**, 290 (1999).

[8] C. Kwon, Q. X. Jia, Y. Fan, M. F. Hundley, and D. W. Reagor, J. Appl. Phys. **83**, 7052 (1998); C. Kwon, Q. X. Jia, Y. Fan, M. F. Hundley, D. W. Reagor, J. Y. Coulter, and D. E. Peterson, Appl. Phys. Lett. **72**, 486 (1998).

[9] D. Koller, M. S. Osofsky, D. B. Chrisey, J. S. Horwitz, R. J. Soulen, Jr., R. M. Stroud, C. R. Eddy, J. Kim, R. C. Y. Auyeung, J. M. Byers, B. F. Woodfield, G. M. Daly, T. W. Clinton, and M. Johnson, Journ. Appl. Phys. **83** 6774 (1998).

[10] R. M. Stroud, J. Kim, C. R. Eddy, D. B. Chrisey, J. S. Horwitz, D. Koller, M. S. Osofsky, R. J. Soulen, Jr., R. C. Y. Auyeung, Journ Appl. Phys. **83**, 7189 (1998).

[11] J. N. Eckstein and I. Bozovic, Annu. Rev. Mater. Sci. **25**, 679 (1995).

[12] J. A. M. Van Roosmalen, P. van Vlaanderen, E. H. P. Cordfunke, W. L. Ijdo, and D. J. W. IJdo, J. Solid State Chem. **93**, 213 (1991).

[13] B. C. Tofield and W. R. Scott, J. Solid State Chem. **10**, 183 (1974).

[14] J. F. Mitchell, D. N. Argyriou, C. D. Potter, D. G. Hinks, J. D. Jorgensen, and S. D. Bader, Phys. Rev. B **54**, 6172 (1996).

[15] A. Urushibara Y. Moritomo T. Arima A. Asamitsu G. Kido Y. Tokura , Phys. Rev. B **51**, 14103 (1995).

[16] B. Altshuler, G. Aronov and P. Lee, Phys. Rev. Lett. **44**, 1288, (1980).

[17] S. Stadler, Y. U. Idzerda, Z. Chen, S. B. Ogale, and T. Venkatesan, Appl. Phys. Lett. **75**, 3384 (1999).

Novel Magnetic Oxides

TRANSPORT PROPERTIES AND MAGNETISM OF β-MnO$_2$

H. SATO [*], T. ENOKI [*], K. WAKIYA [*], M. ISOBE [**], Y. UEDA [**], T. KIYAMA [***], Y. WAKABAYASHI [***][*****], H. NAKAO [***], and Y. MURAKAMI [***]

[*]Department of Chemistry, Tokyo Institute of Technology, Meguro-ku, Tokyo 152-8551, JAPAN
hirohiko@chem.titech.ac.jp
[**]Institute for Solid State Physics, the University of Tokyo, Minato-ku, Tokyo 106-8666, JAPAN
[***]Photon Factory (PF), Institute of Materials Structure Science, High Energy Accelerator Research Organization (KEK), Tsukuba 305-0801, JAPAN
[****]Department of Physics, Faculty of Science and Technology, Keio University, Yokohama 223-8522, JAPAN

ABSTRACT

A comprehensive study of transport properties and magnetism on β-MnO$_2$ reveals the strong coupling between the conduction electrons and the localized spins which are supposed to form a magnetic helix below $T_N \sim 92$ K. We also show the direct evidences of the helical magnetism by means of the measurements of the anisotropy in the magnetic susceptibility and the observation of x ray magnetic scatterings on a single crystal of β-MnO$_2$. These results are consistent with proper-type helix model proposed by Yoshimori [J. Phys. Soc. Jpn. **14**, p. 807 (1959)]. This model also qualitatively agrees with the anisotropy in magnetoresistance that appears below T_N. The pitch of the magnetic helix is not commensurate to the lattice and it is slightly temperature dependent. The intensity of several Bragg peaks drastically changes at T_N suggesting that the magnetic ordering is accompanied by a lattice distortion.

INTRODUCTION

Rutile-type manganese dioxide β-MnO$_2$ has been known as a prototype of helical magnetism for a long time. In addition, relatively high electric conductivity with a large anomaly at the magnetic phase transition temperature has been reported,[1] although the composition MnO$_2$ suggests the existence of only Mn^{4+} and semiconductive electronic state because of the ligand field splitting between the t_{2g} and the e_g levels. For the interest in the origin of the metallic carriers and strong coupling between electronic conduction and magnetism, we have carried out a comprehensive study of electronic transport properties such as resistivity, thermoelectric power (TEP), Hall effect and magnetoresistance (MR). [2]

On the other hand, the experimental evidences of the helical magnetism in β-MnO$_2$ have been insufficient although it is well established theoretically.[3] There has been no publication about the neutron diffraction data, so far, except for the private communication in Ref. 3. The measurement of magnetic susceptibility anisotropy has been also lacking. Bizette already reported the anisotropy of susceptibility.[4] However, strictly speaking, his data is not on a single crystal in a usual sense but on a material obtained by a calcination of γ-MnOOH. We have recently succeeded to obtain a good single crystal of β-MnO$_2$ by a hydrothermal method. Therefore, we have reinvestigated the anisotropy of magnetic susceptibility using a SQUID magnetometer. We have also done x ray magnetic scattering measurements using synchrotron radiation (SR) in order to obtain the precise magnetic structure.

Mat. Res. Soc. Symp. Proc. Vol. 602 © 2000 Materials Research Society

EXPERIMENT

The single crystals of β-MnO$_2$ were synthesized by using a hydrothermal technique almost the same as that reported by Rogers *et al.* [1] The resistivity between room temperature (RT) and 0.3 K was measured by a conventional four-terminal method. Magnetoresistance and Hall coefficient were measured by using a 15 Tesla superconducting magnet system (Teslatron, Oxford Instruments). TEP was measured by a differential method using AuFe-Chromel thermocouples. Magnetic susceptibility was measured by using SQUID magnetometer (Quantum Design MPMS-5) on a single crystal for anisotropy measurement and polycrystalline sample for determination of exact value of susceptibility. The x ray magnetic scattering measurements on a single crystal of β-MnO$_2$ (0.5×0.5×1 mm^3 in size) were done at the experimental station BL-4C in KEK-PF, Tsukuba. The incident beam is monochromatized by a Si(111) double crystal and focused by a bent cylindrical mirror. The energy of the x ray is fixed at 17.46 keV, which corresponds to the energy of conventional x ray source with Mo target.

RESULTS

Resistivity, Thermoelectric Power, and Hall Coefficient

Figure 1(a) shows the resistivity of β-MnO$_2$ with the electric current along the c axis. The value at RT is around 0.3 Ωcm, which is too small for ordinary metal. However, its temperature dependence is almost metallic down to around 200 K. Between 200 K and T_N (92 K), the temperature coefficient becomes negative but the temperature dependence is not so strong as activation-type behavior. Just below T_N, it exhibits a steep drop and becomes metallic and increases again after showing a minimum at around 50 K. Figure 1(b) shows the temperature dependence of the TEP. The TEP also shows a metallic behavior, that is, proportional to temperature. Around T_N, there is an anomaly related to the magnetic phase transition probably because of the enhancement of magnetic scattering. It is to be noted that the slope in $T > T_N$ region and that in $T < T_N$ region are almost the same, indicating that the carrier density does not change at T_N. Figure 1(c) shows the temperature dependence of the Hall coefficient. It shows a negative value around –0.15 cm^3/C, almost independent of temperature except around T_N. This indicates the existence of uncompensated n-type carriers with almost zero excitation energy. From these results, we can estimate the carrier density. The value of Hall coefficient around RT gives the carrier density of 4.8×10^{19} cm^{-3}. The origin of this small density can be explained by the very small amount of oxygen defects like MnO$_{2-\delta}$ (δ = 0.0014). The TEP value at RT is around –80 μV/K, which is explained within a free electron model without any mass-enhancement. The Hall mobility at room temperature is calculated as only 0.48 cm^2/Vs from the experimental value. This extremely small value is far smaller than Mott's limit, therefore, the picture of metallic conduction based on the Boltzmann's equation is no longer correct in this case. The strong scatterings caused by the localized magnetic moments in the t_{2g} cores can account for it, as also seen in perovskite manganates.[5] Below T_N, the mobility increases while the carrier concentration remains constant. This is because the magnetic fluctuation of the magnetic moments in t_{2g} cores disappears below T_N. In the lowest temperature range below 50 K, the mobility decreases again. This is probably due to the localization caused by random potentials.

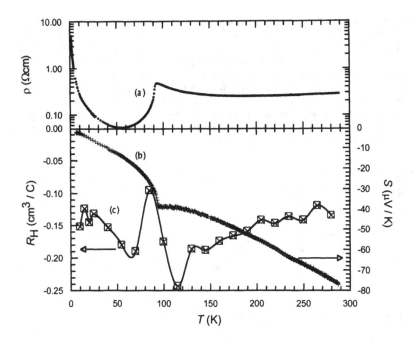

Fig. 1. The temperature dependence of (a)the resistivity, (b)the thermoelectric power, and (c)the Hall coefficient, in β-MnO$_2$.

Magnetic Susceptibility and Magnetoresistance

Magnetic susceptibility is shown in the inset of Fig. 2(a). Above T_N, a Curie-Weiss analysis gives 2.46 emu K/mol and -783 K for Curie constant C and Weiss temperature Θ, respectively. The value of C deviates from the value 1.857 calculated based on the localized spin model with $S = 3/2$. Below T_N, the magnetic susceptibility becomes anisotropic. The susceptibility χ ($H \parallel c$) increases with decreasing temperature while χ ($H \perp c$) decreases. Unlike collinear antiferromagnetism, there is no direction of magnetic field along which the susceptibility goes to zero at $T = 0$ K. This is one of the characteristics of helical magnetism and our data are qualitatively consistent with the proper-type helical spin structure proposed by Yoshimori [3], which is schematically shown in Fig. 2(b). However, the behavior of χ ($H \parallel c$) below T_N is different from his calculation based on molecular field approximation, which predicts temperature-independent behavior. This requires some modification of the theory taking account of spin-wave excitations. The MR, defined as the resistance at 15 T normalized by that at 0 T, is shown in Fig. 2(a). In the measurements, the direction of electric current was always along the c axis, while the direction of the magnetic field was varied. In the paramagnetic region, slightly negative MR appears. This is simply explained by the suppression of the magnetic fluctuation by external magnetic field.[6] The remarkable thing is the appearance of the anisotropy of the MR below T_N. The MR for $H /\!/ c$ is negative but that for $H \perp c$ becomes positive. This should reflect the anisotropy of the response of the helical magnetic structure for the external field. It is predicted that the magnetic field along the c axis simply cants the helix toward the c direction, while the field perpendicular to the c axis deforms the shape of helix as a precursor of the to the transition into fan-like structure. [7, 8] Therefore, in the latter case, the angle between the adjacent spins is no longer identical and the helix becomes including the higher harmonics of the wavenumber of the uniform helix. This can cause additional scatterings and consequently increase resistance. The origin of the strong enhancement of MR in the lowest temperature region has been unsolved. It should be related to the carrier localization, although the simple theory of weak-localization predicts negative MR.

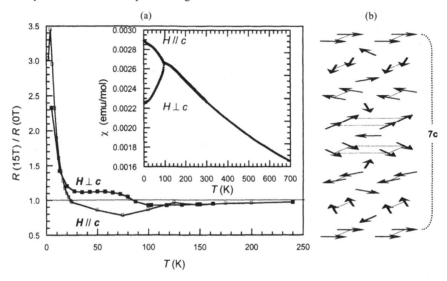

Fig. 2. (a)The magnetoresistance of β-MnO$_2$ with the magnetic field parallel or perpendicular to the c-axis. The inset shows the magnetic susceptibility. (b)The model for helical magnetism of β-MnO$_2$ proposed by Yoshimori.

X Ray Magnetic Scattering

It is surprising that there has been no detailed publication about the experimental evidence of the helical magnetic structure in β-MnO$_2$ while the theory is so established.[3] Several problems remain which we should clarify by experiments. Is the magnetic helix commensurate to the crystal lattice or not? Is the pitch of the helix dependent on temperature? In order to clarify them, we carried out an x ray magnetic scattering experiment. The advantages of this method compared to neutron diffraction are mainly in the following two points. The first is its high resolution in q-space because SR can generate an extremely straight beam. Another advantage is that it is available for a small single crystal with sub-millimeter size. The principle of the magnetic scattering of the x ray is given in the literature.[9] In this measurement, we succeeded to find four satellites at $(1\ 0\ 2+q)$, $(1\ 0\ 2-q)$, $(1\ 1\ 3-q)$, and $(2\ 0\ 1+q)$, where $(h\ k\ l)$ denotes the indices in the q-space. The scattering intensity of these satellites is about 10^{-6} weaker than that of typical fundamental diffractions. Figure 3(a) shows the temperature dependence of the intensity of the (1 0 2+q) satellite. It reminds us the typical temperature dependence of an order parameter of second-order phase transition. The extinction rule for the satellites that $h+k+l = 2n$ are forbidden is consistent with the Yoshimori's model (Fig. 2(b)) in which the magnetic moments on the corner Mn sites and those on the body-center Mn sites are antiparallel. The wavenumber of the helix is slightly dependent on temperature as shown in Fig. 3(b). In Yoshimori's model based on the neutron diffraction data by Erickson, the pitch of the helix is assumed to be $7 / 2\ c$. However, our data of q show a slight deviation from $(2 / 7)\ c^* \approx 0.2857\ c^*$, and there is no evidence of a commensuration to the crystal lattice, as shown in Fig. 3(b).

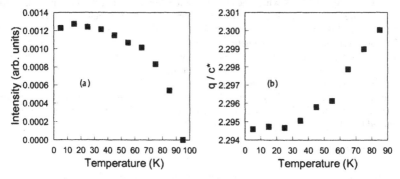

Fig. 3. The temperature dependence of (a)the intensity and (b)the wavenumber of the (1 0 2+q) x ray magnetic scattering in β-MnO$_2$.

In the process of the experiment, we also found that the intensity of some Bragg diffractions shows abrupt change at T_N. Figures 4(a) and 4(b) show the temperature dependence of the intensity of (1 0 2) and (1 0 3) diffraction, exhibiting the drastic change at T_N. Especially, (1 0 2) diffraction undergoes far larger change. It is easily shown that the ($h\ k\ l$) diffraction with $h+k+l = 2n+1$ comes from only oxygen atoms in rutile structure. Therefore, this result shows that the magnetic transition is accompanied by the large displacement of oxygen atoms.

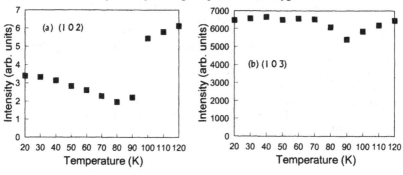

Fig. 4. The temperature dependence of the intensity of (a) (1 0 2) and (b) (1 0 3) Bragg diffractions in β-MnO$_2$.

CONCLUSION

Transport and magnetic properties have been measured on rutile-type manganese dioxide β-MnO_2. The results on resistivity, TEP and Hall effect show metallic conduction with n-type carriers. The origin of the carriers is considered to be e_g electrons doped by oxygen defects, *etc.* Therefore, in this system, conduction e_g electrons and localized magnetic moments in the t_{2g} orbitals coexist like in perovskite manganates. The mobility is extremely small although the effective mass is not so enhanced. This situation is similar to perovskite manganates and strong scattering via Hund coupling between e_g conduction electrons and t_{2g} magnetic moments are responsible for it. Below T_N (~92 K), a transition temperature into a helical magnetic state, the resistivity drastically decreases while the carrier number remains almost constant. This is most likely due to the disappearance of incoherent magnetic scattering below T_N. The magnetic susceptibility and MR becomes anisotropic below T_N, reflecting the formation of a magnetic helix. The negative MR with the magnetic field along the c-axis and the positive MR with the field perpendicular to the c-axis are consistent with the proper-type helix model. X ray diffraction study using SR revealed magnetic scatterings from a helical state. The pitch of the magnetic helix was obtained in high resolution, which shows that the helix is incommensurate to the lattice and the period is slightly temperature-dependent. The intensity of some Bragg peaks drastically changes at T_N. This observation clearly shows that the magnetic transition is accompanied by a large displacement of oxygen atoms.

ACKNOWLEDGMENTS

The authors would like to thank Professor Naoichi Yamamoto for useful advice on the synthetic technique, Mr. Masaya Enomoto for the help in the resistivity measurement at low temperature. The authors also acknowledge Professor Hidekazu Tanaka for useful discussion on the magnetism. This work is partly supported by a Grant-in-Aid for Scientific Research Nos. 10740318 and 10149101 from the Ministry of Education, Science, Sports and Culture, Japanese Government.

REFERENCES

1. D.B. Rogers, R.D. Shannon, A.W. Sleight, and J.L.Gillson, Inorg. Chem. **8**, p. 841 (1969).
2. H. Sato, T. Enoki, M. Isobe, and Y. Ueda, Phys, Rev. B., *in press* (2000).
3. A. Yoshimori, J. Phys. Soc. Jpn. **14**, p. 807 (1959).
4. H. Bizette, J. Phys. Radium **12**, p. 161 (1951).
5. A. Urushibara, Y. Moritomo, T. Arima, A. Asamitsu, G. Kido and T. Tokura, Phys. Rev. B **51**, p. 14103 (1995).
6. T. Kasuya, Prog. Theor. Phys. **16**, p. 45 (1956).
7. T. Nagamiya, K. Nagata, and Y. Kitano, Prog. Theor. Phys. **27**, 1253 (1962).
8. Y. Kitano and T. Nagamiya, Prog. Theor. Phys. **31**, 1 (1964).
9. S.W. Lovesey and S.P. Collins, *X-Ray Scattering and Absorption by Magnetic Materials*, Oxford University Press, New York, 1996.

PROPERTIES OF THE FERRIMAGNETIC DOUBLE-PEROVSKITES A_2FeReO_6 (A=Ba AND Ca)

W. PRELLIER [a,b,*], V. SMOLYANINOVA [a], A. BISWAS [a], C. GALLEY [a], R.L. GREENE [a], K. RAMESHA [c] AND J. GOPALAKRISHNAN [c]

[a] Center for Superconductivity Research and Department of Physics, University of Maryland, College Park, MD 20472, USA.

[b] Current address: Laboratoire CRISMAT-ISMRA, CNRS UMR 6508, 6 Bd. du Maréchal Juin, 14050 Caen, France.

[c] Solid State and Structural Chemistry Unit, Indian Institute of Science, Bangalore 560012, India.

[*] Author to whom correspondence should be addressed : prellier@ismra.fr.

ABSTRACT

We have synthesized ceramics of A_2FeReO_6 double-perovskites A_2FeReO_6 (A=Ba, Ca). Structural characterizations indicate a cubic structure with a=8.0854(1) Å for Ba_2FeReO_6 and a distorted monoclinic symmetry with a=5.396(1) Å, b=5.522(1) Å, c=7.688(2) Å and β=90.4° for Ca_2FeReO_6. The barium compound is metallic from 5K to 385K, i.e. no metal-insulator transition has been seen up to 385K, and the calcium compound is semiconducting from 5K to 385K. Magnetization measurements show a ferrimagnetic behavior for both materials, with T_c=315 K for Ba_2FeReO_6 and above 385K for Ca_2FeReO_6. At 5K we observed, only for Ba_2FeReO_6, a negative magnetoresistance of 10% in a magnetic field of 5T. Electrical, magnetic and thermal properties are discussed and compared to those of the analogous compounds $Sr_2Fe(Mo,Re)O_6$ recently studied.

INTRODUCTION

Perovskite manganites exhibiting a variety of exotic electronic properties [1] that include a spectacular decrease of electrical resistance in a magnetic field [2], the so-called colossal magnetoresistance (CMR) [3], have attracted wide attention in recent years. A significant feature of the electronic structure of the ferromagnetic CMR manganites, revealed by recent experiments [4-6] and theory [7], is that the charge carriers are almost completely spin-polarized at the Fermi level E_F. These materials are half-metallic ferromagnets, where the majority spin states near E_F are delocalized and the minority spin channel is effectively localized. Since half-metallic ferromagnetism and magnetoresistance (MR), especially at low fields, seem to be intimately related to each other [8] - the latter arising from the former- there is an intense search for half-metallic magnets which could be candidate materials for the realization of MR applications. While several double-perovskite oxides of the kind $A_2BB'O_6$ (A being an alkaline earth or rare earth ion and B, B' being d-transition metal ions) have been theoretically predicted to be half-metallic antiferromagnets [9], one such material, Sr_2FeMoO_6 [10], has recently [11-13] been shown to be a half-metallic ferrimagnet exhibiting a significant tunneling-type magnetoresistance at room temperature. More recently, it has also been shown that it is possible to grow by Pulsed Laser Deposition, thin films of Sr_2FeMoO_6 [13-15]. Ba_2FeReO_6 [16] and Ca_2FeReO_6 are ordered double-perovskites whose structure and properties are quite similar to those of Sr_2FeReO_6. In both materials a valence degeneracy between the B-site cation occurs, giving rise to the observed metallic and magnetic properties ; in the Sr_2FeReO_6 case, the valence-degeneracy is between

23

$Fe^{3+}+Mo^{5+} \rightarrow Fe^{2+}+Mo^{6+}$ states, while in the Ba_2FeReO_6 case, the degenerate oxidation states are $Fe^{3+}+Re^{5+} \rightarrow Fe^{2+}+Re^{6+}$. A major difference between the Mo and Re oxides however, is that Mo^{5+} is $4d^1$, whereas Re^{5+} is $5d^2$. This would mean that the conducting charge carrier density of the Re compound would be twice as much as in the Mo compound, while the localized spins centered on Fe remain the same in both oxides. We believe that this difference could have an influence on the magnetotransport behavior, especially in view of the recent report [17] that the bulk low field MR in ferromagnetic metals is mainly determined by the charge carrier density. In view of the foregoing, we have investigated the magnetic and transport properties of A_2FeReO_6 (A=Ba and Ca).

EXPERIMENTAL

Both polycrystalline samples of Ba_2FeReO_6 and Ca_2FeReO_6 were synthesized in the same way by standard solid state technique. First, a precursor oxide of the composition $A_2ReO_{5.5}$ (A=Ba or Ca) was prepared by reacting stoichiometric amounts of ACO_3 and Re_2O_7 at 1000 °C in air for 2 h. Second, this resultant oxide was mixed with required quantities of Fe_2O_3 and Fe powder to obtain the desired composition A_2FeReO_6. Finally, pellets of this mixture were heated in an evacuated sealed silica tube at 910°C during 4 days, followed by another treatment at 960°C during the same time, with intermediate grinding. Since, these two compounds have been prepared in an identical way, their properties can be compared. X-Ray powder diffraction (XRD) patterns were taken using a conventional diffractometer with Cu Kα radiation (λ=1.5406Å). DIAMOND program was used to calculate the lattice parameters, bond lengths and angles. DC resistivity (ρ) was measured by a standard four-probe method, on bars with the approximate dimensions of $(1x3x8)$ mm^3. Magnetization measurements (M) were registered with a SQUID (MPMS Quantum Design) magnetometer. These DC measurements were carried out with increasing temperature after the sample was zero field cooled. The specific heat was measured by relaxation calorimetry in the temperature range 2-16 K. Details of this apparatus have been previously described elsewhere [18].

RESULTS AND DISCUSSION

The XRD patterns of final compounds are shown in Fig.1. Ba_2FeReO_6 (Fig.1a) is indexed on the basis of a cubic cell (*Fm3m*) with a=8.054(1)Å, whereas Ca_2FeReO_6 (Fig.1b) is distorted to a monoclinic symmetry (*P21/n*) with a=5.396(2)Å, b=5.522(2)Å, c=7.688(1)Å and β=90.4°.

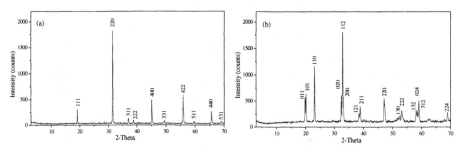

Fig.1 : XRD pattern of A_2FeReO_6. (a) A=Ba and (b) : A= Ca

These results indicate the formation of the double-perovskite for both compounds, as reported earlier [16] and consistent with recent results [11-12]. One also notes that no impurity phase can be detected in the XRD indicating a clean single phase in each case.

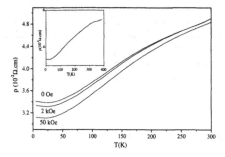 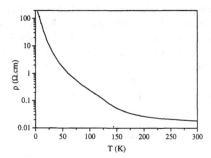

Fig.2a : Resistivity vs T for Ba₂FeReO₆.

Fig.2b : Resistivity vs T for Ca₂FeReO₆

The temperature (T) dependence of ρ is shown for Ba_2FeReO_6 in Fig.2a at various magnetic fields of 0, 0.2 and 5T. The resistivity gradually decreases when the temperature decreases suggesting a metallic behavior below 300K. The zero-field resistivity was also measured up to 385K (inset of Fig.2a) but no change in the metallic behavior was observed in this region. In contrast, Ca_2FeReO_6 shows a semiconducting behavior from 5K to 300K (Fig.2b). While Ba_2FeReO_6 exhibits magnetoresistance at 5K, the resistivity of Ca_2FeReO_6 remains unchanged, even under an applied magnetic field of 8 T. We find, that the latter resistivity does not fit with an activation energy law ($\rho(T) \propto \exp(-E_a/kT)$). As often found in the manganites [19], the resistivity at room temperature is large (50mΩ.cm for Ba_2FeReO_6 and 20mΩ.cm for Ca_2FeReO_6), well above the Mott limit, but perhaps reflecting intergrain resistance in these polycrystalline samples. The resistivity of the Ca_2FeReO_6 compound is ~3.10³ higher than the resistivity of the Ba_2FeReO_6 at low temperature. This difference of magnitude can hardly be explained by the scattering on the grain boundaries since the samples are prepared in the equivalent way (i.e. they have the same density). However, a more detailed understanding of the transport of these materials will require single crystals or oriented thin films. To investigate the MR of the Ba_2FeReO_6, we present the MR at different temperatures in Fig.3. As shown in Fig.2a,

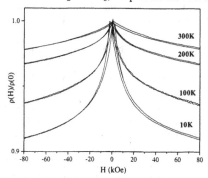

Fig.3 : MR of Ba₂FeReO₆.

the Ba_2FeReO_6 compound exhibits negative MR, with MR defined as MR(T,H)=[R(H)-R(0)]/R(H). The MR is equal to 3% at 10K in a 2000 Oe field. This MR under low field at 10K is smaller than that found in Sr_2FeMoO_6 (10%) [11]. However, the zero field $\rho(T)$ behavior of Sr_2FeMoO_6 is rather different at low temperature than Ba_2FeReO_6 - in Sr_2FeMoO_6 it tends to increase slightly at 10K. This behavior might be due to the preparation procedure of the sample which dramatically affects the $\rho(T)$ [11], but most probably this is a result of the difference in the structure of these compounds. Indeed, in the series Ba_2FeReO_6, Sr_2FeReO_6 [20] and

Ca$_2$FeReO$_6$, the crystal symmetry decreases from cubic to tetragonal to orthorhombic (or distorted monoclinic). This is clearly a manifestation of the decreasing Re-O-Fe bond angle from 180 degrees. Accordingly, the conduction band width would be expected to decrease as we go from Ba to Sr to Ca. Thus, it is not surprising that Ba$_2$FeReO$_6$ is metallic and Ca$_2$FeReO$_6$ is not. A similar conclusion occurs when comparing Ba$_2$FeMoO$_6$ and Sr$_2$FeMoO$_6$ [11,21]; Ba$_2$FeMoO$_6$ is metallic (and cubic) and Sr$_2$FeMoO$_6$, whose structure is tetragonal, has a resistivity which increases when T decreases [11]. The MR for Ba$_2$FeReO$_6$ strongly increases at low field with a slower increase at higher field. This effect occurs mainly at low temperature, since at room temperature the MR is very small. The features are characteristic of intergrain magnetoresistance [12-13]. In fact, Yin et al. Verified recently [13] that spin dependent electrons transfer thought grain boundary [12-13,21] is responsible for the low field MR. At low temperature (Fig.3), a small hysteretic behavior also appears but thus is not of relevance to the issues discussed in this paper.

For Ba$_2$FeReO$_6$, the low temperature magnetization for ferrimagnetic value is 3.04μ_B, which is close to the expected value based on Fe^{3+} and Re^{5+} moments (3μ_B) for a ferrimagnetic state. For the Ca$_2$FeReO$_6$ compound, the low temperature magnetization value for a ferrimagnetic state is calculated to be 3μ_B which is a little higher than the experimental value (2.24μ_B) but in agreement with a previous report [22]. This is probably due to the fact that the full saturation magnetization was not reached even with a 5T magnetic field. The temperature dependence of M under a magnetic field of 10Oe is shown in Fig.4 for Ba$_2$FeReO$_6$. Like several other double-perovskite compounds [201,23], A$_2$FeReO$_6$ (A=Ba and Ca) exhibit ferrimagnetic behavior due to an antiferromagnetic superexchange interaction between spins of Fe^{3+} (S=5/2) and Re^{5+} (S=1). Around 315K, there is a sharp change in the magnetization indicating that a transition from a paramagnetic to a ferrimagnetic state occurs. Thus, the ferrimagnetic transition temperature (T$_c$) of Ba$_2$FeReO$_6$ was estimated to be 315±5 K. The Ca$_2$FeReO$_6$ compound exhibits a higher T$_c$, above 385K (not determined). A detailed study using Arrot plots will be necessary to determine precisely the Curie temperature of theses compounds. However, the exact value is not the issue of the paper. The ferrimagnetic transition temperature of Ba$_2$FeReO$_6$ is lower than the reported value of T$_c$ 410K for Sr$_2$FeReO$_6$ [20,21] but close to the value of Ba$_2$FeMoO$_6$ (T$_c$=340K) [21], whereas the T$_c$ of Ca$_2$FeReO$_6$ seems to be higher (> 385K). This means that these differences are likely caused by the larger ionic radius of Ba^{2+} (1.47Å versus 1.31Å for Sr^{2+} and 1.18Å for Ca^{2+} [24] which leads to a larger Re-O-Fe (or Mo-O-Fe) bond length and therefore a smaller exchange and lower ferrimagnetic transition temperature.

The specific heat data for Ba$_2$FeReO$_6$ and Ca$_2$FeReO$_6$ are shown in Fig.5. In our analysis of the low temperature specific heat we include lattice (C$_{latt}$), metallic (C$_{el}$) and hyperfine (C$_{hyp}$) contributions. When the temperature decreases below K, the specific heat increases due to the hyperfine contribution C$_{hyp}$=A/T^2. Since the nuclear spin of Fe^{56} is zero, the hyperfine contribution in our samples arises from the Re^{186} nuclear spin I=1. The experimental values of A are found to be 135±4 mJ-Kmole and 180±5 mJ-K/mole for Ba$_2$FeReO$_6$ and Ca$_2$FeReO$_6$ respectively. In addition, Ba$_2$FeReO$_6$ has the expected metallic contribution, γT, with γ =23.1±0.2

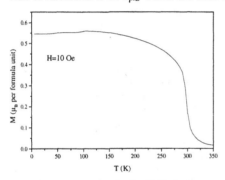

Fig.4 : M vs T for Ba$_2$FeReO$_6$.

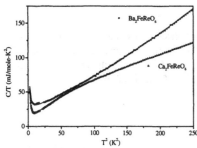

Fig.5 : Specific heat for Ba_2FeReO_6 and Ca_2FeReO_6 plotted as C/T vs T^2.

mJ/mole-K^2. Using $\gamma=\pi k_B N(E_F)/3$, we find the density of states at the Fermi energy $N(E_F)$ to be $5.9 10^{24}$ $eV^{-1}mole^{-1}$. This value is larger than the $N(E_F)$ obtained from the band structure calculation for Sr_2FeMoO_6 ($1.2 10^{24}$ $eV^{-1}mole^{-1}$) [11], probably because Re^{5+} has two electrons on the d orbital $5d^2$, while Mo5+ has only one $4d^1$. To achieve a good fit for our Ba_2FeReO_6 specific heat data two lattice terms are required: $C_{latt}=\beta_3 T^3+=\beta_5 T^5$, where $\beta_3=0.438\pm0.003$ mJ/mole-K^4 and $\beta_5=6.1 \times 10^{-4}$ mJ/mole-K^6. Since $\Theta_D=(12\pi^4 pR/5\beta_3)^{1/3}$, where p=10 is the number of atoms per formula unit, we find the Debye temperature Θ_D to be 354 K, which is similar to Θ_D of other perovskites [18]. The best fit for Ca_2FeReO_6 does not include γT and $\beta_5 T^5$ terms, but requires an additional term $C'\approx\alpha T^a$, where a is close to 2. The absence of the charge carrier term γT is consistent with the insulating resistivity of Ca_2FeReO_6. The $C'\approx\alpha T^a$ term, or more precisely $C'=C'(\Delta,B,T)$, which is the specific heat of excitations with a dispersion relation $\varepsilon=\Delta+B\kappa^2$ of non-magnetic origin, was also found in the charge-ordered perovskite manganites [25]. In the case of insulating Ca_2FeReO_6 a charge of 3+ on the Fe site and a charge of 5+ on the Re site create a situation similar to the Mn^{3+}-Mn^{4+} charge ordering in manganites. The fit to our Ca_2FeReO_6 specific heat data gives $\Delta=7K$ and $B=22meV-\text{Å}^2$ similar in magnitude to the parameters found in the manganites. We believe that the reasons for the presence of the $C'(\Delta,B,T)$ term in Ca_2FeReO_6 are similar to those in $La_{0.5}Ca_{0.5}MnO_3$ [25]. The lattice contribution $\beta_3 T^3$ ($\beta_3=0.25\pm0.006$ mJ/mole-K^4 in Ca_2FeReO_6 is smaller than in Ba_2FeReO_6. This is surprising but perhaps can be understood by the effect of a different crystal structure (cubic for Ba_2FeReO_6 and distorted monoclinic for Ca_2FeReO_6) and smaller Ca mass on the lattice vibrations. A ferrimagnet has a magnetic contribution to the specific heat, $C_{mag}=\Delta T^{3/2}$, which is similar to that of a ferromagnetic one. However, this contribution, can not be resolved from the specific heat data alone, since our data can be fit well without the magnetic term. Since Ca_2FeReO_6 and Ba_2FeReO_6 have a high ferrimagnetic transition temperature, and hence, strong exchange interaction (J), the magnetic term should be small ($\delta\propto J^{-3/2}$).

CONCLUSION

In summary, we have investigated the transport, magnetic and thermal properties of two polycrystalline double-perovskites Ba_2FeReO_6 and Ca_2FeReO_6. Ba_2FeReO_6 displays a metallic behavior below 385 K whereas Ca_2FeReO_6 is insulating below this temperature. The Ba_2FeReO_6 compound exhibits a negative MR at 10K, smaller than the analogous compound Sr_2FeMoO_6. Magnetic measurements indicate a ferrimagnetic behavior, with a Tc=315±5 K for Ba_2FeReO_6 and above 385K for Ba_2FeReO_6. . The specific heat of Ba_2FeReO_6 gives a low temperature metallic contribution with an electron density of states at the Fermi level close to the band structure value. Insulating Ca_2FeReO_6 has no metallic term in the specific heat but rather an extra contribution most likely caused by charge ordering of Fe, Re These data have been explained and compared with those of the analogous compounds $Sr_2Fe(Re,Mo)O_6$.

ACKNOWLEDGMENTS

The authors are grateful to Prof. S. Bhagat for helpful discussions. Partial support of NSF-MRSEC at University of Maryland is acknowledged. The work at Bangalore was supported by the Department of Science and Technology, Government of India. K.R. thanks the Council of Scientific and Industrial Research, New Delhi, for the award of a fellowship.

REFERENCES

[1] A.J. Millis, Nature **392**, 147 (1998).

[2] R. Von Helmot, J. Wecker, B. Holzapfel, L. Schultz and K. Samwer, Phys. Rev. Lett. **71**, 2331 (1993).

[3] S. Jin, T.H. Tiefel, M. McCormack, R.A. Fastnacht, R. Ramesh and L.H. Chen, Science **264**, 413 (1994).

[4] J.-H. Park, E. Vescovo, H.-J. Kim, C. Kwon, R. Ramesh and T. Venkatesan, Nature **392**, 794 (1998).

[5] R.J. Soulen Jr., J.M. Byers, M.S. Osofsky, B. Nadgorny, T. Ambrose, S.F. Cheng, P.R. Broussard, C.T. Tanaka, J. Nowak, J.S. Moodera, A. Barry and J.M.D. Coey, Science **282**, 85 (1998).

[6] J.Y.T. Wei, N.-C Yeh and R.P. Velasquez, Phys. Lett. B **79**, 5150 (1997).

[7] W.E. Pickett and D.J. Singh, Phys. Rev. B **53**, 1146 (1996).

[8] H.Y. Hwang and S.-W. Cheong, Science **278**, 1607 (1997).

[9] W.E. Pickett, Phys. Rev. B **57**, 10613 (1998).

[10] T. Nakagawa, J. Phys. Soc. Japan **24**, 806 (1968).

[11] K.I. Kobayashi, T. Kimura, H. Sawada, K. Terakura and Y. Tokura, Nature **395**, 677 (1998).

[12] T.H. Kim, App. Phys. Lett. 74, 1731 (1999).

[13] H.Q. Yin, J.S. Zhou, R. Dass, J.T. McDevitt and J.B. Goodenough, App. Phys. Lett. **75**, 2812 (1999).

[14] H. Asano, S.B. Ogale, J. Garrison, A. Orozco, E. Li, V. Smolyaninova, C. Galley, M. Downes, M. Rajeswari, R. Ramesh and T. Venkatesan, App. Phys. Lett. **74**, 3696 (1999).

[15] T. Manako, M. Izumi, Y. Konishi, K.I. Kobayashi, M. Kawasaki and Y. Tokura, App. Phys. Lett. **74**, 2215 (1999).

[16] A.W. Sleight and J.F. Weiher, J. Phys. Chem. Solids 33, 679 (1972).

[17] P. Majumdar and P.B. Littlewood, Nature **395**, 479 (1998).

[18] J. J. Hamilton, E.L. Keatley, H.L. Ju, A.K. Raychaudhuri, V.N. Smolyaninova and R.L. Greene, Phys. Rev. B **54**, 14926 (1996).

[19] A. Urushibara, Y. Moritomo, T. Arima, A. Asamitsu, G. Kido, Y. Tokura, Phys. Rev. B **51**, 14103 (1995).

[20] M. Abe, T. Nakagawa and S. Momura, J. Phys. Soc. Jpn. **35**, 1360 (1973).

[21] A. Maignan, B. Raveau, C. Martin and M. Hervieu, J. Sol. State Chem. **144**, 224 (1999).

[22] A.W. Sleight, J. Longo and R. Ward, Inorg. Chem. **1**, 245 (1962).

[23] J. Longo and R. Ward, J. Amer. Chem. Soc. **83**, 2816 (1961).

[24] R.D Shannon, Acta. Cryst. A**32**, 751 (1976).

[25] V.N. Smolyaninova, K. Ghosh and R.L. Greene, Phys. Rev. B **58**, R14725 (1998).

NEW MAGNETIC AND FERROELECTRIC CUBIC PHASE OF THIN-FILM Fe-DOPED BaTiO$_3$

R. MAIER,[1] J. L. COHN,[1] J. J. NEUMEIER,[2] AND L. A. BENDERSKY[3]

[1] Department of Physics, University of Miami, Coral Gables, FL 33124

[2] Department of Physics, Florida Atlantic University, Boca Raton, FL 33431

[3] Materials Science and Engineering Laboratory, National Institute of Standards and Technology, Gaithersburg, MD 20899

ABSTRACT

The properties of a new cubic perovskite phase of thin-film BaFe$_x$Ti$_{1-x}$O$_3$ ($0.5 \leq x \leq 0.75$) are reported. This material is novel because the corresponding bulk compounds have hexagonal structure for comparable x. The films, grown by pulsed laser deposition on MgO and SrTiO$_3$ substrates, are magnetic (ferro- or ferri-, with T$_C$ > 500 °C) and ferroelectric (T$_C$ ~ 200-300 °C).

INTRODUCTION

Compounds in which magnetic and electric polarization exist simultaneously [1] are quite rare in nature, particularly at room temperature and above. Such materials offer the prospect of devices in which, e.g., the magneto- or electrostriction may be used to manipulate the polarization or transition temperature.

The ferroelectric barium titanate (BaTiO$_3$) is currently employed in positive-temperature-coefficient-resistance (PTCR) thermistors and has been widely explored for its potential use in applications such as memories, electro-optic switches and modulators. Iron substitution for titanium in bulk BaTiO$_3$ stabilizes the hexagonal structure [2] rather than the cubic/tetragonal perovskite. Solid solutions of bulk hexagonal BaTi$_{1-x}$Fe$_x$O$_3$ with $0.06 \leq x \leq 0.84$ have been reported [3]. Recently we reported structural characterization of thin-film BaTi$_{1-x}$Fe$_x$O$_3$ (x~0.5) grown by pulsed laser deposition and having the cubic/tetragonal perovskite structure [4]. Here we discuss preliminary measurements of physical properties, establishing the coexistence of magnetic and electric order well above room temperature.

EXPERIMENT

Thin films were grown by pulsed laser deposition, using a 248 nm KrF excimer laser (Lambda Physik, Compex 205), from targets of BaFe$_{0.5}$Ti$_{0.5}$O$_3$, BaFe$_{0.75}$Ti$_{0.25}$O$_3$, and Ba$_4$Fe$_4$Ti$_3$O$_{16}$ (BFTO-E), prepared by standard solid state reaction. The energy density at the target was ~1.5-2.0 J/cm^2, pulse repetition rate 10 Hz, and target-substrate distance 4 cm. The films were deposited on MgO substrates of (100), (110), and (111) orientation, as well as SrTiO$_3$ (100), with substrate temperature 880 °C and oxygen pressure 100 mTorr. Following the depositions, the chamber was filled to 500 Torr oxygen, held at 500 °C for 30 min., and subsequently cooled to room temperature. X-ray diffraction (XRD) was performed using a Philips X'Pert diffractometer with parallel-beam optics. Magnetization was measured with a Quantum Design SQUID magnetometer. Capacitance measurements (LCR meter, HP 4363B) were performed on tri-layer structures composed of films deposited on metallic, Nb-doped SrTiO$_3$ (100) substrates and vapor-deposited gold overlayers. The excitation voltage was 20 mV at 20 kHz. Two-probe dc electrical resistance was investigated by measuring current (Keithley, 6512 electrometer) at various applied dc voltages (Keithley, 230 voltage source).

29

RESULTS

All films have a similar microstructure; a detailed investigation of the films grown from the BFTO-E target on MgO is presented elsewhere [4]. A pseudocubic perovskite phase was observed for all films, with a=4.03Å-4.07Å. Figure 1 shows θ-2θ XRD scans with scattering vector normal to the substrate for films grown from the BFTO-E target on the three orientations of MgO. The appearance of only (*l*00), (*ll*0), and (*lll*) film reflections for each substrate orientation indicates epitaxial growth. XRD pole figures confirm cube-on-cube growth for all substrate orientations; results for MgO (100) are shown in Fig. 2. Asymmetric rocking curve scans of the (303) film reflections were employed to determine the c/a ratio of the tetragonal cell [5]; a weak tetragonality is observed, with the in-plane lattice parameter ~ 0.3% larger than that out-of-plane at room temperature in the as-prepared films. This is somewhat smaller than bulk BaTiO$_3$ (c/a=1.010). Misfit strain between the film and substrates results in interfacial and bulk dislocations and stacking faults evidenced in transmission electron microscopy [4]. {111} twins, typical of bulk and thin-film BaTiO$_3$ [6], were also observed.

The ferroelectric character of the films is somewhat difficult to discern from the capacitance-electric field curves (T=300 K),

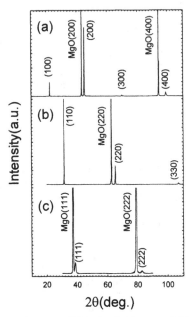

Fig. 1 X-ray diffraction θ-2θ scans of film grown from the BFTO-E target on Mg substrates of orientation (a) (100) (b) (110) and (c) (111).

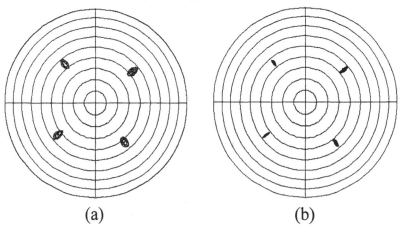

Fig. 2 (a) {110} pole of a film grown from the BFTO-E target on MgO (100) (b) {220} pole of the substrate.

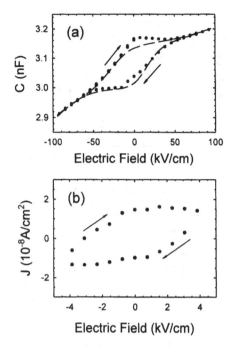

Fig. 3 (a) C-E curve and (b) low-field J-E curve for a Au/ $BaFe_{0.75}Ti_{0.25}O_3$/Nb:SrTiO$_3$ capacitor.

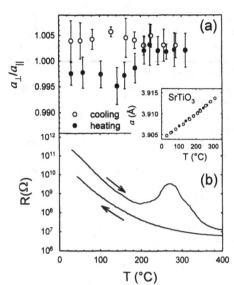

Fig. 4 (a) ratio of out-of-plane and in-plane lattice constants *vs* temperature for a $BaFe_{0.5}Ti_{0.5}O_3$ film grown on SrTiO$_3$ (100). Inset: lattice constant of the substrate. (b) in-plane dc electrical resistance of a film grown from the same target on MgO (100).

as shown in Fig. 3 (a) for an as-prepared film grown from the $BaFe_{0.75}Ti_{0.25}O_3$ target. The capacitance is dominated by losses typical of trap activation, which yield a hysteresis loop (approximated by the dashed curve) that is traversed in a clockwise sense. Such a behavior is not surprising given that the formation of Fe^{3+} and Ba vacancies are associated with oxygen vacancies which may serve as traps, particularly at grain boundaries. The difference between the measured curve and the lossy curve appears to describe a more typical ferroelectric butterfly loop. From these observations we estimate a coercive field, $E_C \approx$ 5-7 kV/cm. The current-voltage characteristic [Fig. 3 (b)] is also hysteretic at low fields with a relatively small leakage current density ~ 10 nA/cm^2 at E=4 kV/cm.

Temperature dependent XRD studies of a film grown from the $BaFe_{0.5}Ti_{0.5}O_3$ target on SrTiO$_3$ (100) indicate a structural transition at ~220 °C [Fig. 4 (a)] and this presumably marks the ferroelectric transition, T_C. Upon heating in air through the transition temperature, the ratio of the out-of-plane lattice parameter to that in-plane, $a_{\perp}/a_{\parallel}$, increases from 0.997 at room-temperature to 1.002, i.e. the tetragonal c axis, initially in the film plane at room temperature in the as-prepared films, reorients normal to the film plane. The high-temperature phase is slightly tetragonal (c/a~1.002) rather than cubic, presumably a consequence of the stress [7] imposed by the SrTiO$_3$ substrate (a=3.910 Å). Upon cooling through T_C the c axis remains normal to the film plane.

This thermal hysteresis in c/a is evident in the in-plane dc electrical resistance [Fig. 4 (b)] of a film grown from the same target on MgO (100) [8]. Upon heating, an increase in resistance, typical of the PTCR effect in bulk BaTiO$_3$ polycrystals [9], is associated with the structural and ferroelectric transition. This phenomenon is understood to be a consequence of a sharp increase in the potential barrier at grain-boundary Schottky contacts; as T increases above T$_C$, the partial compensation of space charge by the ferroelectric depolarizing fields vanishes abruptly. The space charge is determined by the density of acceptor states, which is controlled by grain-boundary impurities and oxygen vacancies. It is known that oxygen vacancies can degrade the PTCR effect. It

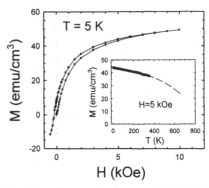

Fig. 5 T=5 K magnetization vs applied field fo a film grown from BFTO-E on SrTiO$_3$ (100) The inset shows M(T, 5 kOe) for the same film The dashed line is a guide.

seems likely that both the reorientation of the c-axis normal to the film plane and possibly oxygen loss at grain boundaries upon heating in air contribute to the absence of PTCR in the in-plane electrical resistance upon cooling. We note that both a-axis and c-axis oriented BaTiO$_3$ films have been grown by PLD on MgO (100) [10] and the reason for the difference in morphology remains unclear. Further investigations are required to determine if the c/a reversal observed here is related to the Fe substitution.

The ferroelectric transition temperatures, inferred from the midpoint temperature of the PTCR, fall in the range 200-300 °C for all films investigated. This is substantially higher than T$_C$=120 °C of bulk and thin-film BaTiO$_3$ grown epitaxially on MgO and SrTiO$_3$. Applied pressure reduces T$_C$ in bulk BaTiO$_3$ [11] and it is plausible that the effective negative pressure associated with the lattice expansion of the present films in comparison to bulk BaTiO$_3$ could account for the higher T$_C$. Thus we tentatively attribute the increase of T$_C$ by ~100 °C to the lattice expansion associated with Fe substitution for Ti.

Magnetization vs field and temperature are shown in Fig. 5 for a film grown from the BFTO-E target on SrTiO$_3$ (100). Extrapolation of M(T) to beyond the measurement range (T $\leq$ 350 K) suggests that the Curie temperature of this material exceeds 500 °C. The T=5 K saturation magnetization corresponds to ~ 0.3 μ_B/f.u. We may compare this value to the value computed for isolated magnetic moments, m=gμ_BS [where g is the Landé g-factor (taken to be 2) and S the ion spin]. Assuming Fe^{3+} (Fe^{4+}) valence states with S=5/2 (S=2) and using the ratio Fe:Ti=4:3 from the target stoichiometry, yields a computed moment of 2.86 (2.29) μ_B/f.u. The substantially smaller value observed for the films suggests that this material is ferrimagnetic rather than ferromagnetic. The coercive field, H$_C$ = 150-200 Oe.

CONCLUSIONS

We have reported on the physical properties of epitaxial, pseudocubic thin-film BaFe$_x$Ti$_{1-x}$O$_3$ with high Fe concentration, a metastable material that is not found in bulk. The material is both magnetic (likely ferrimagnetic) and ferroelectric well above room temperature. The crystallographic orientation of the ferroelectric tetragonal phase can be manipulated by annealing, and further investigations are underway to determine the origin of this phenomenon.

ACKNOWLEDGEMENTS

The work at the University of Miami was supported by NSF Grant DMR-9504213.

REFERENCES

1. G. A. Smolenskii and I. Chupis, Sov. Phys. Usp. **25**, 475 (1982).
2. J. Akimoto, Y. Gotoh, and Y. Osawa, Acta Crystallogr. **C50**, 160 (1994); H. D. Megaw, Proc. Phys. Soc. London **58**, 133 (1946).
3. T. A. Vanderah, J. M. Loezos, and R. S. Roth, J. Sol. St. Chem., **121**, 38 (1996).
4. L.A. Bendersky, R. Maier, J. L. Cohn, and J. J. Neumeier, J. Mater. Res., in press.
5. S. Kim, S. Hishita, Y. M. Kang, and S. Baik, J. Appl. Phys. **78**, 5604 (1995).
6. K. P. Fahey, B. M. Clemens, and L. A. Wills, Appl. Phys. Lett. **67**, 2480 (1995); C. Jia, K. Urban, M. Mertin, S. Hoffmann, and R. Waser, Phil. Mag. A **77**, 923 (1998).
7. N. A. Pertsev, A. G. Zembilgotov, and A. K. Tagantsev, Phys. Rev. Lett. **80**, 1988 (1998); N. A. Pertsev, A. G. Zembilgotov, S. Hoffmann, R. Waser, and A. K. Tagantsev, J. Appl. Phys. **85**, 1698 (1999).
8. Shorting of the current through the more conductive $SrTiO_3$ substrates precludes measurement of the in-plane electrical resistance of films grown on $SrTiO_3$.
9. W. Heywang, Sol. St. Electron. **3**, 51 (1961); G. H. Jonker, *ibid.*, **7**, 895 (1964); G. V. Lewis, C. R. A. Catlow, and R. E. W. Casselton, J. Am. Ceram. Soc. **68**, 555 (1985); D. Y. Wang, *ibid.*, **73**, 669 (1990); A. B. Alles and V. L. Burdick, *ibid.*, **76**, 401 (1993); T. Miki, A. Fujimoto, and S. Jida, J. Appl. Phys. **83**, 1592 (1998).
10. M. G. Norton, K. P. B. Crackle, C. B. Carter, J. Am. Ceram. Soc. **75**, 1999 (1992); K. N. Nashimoto, D. K. Fork, and T. H. Geballe, Appl. Phys. Lett. **60**, 1199 (1992); J. Gong, D. H. Kim and H. S. Kwok, Appl. Phys. Lett. **67**, 1803 (1995).
11. G. A. Samara, Phys. Rev. **151**, 378 (1966).

POLARIZATION DYNAMICS IN $La_{5/3}Sr_{1/3}NiO_4$

N. Hakim*, Z. Zhai*, C. Kusko*, P.V.Parimi*, S-W. Cheong**, and S. Sridhar*†.

*Physics Department, Northeastern University, 360 Huntington Avenue,Boston,MA 02115

**Department of Physics and Astronomy, Rutgers University, Piscataway, NJ 08855.

ABSTRACT

Dynamic susceptibility measurements at microwave frequencies $(2 - 10GHz)$ are a sensitive probe of charge dynamcis in $La_{5/3}Sr_{1/3}NiO_4$. Below the charge ordering temperature of $240K$, a dielectric loss peak due to a relaxation mode with a large dielectric susceptibility is observed, and is associated with charge stripe formation. The dielectric response for $H_\omega||b$ ($E_\omega \perp b$) is well represented by $\bar{\varepsilon}(T) = \varepsilon_o/(1 - i\omega\tau(T))$, with $\varepsilon_o \sim 50$, and $\tau(T) = 2 \times 10^{-9}(sec)\exp(-T/37K)$. Parallel conductivity $\sigma(T)$ contributions dominate at higher temperatures and for $H_\omega||c$ ($E_w \perp c$). The dielectric loss peak observed indicates that the charge relaxation rates lie in the GHz frequency ranges.

INTRODUCTION

The carrier doping in Mott insulators such as $La_2MO_{4+\delta}$ ($M = Ni, Cu$) has attracted much attention bacause of the existence of the high T_c superconductivity in hole doped La_2CuO_4. These quasi-two dimensional electronic systems exhibit charge and spin correlations in which doped holes tend to undergo stripe like ordering on the domain walls of antiferromagnetic stripes[1, 2, 6]. The evidence of dynamic stripes (spin fluctuations) in superconducting $La_{2-x}Sr_xCuO_{4-\delta}$ and static stripe correlations in nonsuperconducting $La_{2-x}Sr_xNiO_4$ has generated compelling interest about the interplay of stripe dynamics and superconductivity.

A variety of measurements have revealed three successive transitions associated with quasi-two dimensional commensurate charge ordering (at $\sim 240K$) and spin stripe ordering ($\sim 190 - 160K$) [1, 10, 9, 8, 2] in $La_{5/3}Sr_{1/3}NiO_4$. The spin ordering at $190K$ is driven by charge ordering when the ordering between charge stripes takes place. The existence of stripe glass is proposed in the temperature regime, $240 - 190K$[2, 9, 8, 11], but the details of orientational order is missing probably due to the dominant effect of short stripes. A key issue that has arisen is the role of the measurement time-scale since stripes, and more generally magnetic and charge correlations, are now believed to have strong dynamical properties, and previous measurements have been principally carried out with probes having very different time scales such as Neutron scattering ($\sim 10^{-13}$ sec) and NMR ($\sim 10^{-7}$ sec).

In this paper, we present measurements of the dynamic (microwave) response of $La_{5/3}Sr_{1/3}NiO_4$ using a precision superconducting microwave cavity at $10GHz$, supplemented by measurements at $2GHz$ using a normal Cu resonator. Our measurements probe short time ($\sim 10^{-11}$ sec) or high frequency dynamics of charge correlations in this material. We find that charge ordering at $\sim 240K$ suppresses the dia-electromagnetic contribution caused by eddy currents due to the conductivity. Instead, charge ordering is accompanied by the onset of a dynamic dielectric susceptibility, which freezes out (quasi-statically at the finite measuring frequency) as the temperature is lowered due to rapid increase of the relaxation time. Our results can be succinctly summarized in terms of a T-dependent conductivity $\sigma(T)$ and a dielectric constant $\tilde{\varepsilon}(T) = \varepsilon_o/(1 - i\omega\tau(T))$. The charge relaxation time $\tau(T)$ increases exponentially with decreasing temperature T. A quantitative fit to the data is obtained with the form $\tau(T) = \tau_o \exp(-T/T_o)$.

The $La_{5/3}Sr_{1/3}NiO_4$ single crystals were prepared using a floating zone technique. Details of the crystal growth are given elsewhere[2]. The high quality of these crystals is well established by thorough characterization by several techniques[2].

EXPERIMENT

The principal measurements reported here are carried out using a superconducting microwave cavity. The superconducting cavity is made of Niobium,which is a superconductor below $T_c = 8.9K$. The dimension of the cylindrical cavity are: radius R=2.22 cm, and a length of L = 2.54cm. The TE_{011} mode resonates at $10GHz$. The sample supported on a sapphire rod is inserted and centered through a hole made at the bottom of the cavity. The entire cavity and assembly is vacuum tight, and in turn is placed in a bath of liquid $_4He$. To heat up the sample to higher temperature, the sapphire rod is thermally isolated from the cavity walls and a 50Ω heating coil is used to control the sample temperature from $4K - 300K$. The high quality factor $Q \sim 2 \times 10^8$ enables us to perform high precision microwave measurements. These experiments have been extensively utilized previously for measuring a variety of materials, including superconducting cuprate, manganate and borocarbide crystals [3, 4]. In all of the measurements, the sample (typically $2 \times 2 \times 1\ mm^3$) was placed at the center of the cavity where the H_ω is maximun and $E_\omega = 0$ fot the TE_{011} mode.

The copper split ring resonator has a cylindrical shape with a split along the side. The inner radius $R_{in} = 0.395cm$, the outer radius $R_{out} = 1.037cm$, and the split gap is of thickness = 0.07cm. These dimensions give a resonance frequency of $2GHz$, and the quality factor $Q \sim 2000$. The resonator is placed within a conducting cylinder to maintain highest Q factor. Heating and supporting the sample inside the ring is done in a way similar to that supperconducting cavity discussed above.

We define an electromagnetic susceptibility $\tilde{\zeta}_H$ which is obtained from the measured cavity resonance parameters by : $f(0) - f(T) + i\Delta f(T) = g[\zeta_H'(T) + i\zeta_H''(T)]$. Here $f(T)$ is the resonant frequency, $\Delta f(T)$ is the width of the resonance, and g is a sample geometric factor, assuming the sample is in the shape of a sphere. When the sample is placed in the center of the cavity and the TE_{011} mode is used, so that the sample is at a region of maximum microwave magnetic field, careful analysis of the cavity perturbation equations shows that the measured EM susceptibility $\tilde{\zeta}_H$ (we use the subscript H to denote that the sample is placed in an H field region), is related to the magnetic ($\tilde{\chi}_M$) and dielectric ($\tilde{\chi}_P$) susceptibilities in the following ways [5]:

$$
\begin{aligned}
\tilde{\zeta}_H &= \tilde{\chi}_M & (k_oa)^2\tilde{\chi}_E \ll \chi_M & \quad (1) \\
&= \frac{1}{10}(k_oa)^2\tilde{\chi}_P & (k_oa)^2\tilde{\chi}_P \gg \chi_M & \quad (2) \\
&= \frac{3}{2}\left(1 - \frac{3}{(ka)^2} + \frac{3\cot ka}{ka}\right) & \text{arbitrary } ka & \quad (3)
\end{aligned}
$$

where $k^2 = k_o^2 \left(\tilde{\varepsilon} + i\frac{\sigma}{\omega\varepsilon_o} \right)$, $\tilde{\varepsilon} = \varepsilon' + i\varepsilon'' = 1 + \chi'_P + i\chi''_P$ is the complex dielectric constant, a is the sample diameter. Note that $\tilde{\zeta}_H$ represents the effective susceptibility which can include eddy current (or conductivity) contributions in addition to dynamic dielectric and magnetic response. The experiment measures the magnetic susceptibility $\tilde{\chi}_M$ only if the dielectric and conductivity contributions are negligible. Since the sample size is typically $2mm$, and hence $k_0 a \sim 0.2$, the dielectric contribution dominates if $\chi_P/\chi_M > 250$. This condition appears to be met in most of the oxides which are even slightly doped and/or weakly conducting, and certainly at high temperature. Thus although the sample is placed in a magnetic field region, at these high frequencies we mostly measure the dielectric (polarization) and/or conductivity dynamics rather than the spin dynamics. Only in very high resistance insulators, such as Sr_2CuO_3 and $ZnCr_2O_4$, and possibly at very low $T < 20K$ in $La_{5/3}Sr_{1/3}NiO_4$, we are possibly measuring the magnetic susceptibility at these high frequencies. The results for $\tilde{\zeta}_H$ can be used to extract information regarding the dielectric permittivity $\tilde{\varepsilon}$ and the conductivity σ.

While the loss term $\zeta''(T)$ is measured absolutely, the technique yields changes $\delta\zeta'_H(T) = \zeta'_H(T) - \zeta'_H(3K)$ in susceptibility with very high precision. In the present measurements since $\zeta'_H(T \to 0) \to 0$, $\delta\zeta'_H(T) \sim \zeta'_H(T)$ does represent an absolute measure of $\zeta'_H(T)$ for most of the temperature range in this work. Comparison of absolute values of the present microwave susceptibility with dc magnetic susceptibility $\chi_M(f = 0, T)$ reveals that we are observing completely new phenomena at these frequencies. Our experiments also enable us to measure the anisotropy by varying the microwave magnetic field direction (H_ω) with respect to the crystal axes $(a, b$ or $c)$.

RESULTS

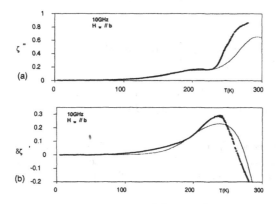

Figure 1: Microwave susceptibility $\zeta''_{Hb}(10GHz, T)(a)$ and $\delta\zeta'_{Hb}(10GHz, (T)(b)$ versus T for $La5/3Sr1/3NiO4$. Solid lines represent he susceptibility calculated from the dielectric and eddy current conributions. The dielectric loss peak at around 210K is visible in $\zeta''_{Hb}(10GHz, T)$ data and the model calculations.

The $10GHz$ susceptibility, $\zeta''_{H,b}(10GHz, T)$ for $H_\omega // b$ is shown in Fig. 1(a). A sharp drop

37

is seen from $300K$ which is arrested around $240K$ and followed by a peak in the absorption at around $210K$. Further decrease in T results in monotonic decrease of $\zeta''_{H,b}$. The features observed in $\zeta''_{H,b}$ are reflected in $\delta\zeta'_{H,b}$. As can be seen from Fig. 1(b), the high temperature conductivity response which is dia-electromagnetic, is dominant above $240K$, and at lower T the dielectric response takes over, resulting in a peak at $240K$.

We have also carried out measurements at $2GHz$ to investigate the frequency dependence and anisotropy of the microwave features. Overall the features observed at $10GHz$ are reproduced at this frequency as seen in Fig. 2, in the $\zeta''_{H,b}(2GHz,T)$, $H_w//b$ data. However, the absorption peak which appeared in the $10GHz$ data around $210K$ has moved down to $150K$, indicating a clear frequency dependance to this process.

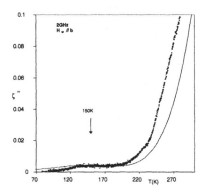

Figure 2: The microwave susceptibility, ζ''_{Hc} at 2 GHz as a function of temperature. Here the dielectric peak loss moved down to 150K at 2 GHz.

A quantitative fit of the data to Eq(3) is obtained using two contributions which can be represented as $\sigma(T) - i\omega\tilde{\varepsilon}(\omega, T)$, which are:

1. a complex T-dependent dielectric function $\tilde{\varepsilon} = \varepsilon' + i\varepsilon'' = \varepsilon_o/(1 - i\omega\tau(T))$. Best fits to the data below $240K$ give $\varepsilon_o = 50$, $\tau(T) = \tau_o \exp(-T/T_{\tau o})$, with $\tau_o = 2 \times 10^{-9}$ sec, and $T_{\tau o} = 40K$.

2. a T dependent conductivity $\sigma(T) = \sigma_o \exp(-T_{\sigma o}/T)$, with $\sigma_o = 5 \times 10^6 (\Omega - m)^{-1}$ and $T_{\sigma o} = 3160K$. This conductivity value is intermidiate between the measured dc conductivity along the c-axis and the a-b (Ni-O) plane.

The dielectric loss peak occurs at a peak temperature T_p where $\omega\tau(T_p) = 1$. For our data, $T_p = 210K$ at $10GHz$, and $T_p = 150K$ for the $2GHz$. The comparison of this model using eq.1 and including the above conductivity and dielectric contributions is shown in Fig 1 &2, and is seen to describe all the essential features of the data.

When $H_w \| c$ at $2GHz$, Fig. 3a shows an absorption peak around $230K$ in $\zeta''_{H,c}(2GHz,T)$. This feature is reflected as a decrease in $\zeta'_{H,c}$ as a change of state to dia-electromagnetism. Knowing that the conductivity [12] is larger along the a-b plane, the nickel-oxygen plane,

38

indicates that this response is mainly due to in-plane conductivity contributions. There is a small dielectric contribution also in this configuration at lower T. In the $H_w//b, 10GHz$ & $2GHz$, the dielectric contribution below $230K$ is more clearly visible since the conductivity is less along the c-axis. Above $230K$, where the charges are free, eddy current response dominates which explains the increase in absorption at high temperature.

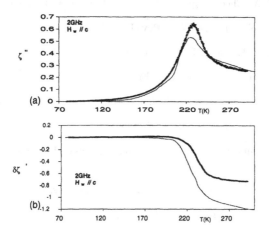

Figure 3: The microwave susceptibility, ζ''_{HC} (a), $\delta\zeta'_{Hc}$ (b) at 2 GHz as a function of tempreture. Here the conductivity term dominates and results in the large peak in ζ''_{HC}. The dielectric loss peak is present but much smaller.

A clear theme is emerging from the present measurements on $La_{5/3}Sr_{1/3}NiO_4$ when viewed with an extensive set of data taken by us on other oxides[7], including the spin chain/ladder compounds of the $Sr - Cu - O$ family, the superconducting cuprates such as $Y:123$ and the $Hg-Ba-Cu-O$ family, and the CMR manganites $La-Sr-Mn-O$. New strong dielectric contributions appear at high frequencies both as dispersion-like changes in ζ'_H accompanied by absorption peaks in ζ''_H below charge ordering transitions.

CONCLUSION

In conclusion we have performed microwave measurements on $La_{5/3}Sr_{1/3}NiO_4$ and observed signatures of charge dynamics which are not observed in other measurements. Our results show that charge ordering is accompanied by the occurrence of dielectric relaxation modes with a large dynamic susceptibility. The present results have important implications for microwave measurements on other oxide materials, such as the spin ladder and superconducting materials, since they demonstrate the phenomenology associated with stripe formation and charge ordering.

ACKNOWLEDGMENT

This work was supported by NSF-ECS-9711910andAFOSR.

References

[1] C. H. Chen, S-W. Cheong, and A. S. Cooper, Phys. Rev. Lett. 71, 2461 (1993).

[2] S-H. Lee, and S-W. Cheong, Phys. Rev. Lett. 79, 2514 (1997).

[3] S. Sridhar and W.Kennedy, Rev. Sci. Instr., 59, 71 (1983)

[4] H. Srikanth, Z.Zhai, S.Sridhar and A.Erb, Phys. Rev. B, 57, 7986 (1998).

[5] Z. Zhai, C. Kusko, N. Hakim, S. Sridhar, (to be published).

[6] C. Hess et al., Phys. Rev. B 59, 10397 (1999).

[7] Z. Zhai, P.V.Parimi, N.Hakim, J.B.Sokoloff, S.Sridhar, U. Ammerahl, A. Vietkine, A. Revcolevschi, cond-mat/9903198.

[8] G. Blumberg, M. V. Klein, and S-W. Cheong, Phys. Rev. Lett. 80, 564 (1998).

[9] A. P. Ramirez, P. L. Gammel, S-W. Cheong, and D. J. Bishop, Phys. Rev. Lett. 76, 447 (1996).

[10] S- W. Cheong, H. Y. Hwang, C. H. Chen, B, Batlogg, L. W. Rupp, Jr., and S. A. Carter. Phys. Rev. B 49, 7088 (1994).

[11] Y. Yoshinari, P. C. Hammel, and S-W. Cheong, Phys. Rev. Lett. 82, 3532 (1999)

[12] T. Katsufuji, T. Tanabe, T. Ishikawa, Y. Fukuda, T. Arima, and Y. Tokura, Phys. Rev. B 54, 14230 (1996).

HIGH PRESSURE MEASUREMENTS ON Tl$_2$Mn$_2$O$_7$

M. Núñez-Regueiro[*], R. Senis, W. Cheikh-Rouhou, P. Strobel, P. Bordet, M. Pernet[**], M. Hanfland[§], B. Martinez and J. Fontcuberta[#].

[*]   Centre de Recherches sur les Très Basses Températures., C.N.R.S., BP166 cedex 09, 38042 Grenoble, France
[**]   Laboratoire de Cristallographie, C.N.R.S., BP166 cedex 09, 38042 Grenoble, France
[§]   European Synchrotron Radiation Facility, BP 220, 38043 Grenoble, France
[#]   ICMAB-CSIC,Campus Universitat Autonoma de Barcelona, Bellaterra 08193, Spain

ABSTRACT

Transport and structural properties of the colossal magnetoresistance pyrochlore Tl$_2$Mn$_2$O$_7$ are studied as a function of applied pressure up to ~20GPa. This allows us to probe the effect of structural changes on the ferromagnetic transition and the transport properties. We observe a non-monotonous pressure dependence of the ferromagnetic transition temperature. We correlate this unusual variation with the structural parameters that, according to electronic band calculations, are key in controlling the properties of these materials.

INTRODUCTION

Colossal magnetoresistance (CMR) materials are largely studied due to its potential technological applications. The fundamental interaction at the origin of CMR for the perovskite manganites is well established : double exchange (DE) [1] where electron hopping between mixed valence Mn^{3+}/Mn^{4+} ions mediates ferromagnetic interactions leading to a ferromagnetic ordering with metallic conduction below T$_c$. Near T$_c$ the applied magnetic field tends to align the local spins and the electron transfer increases. As a result, a sharp resistivity decrease is observed. The high resistivity above T$_c$ and the CMR are explained by the combination of both Mn^{3+}/Mn^{4+} DE and transport via Jahn-Teller polarons.[2]. The pyrochlore manganites[3] are interesting because the above scenario cannot be easily applied in Tl$_2$Mn$_2$O$_7$ compounds in which the high temperature state is metallic (dρ/dT > 0) and there are no traces of significant Mn^{3+}/Mn^{4+} mixed valence since Mn^{4+} is the only Mn ion present[4]. Therefore, neither the DE mechanism nor any significant Jahn-Teller distortions seem to be operative in Mn pyrochlores. Strong carrier scattering by spin fluctuations (SF), associated with the ferromagnetic ordering of the Mn-O lattice, has been invoked to explain CMR in Mn pyrochlores. The mechanism would be similar to that of ferromagnetic transition metals[5], itinerant charge carriers (Tl 6s electrons) scattering against localized moments (Mn 3d electrons) with the small carrier density of the pyrochlores enhancing the otherwise well-known effect[6].

Recently, supplementary evidence for this picture has been put forward [7] through the compound Tl$_2$Mn$_{2-x}$Ru$_x$O$_7$. Ru 4d electrons seem to have an itinerant character for x≤1, and both

41

the ferromagnetic transition temperature T_c and the magnetoresistance decrease with x. In other words, Ru substitution increases the Tl bands carrier concentration without disrupting Tl-O coordination. Another way of modifying charge concentration is by application of strong pressure, e.g. the increase of hole concentration of the superconducting Cu-O layers in cuprates under pressure [8]. We have thus performed electrical resistivity and structural measurements up to 20GPa on the $Tl_2Mn_{2-x}Ru_xO_7$ system.

EXPERIMENT

The electrical resistivity measurements were performed in a sintered diamond Bridgman anvil apparatus using a pyrophillite gasket and two steatite disks as the pressure medium. The Cu-Be device that locked the anvils can be cycled between 1.2K and 300K in a sealed dewar. Pressure was calibrated against the various phase transitions of Bi under pressure at room temperature and by superconducting Pb manometer at low temperature. The temperature was determined using a calibrated carbon glass thermometer with a maximum uncertainty (due mainly to temperature gradients across the Cu-Be clamp) of 0.5K. Four probes electrical resistivity d.c. measurements were carried out using a Keithley 182 nanovoltmeter combined with a Keithley 238 current source.

High-pressure diffraction experiments were performed at the ID09 beam line of the *European Synchrotron Radiation Facility* in Grenoble. The high-pressure conditions were obtained by using a diamond anvil cell in combination with a ruby luminescence method [9] for pressure measurements. Nitrogen was used as pressure transmitting medium. X rays from an undulator source were focused vertically by a Pt coated Si mirror and horizontally by an asymmetrically cut bent Si(111) Laue monochromator. Monochromatic radiation with wavelength of 0.41318Å was used and diffraction patterns were collected, after typical exposure time of 1-2minutes, on an A3-size Fuji image plate located at $\approx$40cm from the sample. The plates were scanned on a Molecular Dynamics image plate reader, and processed by using the Fit2D software[10]. The images were corrected for spatial distortion effects, and tilt corrections and wavelength/distance calibrations were obtained from standard silicon powder images. After removal of spurious peaks the corrected images were averaged over 360° about the direct beam position, yielding Intensity vs. 2θ spectra. These data were then analyzed by the Rietveld technique, using the Fullprof[10] software.

RESULTS

The results of our electrical resistance measurements are shown in Fig. 1. The metallic dependence that is observed in the compound at room pressure has changed into an activated behavior. In quasihydrostatic pressure measurements, the stresses that appear on stabilization of the cell can be strong and normally introduce defects and can even crumb the sample. Although the smaller grains are in electrical contact, if we assume that the conducting mechanism

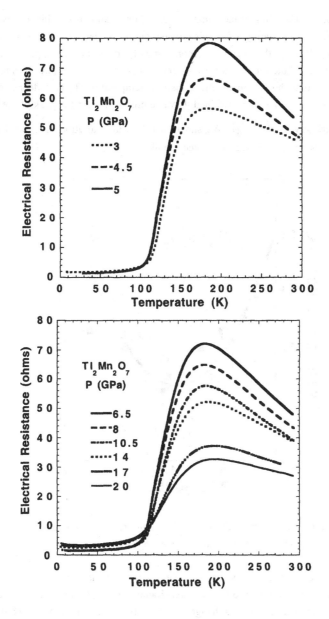

Figure 1: Evolution of the electrical resistance with pressure for $Tl_2Mn_2O_7$. Upper panel : low pressures; lower panel : high pressures.

is by spin-polarized tunneling of electrical carriers[11],we can expect a higher resistance, as a smaller grain size will certainly affect this conduction mechanism. The resistance continues to increase up to 5GPa and then decreases monotonically to the highest applied pressure. This decrease may be attributed to an increase of carrier density with pressure. It is also apparent from our results that the ferromagnetic transition temperature T_c first decreases and then increases with pressure. We have obtained its value precisely from the peak of the logarithmic temperature derivative, $^{-1}/_R\,^{dR}/_{dT}$. We show on Fig. 2 the variation of T_c with pressure. A similar behavior is observed in Ru substituted samples.

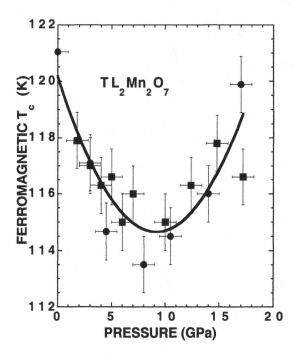

Figure 2: Evolution with pressure of the ferromagnetic transition temperature for two samples (circles and squares). The line is just a guide to the eyes.

We have also followed the pressure dependence of the structural parameters of the pyrochlore structure. According to Goodenough-Kanamori's rules [12], the Mn-O(2)-Mn angle can determine the type of superexchange interactions, from ferromagnetic (FSE) at 90° to antiferromagnetic (AFSE) at 180°. We show on Fig. 3 its variation with pressure. The ambient pressure value is empirically reported to correspond to FSE. If the FSE were the only interaction responsible for ferromagnetism in $Tl_2Mn_2O_7$, a decrease of this angle under pressure, as we observe for P<8GPa and was also previously reported for measurements below 1.5GPa [13], should be associated with an increase of T_c. We observe, though, initially a decrease of T_c and only at high pressures we recover the increase of T_c expected by FSE. All other structural parameters, and in particular, the Tl-O(1) and Mn-O(2) distances, also change monotonically with pressure..

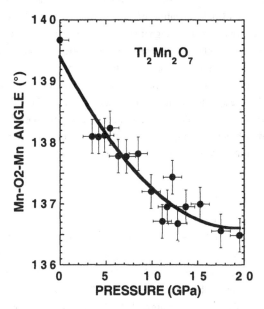

Figure 3: Variation with pressure of the angle Mn-0(2)-Mn, important for superexchange interactions.

DISCUSSION

It was initially supposed that FSE was the driving interaction to explain ferromagnetism in $Tl_2Mn_2O_7$ [4,6], as there was no evidence for DE due to a Mn^{3+}/Mn^{4+} mixed valence. However, as we gather from Fig. 2 and 3, this cannot be the only interaction present, as only at high pressures does a decrease of the angle correlate with an increasing T_c. Other theories have been put forward to explain ferromagnetism in this materials. A more subtle DE mechanism has been used to explain ferromagnetism in this semi-metallic material. Band structure calculations [14] have shown a strong anti-polarization of the Tl-O Γ_1 electrons with respect to Mn^{4+} moments. The itinerant electrons from the Tl-O bands would transport the information in a way similar as in the RKKY interaction, but enhanced by the complete polarization of the conducting electrons that would lead to a ferromagnetic alignment of Mn^{4+} spins. In fact this is a double-exchange interaction where the conduction electrons originate from atoms different to those responsible for the localized magnetic moments, in contrast to perovskite manganites, where both carriers and moments are issued from Mn orbitals. Within this explanation, the ferromagnetic transition temperature would depend directly on $N(E_F)$. However, the measurements [7] on $Tl_2Mn_{2-x}Ru_xO_7$ indicate that in the case of Ru substitution T_c decreases with increasing x and increasing carrier concentration. We must take into account, though, that Ru substitution, even in the case of delocalization of its electrons, decreases the number of Mn^{4+} moments, and should probably affect ferromagnetism in a stronger manner than the rather weak ferromagnetic interaction through electrical carriers. We also observe that the Tl-O(1) distance that decreases monotonically with pressure may point towards an increase of the carrier density and of the mobility of the charges, thus inducing a decrease of the resistivity, as we observe at higher pressures. Within the reservation that this increase must be confirmed by carrier density measurements, such as Hall effect, our data would indicate the importance of carrier density for the ferromagnetism of pyrochlores. In any way, the origin of the decrease of T_c at low pressures is still not understood, and further work is needed to clarify this issue.

REFERENCES

1. C. Zener, Phys. Rev. **82**, 403(1951)

2. A.J. Millis, P.B. Littlewood and B.I. Shraiman, Phys. Rev. Lett. **74**, 5144 (1995) ; H., Roëder, J. Zhang and A.R. Bishop, Phys. Rev. Lett. **76**, 1356 (1996)

3. M.A. Subramanian, G. Aravamudan and G.V. Subba Rao, Prog. Solid St. Chem. **15**, 55 (1983) ; Y. Shimakawa, Y. Kubo and T. Manako, Nature **379**, 53 (1996)

4. M.A. Subramanian, B.H. Toby, A.P. Ramirez, W.J. Marshall, A.W. Sleight and G.H. Kwei, Science **273**, 81 (1996) ;J.W. Lynn, L. Vasiliu-Doloc and M.A. Subramanian, Phys. Rev. Lett. **80**, 4582 (1998)

5. M.E. Fischer and J.S. Langer, Phys. Rev. Lett. **20**, 665 (1968)

6. P. Majumdar and P. Littlewood, Phys. Rev. Lett. **81**, 1314 (1998) ; Nature **395**, 479(1998)

7. B. Martínez, R. Senis, J. Fontcuberta, X. Obradors, W. Cheikh-Rouhou, P; Strobel, C. Bougerol-Chaillout and M. Pernet, Phys. Rev. Lett. **83**, 2022 (1999)

8. See for example M. Núñez-Regueiro and C.Acha "Studies of High Temperature Superconductors", **24**, 203, edited by A. Narlikar, Nova science New York (1997) and references therein.

9. H.K. Mao, J. Xu and P.M. Bell, J. Geophys. Res. **91**, 4673 (1986).

10. A.P. Hammersley *et al.* High Pressure Research **14**, 235 (1996).

11. H.Y. Hwang and S-W. Cheong, Nature **389**, 942 (1997)

12. J.B. Goodenough, Phys. Rev. **100**, 564 (1955)

13. Yu.V. Sushko; Y Kubo; Y. Shimakawa and T. Manako, Physica-**B259-261**, 831 (1999)

14. S.K. Mishra and S. Satpathy, Phys. Rev. **B58**, 7585 (1998)

MAGNETORESISTANCE IN THIN FILMS OF SILVER CHALCOGENIDES

I.S. CHUPRAKOV *, V.B. LYALIKOV *, K.-H. DAHMEN *, P. XIONG **
* Chemistry Department and MARTECH, Florida State University, Tallahassee, FL 32306
** Physics Department and MARTECH, Florida State University, Tallahassee, FL 32306

ABSTRACT

Oriented and non-oriented thin films of silver(I) telluride, Ag_2Te, were prepared by e-beam evaporation, vapor transport technique and Chemical Vapor Deposition (CVD). Crystallinity and orientation of the films were studied by $\Theta-2\Theta$ XRD, rocking curve and pole figure measurements. The origin and conditions for the oriented growth are discussed. Special microdevice was prepared by photolitography from the oriented films of Ag_2Te in order to investigate magnetoresistance (MR) in this material. It was proved that the reported earlier negative MR in Ag_2Te films is a completely geometrical effect, which can be observed using non-linear arrangement of current and voltage contacts.

INTRODUCTION

A large variety of magnetoresistive materials have been intensively studied for the last decade because of the potential applications in magnetic recording and as magnetic field sensors. Both intrinsic (narrow bandgap semiconductors [1], manganites [2,3]) and artificial (magnetic multilayers [4], magnetic heterogeneous alloys [5,6]) magnetoresistive materials demonstrate high MR values. Non-magnetic narrow-gap semiconductors with high electron mobility, like InSb or InAs, are currently used for manufacturing of magnetoresistors [1].

Silver(I) telluride, Ag_2Te, has two phases: low-temperature monoclinic $\alpha-Ag_2Te$ and high-temperature cubic $\beta-Ag_2Te$ with transition temperature of $140°C$. The high-temperature phase is metallic with high mobility of Ag atoms and superionic conductivity, while the low-temperature phase is a semiconductor with a band gap of 0.1 eV and small effective mass of the electron carriers of $0.050-0.069m_o$, with m_o the free-electron mass [7]. Having resistivity in the order of milliohms·cm, $\alpha-Ag_2Te$ is a promising material for different MR devices.

Peculiar MR properties were found recently in slightly non-stoichiometric silver telluride, $Ag_{2\pm\delta}Te$. High MR values were achieved in both bulk material [8] and thin films [9] of this compound. Unusual linear field dependence of MR down to several Oe was observed in the bulk Ag_2Te [8] and drew attention of some theoreticians [10]. Recently we reported another interesting feature of silver telluride: MR of oriented films has a clear peak at about 100K, whereas MR of non-oriented films is almost temperature independent below 120K [9]. This effect is similar to the one for GMR manganites [11-13]. Additionally, we observed unusual for non-magnetic materials negative MR at parallel field-current orientation in Ag_2Te films. At transverse field-current orientation MR turns positive [14]. All these effects still require elaborate experimental and theoretical study.

EXPERIMENT

Oriented thin films of silver telluride were grown by electron-beam evaporation from a mixture of Ag_2Te (Aldrich Chemicals, purity 99.999%) and Ag (Strem Chemicals, purity 99.99%) powders. Ag_2Te was ground, thoroughly mixed with Ag powder at a 16:1 ratio in weight and then pressed into a pellet. This pellet was used as the deposition source and was melted prior to the evaporation in a vacuum chamber. The pressure in the chamber was ~5·10^{-6} torr during the deposition processes. Fused quartz plates were used as substrates, which were heated by halogen lamps to 490°C. The source-substrate distance was 30 cm and the deposition rate was maintained at ~5 Å/min. The deposited films had a thickness of 275±5 nm. The film composition was found to be $Ag_{2.19}Te$ by X-ray photoelectron spectroscopy (XPS).

Non-oriented films of silver telluride were grown by a vapor transport technique described earlier [9]. It includes plasma sputtering of Ag on a quartz substrate followed by tellurium vapor treatment. The obtained films were rich in Te with a composition of $Ag_{1.92}Te$ according to XPS analysis. The film thickness was 280±5 nm.

CVD experiments were conducted in a horizontal cold-wall CVD reactor. Graphite substrate holder was heated by halogen lamps. Helium or hydrogen were used as a carrier gas.

X-ray diffraction (XRD) analysis was performed on D-5000 diffractometer by Siemens, Inc., equipped with HTK 10 high temperature attachment using Cu K_α radiation. Resistivity measurements were carried out with AC phase-sensitive detection in an in-plane field up to 8 T in the standard 4-probe and the van der Pauw geometries.

RESULTS AND DISCUSSION

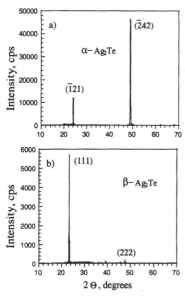

Figure 1. XRD pattern of oriented Ag_2Te film at a) room temperature and b) 200°C.

The films obtained by e-beam evaporation technique were highly oriented. Strong ($\bar{1}21$) orientation of low-temperature α-Ag_2Te monoclinic phase [15a] and (111) orientation of high-temperature β-Ag_2Te cubic phase [15b] was found by Θ-2Θ XRD measurement (see Fig.1) and was confirmed by a rocking curve measurement of (242) reflection of α-Ag_2Te (Fig.2). The full width at half maximum (FWHM) of this peak was about 1.5°. The pole figures of ($\bar{2}42$) and (040) reflections shown in Figure 3 demonstrate that the films have only out-of-plane orientation. So strong orientation of the film deposited on an amorphous substrate can be attributed to the high mobility of Ag atoms in β-Ag_2Te. It is known [7] that silver atoms don't have any definite position in the crystal structure of the high-temperature Ag_2Te phase. The cubic face-centered structure of the tellurium lattice adopts the most closely packed (and the lowest energy) (111) orientation during the deposition at high temperature. The low-temperature phase formed upon cooling inherits similar orientation of the tellurium sub-lattice with ($\bar{1}21$) plane of the monoclinic lattice parallel to the surface of the film.

Films obtained by the vapor transport technique were polycrystalline according to XRD analysis (Fig.4). ($\bar{1}21$) and ($\bar{2}42$) reflections of the low-temperature phase and (111) reflection of the high-temperature phase are slightly increased with respect to the standard. However no clear orientation was observed. Safran, et al. [16] showed recently that strong texture in Ag_2Se films could be achieved by slow cooling of polycrystalline films. The cooling rate of about 1 °/min during the phase transition was required in order to get strong (001) orientation of low-temperature Ag_2Se phase on an amorphous substrate. It seems that this effect does not take place

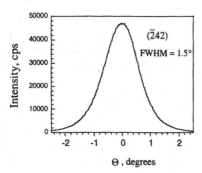

Figure 2. Rocking curve of $(\bar{2}42)$ reflection of α-Ag$_2$Te.

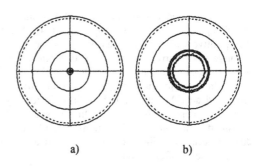

a)   b)

Figure 3. Pole figures of a) $(\bar{2}42)$ and b) (040) reflections of α-Ag$_2$Te.

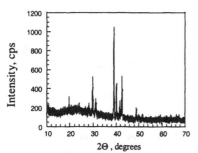

Figure 4. XRD pattern of non-oriented α-Ag$_2$Te film.

in case of Ag$_2$Te film. XRD pattern of the film cooled at 1 $^\circ$/min rate was essentially the same as the one for the as-prepared polycrystalline film. Nevertheless, relatively strong $(\bar{1}21)$ orientation appears in the Ag$_2$Te films during high-temperature (400°C) annealing for several hours. This also can be explain by the superionic mobility of silver atoms in β-Ag$_2$Te.

Ag$_2$Te films were obtained also by Chemical Vapor Deposition (CVD) technique. Initial experiments using [Ag(hfa)PMe$_3$] as silver precursor resulted either in precipitation of non-volatile Ag-Te complex in case of using [Te$_2$Ph$_2$] as a Te-precursor, or in pure Ag deposition in case of using [(Ph—CH=CH)$_2$Te] as a Te-precursor. However, polycrystalline Ag$_2$Te films can be obtained by using [(n-Bu—CH=CH)$_2$Te] precursor. Essential parameters of the CVD process are summarized in the Table I. The XRD pattern of the films were similar to the ones for the films prepared by the vapor transport technique. However, the films were not uniform enough to be acceptable for the MR measurements.

Table I. Parameters of Ag$_2$Te Chemical Vapor Deposition.

Ag:Te atomic ratio in the precursor mixture	1:2
Evaporation temperature	120°C
Hydrogen flow	40 sccm
Pressure	1 mbar
Deposition temperature	500°C
Substrates	SiO$_2$, Si, Stainless steel

As we reported earlier, unusual MR sign reversal was observed in thin films of silver telluride prepared by e-beam evaporation [14]. In a square van der Pauw contacts geometry MR was positive when current was perpendicular to the magnetic field (transverse geometry) and negative when current was parallel to the field (parallel geometry). The applied magnetic field was always parallel to the film surface.

In order to study the anisotropy of MR and the sign reversal in more detail we have fabricated an L-bar device out of the oriented Ag_2Te film by photolitography and ion beam milling, as depicted in Figure 5. In this geometry both transverse and parallel MR could be measured simultaneously. The transverse (T) and parallel (II) MR again exhibit qualitatively different behavior (Figure 6), but dissimilar to the previous measurements. Slight negative MR effect at the parallel field-current orientation still can be seen at about 1 T. However, at higher fields MR turns positive.

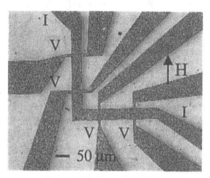

Figure 5. L-bar microdevice.

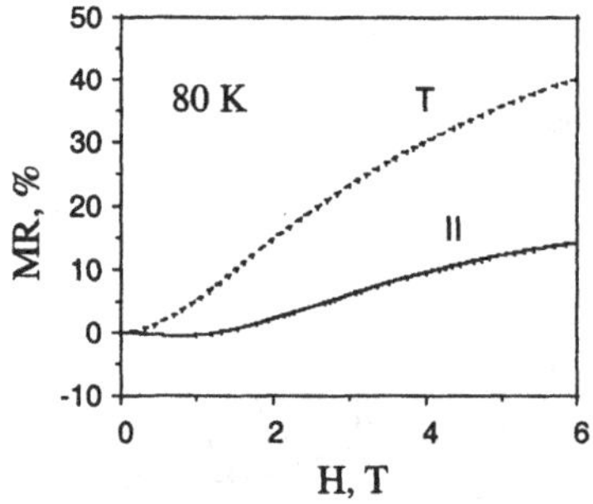

Figure 6. Field dependence of MR for the L-bar microdevice.

In order to understand the difference in the obtained results and further explore the origin of the anomalous MR behavior observed in the square van der Pauw configuration, we performed MR measurements on the oriented $Ag_{2.19}Te$ film with different arrangements of the current and voltage leads as depicted in Fig. 7.

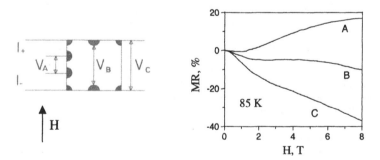

Figure 7. Contacts arrangement and field dependence of parallel MR for different measurement configurations.

The electrode arrangement A corresponds to the conventional 4-point resistivity measurement geometry. MR field dependence for this arrangement is very similar to the one obtained for the L-bar structure. As we moved the voltage leads further away from the current leads, the MR took on increasingly larger negative values. The electrode arrangement C is the square van der Pauw geometry we used in the previous measurements. MR is negative in this case and reproduces the results we obtained before. This negative MR also exhibited strong temperature dependence with maximum of –57% at 150K and –22% at room temperature. Detailed description of the negative MR measurements on the oriented and non-oriented samples of Ag_2Te thin films at different temperatures will be published soon [17].

CONCLUSIONS

Oriented films of Ag_2Te can be obtained on amorphous substrates during e-beam evaporation at high temperature or high-temperature annealing of polycrystalline films. High mobility of Ag atoms in the β-Ag_2Te phase favors the formation of the low-energy (111) orientation of the high-temperature phase and, upon cooling, ($\bar{1}21$) orientation of the low-temperature phase. Non-oriented polycrystalline Ag_2Te films can be prepared by vapor transport technique and CVD. Slow cooling of the polycrystalline films does not induce any preferred orientation of the films.

From the results obtained on special microdevices it can be concluded that the reported previously negative MR in the films of Ag_2Te is completely geometrical effect. It can be observed by using non-linear arrangement of current and voltage contacts similar to the van der Pauw resistivity measurement technique.

ACKNOWLEDGEMENTS

We thank S. von Molnár for many useful discussions and S. Watts, S. Wirth, P. Barber, and E. Lochner for technical assistance. This work was supported in part by DARPA and the Office of Naval Research under contract ONR-N00014-96-1-0767.

REFERENCES

1. J. Heremans, J. Phys. D-Appl. Phys. **26**, p. 1149 (1993).

2. R. von Helmolt, J. Wecker, B. Holzapfel, L. Schultz, and K. Samwer, Phys. Rev. Lett. **71**, p. 2331 (1993).

3. S. Jin, T. H. Tiefel, M. McCormack, R. A. Fastnacht, R. Ramesh, and L. H. Chen, Science **264**, p. 413 (1994).

4. M. N. Baibich, J. M. Broto, A. Fert, F. N. Vandau, F. Petroff, P. Eitenne, G. Creuzet, A. Friederich, and J. Chazelas, Phys. Rev. Lett. **61**, p. 2472 (1988).

5. A. E. Berkowitz, J. R. Mitchell, M. J. Carey, A. P. Young, S. Zhang, F. E. Spada, F. T. Parker, A. Hutten, and G. Thomas, Phys. Rev. Lett. **68**, p. 3745 (1992).

6. J. Q. Xiao, J. S. Jiang, and C. L. Chien, Phys. Rev. Lett. **68**, p. 3749 (1992).

7. H. Kikuchi, H. Iyetomi, and A. Hasegawa, J. Phys.-Condes. Matter **9**, p. 6031 (1997).

8. R. Xu, A. Husmann, T. F. Rosenbaum, M. L. Saboungi, J. E. Enderby, and P. B. Littlewood, Nature **390**, p. 57 (1997).

9. I. S. Chuprakov, and K. H. Dahmen, Appl. Phys. Lett. **72**, p. 2165 (1998).

10. A. A. Abrikosov, Phys. Rev. B-Condens Matter **58**, p. 2788 (1998).

11. A. Gupta, G. Q. Gong, G. Xiao, P. R. Duncombe, P. Lecoeur, P. Trouilloud, Y. Y. Wang, V. P. Dravid, and J. Z. Sun, Phys. Rev. B-Condens Matter **54**, p. 15629 (1996).

12. H. Y. Hwang, S. W. Cheong, N. P. Ong, and B. Batlogg, Phys. Rev. Lett. **77**, p. 2041 (1996).

13. R. Shreekala, M. Rajeswari, K. Ghosh, A. Goyal, J. Y. Gu, C. Kwon, Z. Trajanovic, T. Boettcher, R. L. Greene, R. Ramesh, and T. Venkatesan, Appl. Phys. Lett. **71**, p. 282 (1997).

14. I. S. Chuprakov, and K. H. Dahmen, J. Phys. IV **9**, p. 313 (1999).

15. JCPDS database, files a) 34-142; b) 6-575.

16. G. Safran, L. Malicsko, and G. Radnoczi, J. Cryst. Growth **205**, p. 153 (1999).

17. I. S. Chuprakov, K. H. Dahmen, and P. Xiong, submitted to Appl. Phys. Lett.

Poster Session I

Synthesis and Properties of Epitaxial Thin Films of c-axis oriented Metastable Four-Layered Hexagonal BaRuO$_3$

M. K. Lee, C. B. Eom*, W. Tian, X. Q. Pan **, M. C. Smoak, F. Tsui ***, J. J. Krajewski****
*Department of Mechanical Engineering & Materials Science, Duke University, Durham, NC 27708
** Department of Materials Science & Engineering, University of Michigan, Ann Arbor, MI 48109
*** Department of Physics & Astronomy, University of North Carolina, Chapel Hill, NC 27599
****Bell Laboratories, Lucent Technologies, Murray Hill, NJ 07974

ABSTRACT

We have grown epitaxial thin films of metastable four-layered hexagonal (4H) BaRuO$_3$ on (111) SrTiO$_3$ by 90° off-axis sputtering techniques. X-ray diffraction and transmission electron microscopy experiments reveal that the films are single crystals of c-axis 4H structures with an in-plane epitaxial arrangement of BaRuO$_3$ [2$\bar{1}$$\bar{1}$0] // SrTiO$_3$ [110]. Smooth multilayer growth has been observed in these films with a step height equaling the size of half unit cell. In-plane resistivity of the films is metallic, with a room temperature value of about 810 μΩ-cm and slightly curved temperature dependence. Their magnetic susceptibility is paramagnetic. The metastable layered compounds can be very useful for understanding new solid-state phenomena and novel device applications.

INTRODUCTION

The recent discovery of superconductivity in Sr$_2$RuO$_4$ compound without any copper and doping[1] has led to renewed interest in the physical properties, particularly magnetism, and structural chemistry of the ruthenium based oxides (ARuO$_3$: A = Ba, Sr, Ca). Among all ruthenates, only BaRuO$_3$, which has the largest A-site cation in the series, possesses a hexagonal structure in the bulk, while others have a GdFeO$_3$-typed orthorhombic structure instead.[2]

The structural chemistry of ruthenates can be described in terms of hexagonal and cubic close packing of AO$_3$ layers. If all AO$_3$ layers are cubic close packed, the RuO$_6$ octahedra form a three dimensional array by sharing only one oxygen to give the cubic-like structure, such as cubic, tetragonal and orthorhombic. In contrast, if AO$_3$ layers are entirely hexagonal close packed, the RuO$_6$ octahedra are shared by three oxygens to form the hexagonal structure. Owing to the two basic packing forms, it is recognized that the bulk BaRuO$_3$ ceramic has three different crystal structures. They are (1) the nine-layered rhombohedral structure (9R) with a = 5.75 Å and c = 21.6 Å,[3] (2) the six-layered hexagonal structure (6H) with a = 5.71 Å and c = 14 Å,[4] and (3) the four-layered hexagonal structure (4H) with a = 5.73Å and c = 9.5 Å,[4] depending on the amount of hexagonal and cubic close packing of the BaO$_3$ layers. The 9R phase, which has been considered as the most stable, corresponds to a stacking sequence of CHHCHHCHH, where "C" and "H" represent cubic and hexagonal close packing, respectively. The 6H phase, which has been found mainly as a result of high-pressure bulk synthesis, possesses a stacking sequence of CCHCCH with more cubic close packing. Finally the 4H stacking sequence is CHCH, which contains equal amount of hexagonal and cubic close packing. Like its 6H counterpart, the 4H structure containing pure Ba-Ru-O has not been synthesized reproducibly under ordinary conditions, so it has been considered as *metastable*. The first bulk 4H Ba-Ru-O ceramic was synthesized under

Mat. Res. Soc. Symp. Proc. Vol. 602 © 2000 Materials Research Society

high pressure (between 15 and 30 kbars) by Longo *et al.*,[4] and its single crystal form was grown using a BaCl$_2$ flux with first refinement of crystal structure.[5] More recently, the electrical and magnetic properties of bulk single crystal form have been characterized,[6] although this phase is still regarded as not reproducible. In this letter, we report the synthesis, electrical transport, and magnetic properties of epitaxial thin films of metastable 4H BaRuO$_3$.

EXPERIMENT

In order to match the in-plane symmetry and lattice of (0001) 4H BaRuO$_3$, the following substrates have been chosen for the study: (1) (0001) Al$_2$O$_3$ with a lattice parameter d_{100} of 4.121 Å and a lattice mismatch ε_{100} of +20.41 %, (2) (111) SrTiO$_3$ with d_{110} = 5.523 Å and ε_{110} = +3.73%, (3) (111) MgAl$_2$O$_4$ with d_{220} = 5.715 Å and ε_{220} = +0.24%, and (4) (111) MgO with d_{110} = 5.962 Å and ε_{110} = −3.91%. The films were grown by the 90° off-axis sputtering techniques[7] from a stoichiometric BaRuO$_3$ ceramic target.

RESULTS

Their crystal structures and epitaxial arrangements were studied by x-ray diffraction and transmission electron microscopy (TEM) experiments. All films studied exhibit predominantly c-axis texture. The films grown on (111) MgAl$_2$O$_4$ substrates exhibit two hexagonal domains, rotated by about 4° with respect to each other in the growth plane, and they also contain a small amount of impurity phases. It is considered that the spinel structure of MgAl$_2$O$_4$ substrate does not match structurally with the 4H BaRuO$_3$ phase, even though the $d_{(220)}$ of (111) MgAl$_2$O$_4$ plane has a good match with the in-plane 4H BaRuO$_3$. The films grown on (0001) Al$_2$O$_3$ substrates also contain considerable amount of (110) 4H texture. These films and their counterparts grown on (111) MgO having Rochelle salt structure also contain impurity phases, which have not yet been determined. In contrast, the films grown on (111) SrTiO$_3$ do not show any impurity phases, and they are single crystals of (0001) 4H phase, as illustrated in Fig. 1 (a). The observed behavior is consistent with the notion that among the substrates chosen for the study, (111) SrTiO$_3$ matches with (0001) 4H BaRuO$_3$ the best, in terms of both chemical structure and lattice parameter, particularly for being a perovskite also. In what follows, we describe the structural, electrical, and magnetic properties of the films grown on (111) SrTiO$_3$.

Lattice parameters of the films grown on (111) SrTiO$_3$ are measured by normal θ-2θ scan and grazing incidence diffraction (GID) experiments. For a 4600 Å thick film, they are 9.50 ± 0.01 Å out-of-plane, and 5.728 ± 0.002 Å in-plane, both of which are the same as the bulk values. The epitaxial arrangement is determined by the azimuthal φ-scans of both BaRuO$_3$ (10$\bar{1}$1) and SrTiO$_3$ (002) reflections, as shown in Fig. 1 (b). The observed six-fold intensities of the azimuthal BaRuO$_3$ (10$\bar{1}$1) reflection [Fig. 1 (b)] confirm the presence of a single hexagonal domain in the growth plane. The azimuthal scans also indicate that BaRuO$_3$ [2$\bar{1}$$\bar{1}$0] is parallel to SrTiO$_3$ [110] [see Fig. 1 (b)]. The measured full width at half maximum (FWHM) of the rocking curve across the (0004) reflection is 0.41°, and that of the off-axis φ scan across (10$\bar{1}$1) is 0.54°. These are slightly wider than the corresponding values for the SrTiO$_3$ substrate of 0.25° across the (111) reflection and 0.41° across (002), respectively. The x-ray diffraction results indicate that the RaRuO$_3$ films are high quality single crystals.

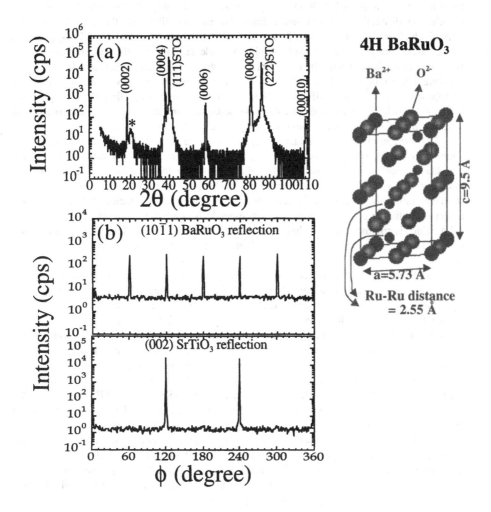

Fig. 1. X-ray diffraction patterns of a 4600 Å BaRuO$_3$ thin film grown on (111) SrTiO$_3$ substrate. (a) A normal θ-2θ scan, and (b) an azimuthal off-axis φ scan across the BaRuO$_3$ (10$\bar{1}$1) reflection in the (0001) plane (upper panel), and an azimuthal scan across SrTiO$_3$ (002) reflection in the (111) plane (lower panel). The Bremsstrahlung peak from the substrate is labeled by "*".

The structural properties have been studied further by cross-section TEM including selected area electron diffraction and imaging experiments. Fig. 2 (a) is a bright field TEM image taken from the 4600 Å thick $BaRuO_3$ film grown on (111) $SrTiO_3$ substrate. Fig. 2 (b) and 2 (c) are electron diffraction patterns taken from the $BaRuO_3$ film and the $SrTiO_3$ substrate, respectively. It is identified that Fig. 2 (b) is the diffraction pattern of the 4H $BaRuO_3$ structure. These studies revealed that the $BaRuO_3$ film is single crystal with the 4H structure, grown epitaxially along the c-axis on the (111) $SrTiO_3$ substrate. The lattice constants determined by TEM are 5.72 ± 0.08 Å in-plane and 9.53 ± 0.07 Å out-of-plane. The TEM results are consistent with those obtained by x-ray diffraction. The dark bands in the cross-section TEM image shown in Fig. 2 (c) indicate the presence of anti-phase boundaries and stress-induced dislocations. The latter is typical for an epitaxial system with large lattice mismatch.

Growth mechanism and surface morphologies have been studied by scanning tunneling microscopy (STM). The films grown on (111) $SrTiO_3$ substrate exhibit large concentric "birthday cake-like" terraces about 1000 to 2500 Å in diameter, leading to a relatively small roughness, as illustrated in Fig. 3 (a). The root-mean-squared (RMS) roughness is about 8 to 10 Å. The observed uniformly distributed 3-dimensional surface morphology indicates a 2-dimensional isotropic multilayer growth. The measured step-height between the concentric terraces is about 4.7 Å, which is approximately one half of the c-axis lattice parameter of 4H $BaRuO_3$, as illustrate in the STM line-scan shown in Fig. 3 (b). Therefore, each terrace is terminated by a BaO_3 unit cell.

Unlike $CaRuO_3$ and $SrRuO_3$, the conduction mechanism of $BaRuO_3$ can be explained in terms of Ru metal-metal interactions, because the face sharing octahedra brings the Ru ions into very close proximity.[8] In fact, the distance between Ru ions of the face shared octahedra is 2.55 Å [see Fig. 1 (a)], which is even shorter than Ru metal (2.65 Å). Since the direct Ru-Ru interactions in the face-shared Ru octahedra are different from those of Ru-O-Ru in the corner-shared octahedra, one should also expect a different electrical transport behavior for samples with 4H stacking from those of other types of stacking. The electrical resistivity of 4H $BaRuO_3$ thin film has been examined. Figure 4 illustrates the normalized in-plane resistivity as a function of temperature for a 4600 Å film. The observed resistivity at 295 K is about 810 $\mu\Omega$-cm, which is nearly twice the value for a bulk single crystal.[6] The observed higher resistivity in the film arises evidently from the presence of anti-phase boundaries and misfit dislocations described above [Fig. 2 (c)]. The temperature dependence of the measured resistivity is metallic over the whole range of the measurement, and it is slightly curved at both the high and low temperature regimes. It is nearly quadratic below 80 K, as illustrated in the nearly linear T^2 dependence with a slope of 3.0×10^{-2} $\mu\Omega$-cm/K^2 in Fig. 4 inset. The observed behavior is similar to that of the bulk 4H single crystal.[6] The residual resistivity is determined to be 220 $\mu\Omega$–cm.

Magnetic properties of the 4H $BaRuO_3$ films have been studied by SQUID magnetometry. The observed susceptibility is paramagnetic and nearly temperature independent, indicating that it is dominated by Pauli paramagnetism. The measured susceptibility is $(8 \pm 2) \times 10^{-5}$ emu/cm^3. These are also comparable to those observed in the bulk.[6]

CONCLUSIONS

In summary, we have synthesized metastable epitaxial 4H $BaRuO_3$ thin films, and have subsequently studied their structural, electrical transport and magnetic properties. The $BaRuO_3$

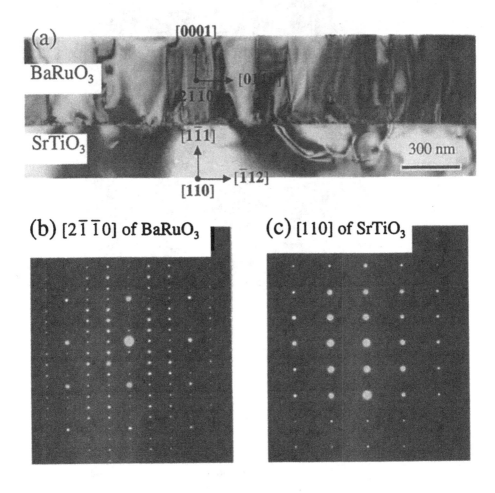

Fig. 2. Cross-section TEM of a 4600 Å BaRuO₃ thin film grown on (111) SrTiO₃ substrate: (a) cross-section bright field image of the film and selected area electron diffraction patterns of (b) the film and (c) the substrate. The 4-layered stacking of BaRuO₃ is shown clearly in (b).

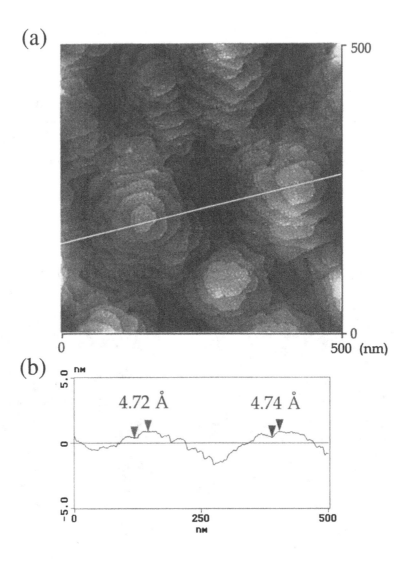

Fig. 3. (a) STM image of a 4H (0001) BaRuO$_3$ thin film grown on (111) SrTiO$_3$ and (b) STM line scan along the line drawn in (a) to indicate the observed step-height.

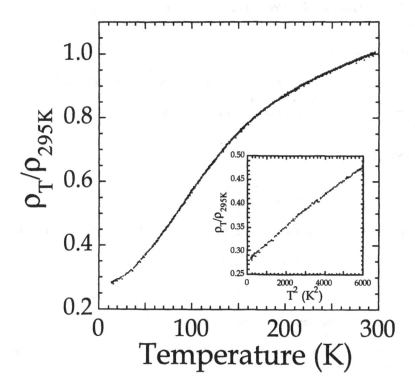

Fig. 4. In-plane normalized electrical resistivity as a function of temperature of a 4600 Å 4H (0001) BaRuO$_3$ thin film grown on (111) SrTiO$_3$. The inset illustrates the quadratic temperature dependence of the resistivity at low temperatures.

films grown on (111) $SrTiO_3$ exhibit a purely c-axis single crystal 4H structure with good crystalline qualities and smooth surfaces. These metastable epitaxial thin films are very important both scientifically and technologically. They can help to understand new phenomena such as superconductivity in layered compounds without any copper oxide layers.[1,6] They also show promises as metallic templates for the synthesis and characterization of epitaxial heterostructures for the novel electronics applications, particularly for growth of (111) perovskites and c-axis hexagonal oxides. For instance, a (0001) $BaRuO_3$ electrode layer with a smooth surface will allow the growth of (111) oriented ferroelectric heterostructures that require a high crystalline quality and clean interfaces.

ACKNOWLEDGMENTS

This work was supported by ONR N00014-95-1-0513, NSF DMR-9421947, a NSF Young Investigator Award (CBE), a David and Lucile Packard Fellowship (CBE), College of Engineering at the University of Michigan (XQP), NSF DMR-9703419 (FT) and NSF DMR-9601825 (FT).

REFERENCES

1. Y. Maeno, H. Hashimoto, K. Yoshida, S. Nishizaki, T. Fujita, J. G. Bednorz, and F. Lichtenberg, Nature (London) **372**, p. 532 (1994).

2. H. Kobayashi, M. nagata, R. Kanno, and Y. Kawamoto, Mat. Res. Bull. **29**, p. 1271 (1994).

3. P. C. Donohue, L. Katz, and R. Ward, Inorg. Chem. **4**, p. 306 (1965).

4. J. M. Longo and J. A. Kafalas, Mat. Res. Bull. **3**, p. 687 (1968).

5. S. T. Hong and A. W. Sleight, J. Solid State Chem. **128**, p. 251 (1997).

6. J. T. Rijssenbeek, R. Jin, Y. Zadorozhny, Y. Liu, B. Batlogg, and R. J. Cava, Phys. Rev. B **59**, p. 4561 (1999).

7. C. B. Eom, R. J. Cava, R. M. Fleming, J. M. Phillips, R. B. van Dover, J. H. Marshall, J. W. P. Hsu, J. J. Krajewski, and W. F. Peck, Jr., Science **258**, p. 1766 (1992).

8. P. R. V. Loan, Ceram. Bull. **51**, p. 231 (1972).

For further information, contact:

Chang-Beom Eom
Dept. of Mech. Eng. & Mater. Sci.
Duke university
Durham, NC 27708

Telephone: (919) 660-5329
Fax: (919) 660-8963
e-mail: eom@acpub.duke.edu

The role of strain in low-field magnetotransport properties of manganite thin films

Y. F. Hu, H. S. Wang, E. Wertz and Qi Li
Department of Physics, Pennsylvania State University, University Park, PA 16802

ABSTRACT

Strain-induced large low-field magnetoresistance has been observed in very thin $Pr_{0.67}Sr_{0.33}MnO_3$ films[1]. To better understand the role of strain in the low-field magnetotransport properties of manganite thin films, we have studied and compared very thin (3-20 nm) $Pr_{0.67}Sr_{0.33}MnO_3$ (PSMO), $La_{0.67}Ba_{0.33}MnO_3$ (LBMO), $La_{0.67}Sr_{0.33}MnO_3$ (LSMO) and $La_{0.67}Ca_{0.33}MnO_3$ (LCMO) films grown on different substrates, such as $LaAlO_3(001)$ (LAO), $NdGaO_3(110)$ (NGO), and $SrTiO_3(001)$ (STO). Due to the lattice mismatch between the films and the substrates ranging from -2.6% to +1%, different strains can be imposed to the films. We have found that: (1) large low-field magnetoresistance(LFMR) behaviors are observed in PSMO, LCMO and LSMO thin films on LAO substrates when a magnetic field is applied perpendicular to the film plane, but the maximum LFMR is the largest in PSMO and LCMO samples; (2) most of the films grown on STO substrates show positive MR when a magnetic field is applied perpendicular to the film plane, and when the field is parallel to the film plane all films show negative MR regardless of the substrates; (3) the large low-field MR is strongly dependent on the film thickness and the composition of the manganites. The anomalous low-field MR effect will be discussed based on strain-induced magnetic anisotropy and domain rotation and movement.

INTRODUCTION

During the last decade, tremendous efforts have been devoted to heterogeneous ferromagnetic materials, such as thin-film multilayers and cluster-alloy compounds which display so-called giant magnetoresistance effect (GMR). The discovery of "colossal" magnetoresistance (CMR) in perovskites has renewed interest in these materials. It launched a frenetic scientific race to understand the cause of the effect and raised expectations on potential applications. The crucial problem for developing practical devices based on colossal magnetoresistance is that a large response is only observed in relatively large magnetic fields -- typically a few tesla. However, for magnetic field-sensing applications, low-field responsivity is necessary.

On the other hand, one important issue both for physics and for any possible application is the strain dependence of material properties. In particular, many proposed applications involve films, and films typically have large biaxial strains since lattice distortion can be easily introduced in thin films due to lattice mismatch between the film and the substrate. Therefore, the thin film samples provide good candidates for the study of the biaxial strain effect on the electrical and magnetic properties of the materials.

We have recently reported the strain-induced large low-field magnetoresistance in very thin $Pr_{0.67}Sr_{0.33}MnO_3$ (PSMO) films grown on $LaAlO_3(001)$ (LAO) substrates[1]. The films are under compressive strain imposed by the lattice mismatch with the substrate. We obtained MR $\approx$ -92% at H=800 Oe (MR is defined as (R(0)-R(H))/R(0)) and T=70 K. We also observed large low-field

Mat. Res. Soc. Symp. Proc. Vol. 602 © 2000 Materials Research Society

magnetoresistance anisotropy between magnetic fields applied parallel and perpendicular to the film surfaces[2]. In this paper, we report the comparison of the low-field magnetotransport properties between different manganite thin films.

EXPERIMENT

To study and compare the role of strain in low-field magnetotransport properties, we have grown very thin (3-20 nm) PSMO, $La_{0.67}Ba_{0.33}MnO_3$ (LBMO), $La_{0.67}Sr_{0.33}MnO_3$ (LSMO) and $La_{0.67}Ca_{0.33}MnO_3$ (LCMO) films on different substrates, such as LAO, $NdGaO_3$ (110) (NGO) and $SrTiO_3(001)$ (STO), where the lattice mismatch between the films and the substrates ranges from -2.6% to +1%.

The films were epitaxially grown by pulsed-laser deposition (PLD). The ceramic targets were synthesized by standard solid state reaction at 1300 °C starting from the high purity Pr_6O_{11}, La_2O_3, $SrCO_3$, CaO and MnO_2 powders. The powders were thoroughly mixed according to desired stoichiometry, and then sintered for four days in an oxygen atmosphere with three intermediate careful grinding and mixings for homogenization. Finally, the powders were pressed into the pellet form, and sintered for two days to become the targets. An excimer laser with the wavelength of 248 nm, the energy density of ~2 J/cm², and the repetition rate of 5 Hz was used. The growth rates were around 0.5/pulse and the thickness of the film was determined by the nominal value. The substrate temperatures were 750-800 °C and oxygen pressures were 350-800 mTorr during the deposition.

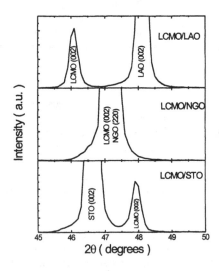

Fig.1 . X-ray diffraction results of the LCMO films grown on different substratyes.

The table 1 shows the lattice constants of the bulk manganite materials and the substrates, and the lattice mismatch between them. X-ray diffraction (XRD) experiments showed that all the films were epitaxially grown, and confirmed the existence of different strains(see figure 1). The MR measurements were carried out using a Quantum Design PPMS 6000 system with standard four-terminal method.

Table 1. The lattice constants of the bulk manganite materials and the substrates, and the lattice mismatch between them.

	PSMO (a=3.866 Å)	LBMO (a=3.897 Å)	LSMO (a=3.88 Å)	LCMO (a=3.867 Å)
STO (a=3.905 Å)	1%	0.2%	0.6%	1%
NGO (a=3.862 Å)	-0.1%	-0.9%	-0.5%	-0.1%
LAO (a=3.794 Å)	-1.9%	-2.6%	-2.2%	-1.9%

Results

Figure 2(a) shows the LFMR effect of a 7.5 nm thick PSMO film with the compressive strain. The magnetic field was scanned from –5 kOe to 5 kOe, and the field was applied perpendicular to the film plane. As seen in the figure, large LFMR and MR hysteresis were observed. Such large LFMR effect was only observed in the compressively-strained films, and only when the field was applied perpendicular to the film plane. When the field is parallel to the film plane, the effect is much smaller.

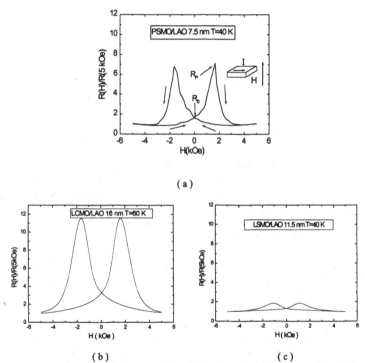

Fig. 2. LFMR and hysteresis loops of (a) PSMO, (b) LCMO, and (c) LSMO films on LAO substrates. The arrows in (a) show the field scanning directions.

Similar behaviors were also observed in the LCMO and LSMO films on LAO substrates. Figure 2(b) and 2(c) show the MR and hysteresis loops of a 16 nm-thick LCMO film and 11.5 nm-thick LSMO film. Large LFMR were obtained in both samples, but the LFMR ratio is smaller in LSMO samples than that in PSMO and LCMO samples. This type of LFMR effect has not been observed so far in the LBMO films.

Positive MR curves were observed in the tensile-strain samples. Figure 3 shows the negative and positive MR in the 6 nm-thick LCMO films with different strains. Left curve shows the negative MR and the hysteresis loop for the film on the LAO substrate with compressive strain, and right curve shows the positive MR and the hysteresis loop for the film on the STO substrate with tensile strain. It is seen that, for the compressively-strained sample, MR is much larger than that of the tensile-strained sample, but with opposite signs. In our study, this kind of behavior is only observed when the field is applied perpendicular to the film plane. When the field is applied parallel to the film plane, only *negative* MR has been observed in the films on all substrates.

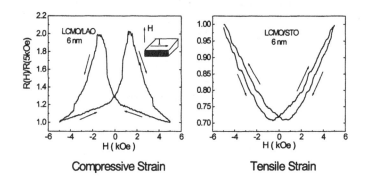

Compressive Strain   Tensile Strain

Fig. 3. Negative and positive MR in the 6 nm-thick LCMO films with different strains.

The large LFMR we have obtained depends strongly on the film thickness. Figure 4 shows the maximum LFMR as a function of film thickness of the LSMO films on LAO substrates when the field is applied perpendicular to the film plane. The LFMR was defined as $(R_P-R_0)/R_0$, where R_P and R_0 were defined in figure 2(a). Only in a vary narrow thickness range, the sample shows the large LFMR effect. Below the critical thickness, the sample is insulating due to surface disorders. When the film is thicker than 15 nm, the LFMR is very small and similar to bulk materials.

We proposed that the LFMR properties of the differently strained samples can be explained by the magnetic anisotropy and the domain rotation. The magnetization measurement of PSMO films[3] shows that, only a small amount of compressive strain will produce a spontaneous out-of-plane magnetization. Our results[3,4] show that a compressive strain induced by LAO substrate produces a magnetically easy axis normal to the film plane. This anisotropy is strong enough to overcome the effects of thin film demagnetization, resulting in a spontaneous out-of-plane magnetization. Under a large perpendicular field, all magnetic domains are aligned along the field direction, and hence the resistance is low. When the field is near the coercive field, opposite aligned domains are present. The resulting high resistance is most likely due to spin dependent scattering from the opposite aligned domains. When the field is applied parallel to the film plane, a much higher field is required to align all the domains along the magnetic hard axis, and the MR ratio is relatively small. Since a large MR due to spin-dependent scattering relies on unaligned ferromagnetic entities, we expect narrow domain walls in the compressively-strained films.

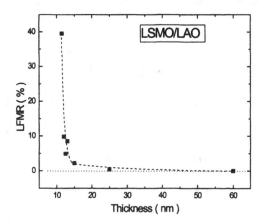

Fig.4. The maximum LFMR as a function of film thickness of the LSMO films on LAO substrates.

On the other hand, it has previously been shown that an in-plane biaxial tensile strain induced by an STO substrate is sufficient to make a magnetically hard axis normal to the film plane[5]. When a perpendicular field is applied, the magnetization will rotate out of the plane and be perpendicular to the film plane. As discussed in detail by Eckstein et al.[6], the resistance increases with field due to the increasing angle between the measuring current and the magnetization, resulting in a positive MR. When the field is applied parallel to the film plane, the LFMR hysteresis is due to the domain rotation and movement within the film plane and the negative MR is due to the increased magnetization in the field direction.

The reason why the compressively-strained LBMO films do not show the similar LFMR effect as LCMO and LSMO films is not well understood yet. But it could possibly be due to the difference of the domain wall and domain structures. The fact that LSMO samples also show smaller effect and only very narrow thickness range samples exhibit the effect may be due to the same reasons. Further study on the domain wall and domain structures is required to understand the mechanism of the LFMR.

CONCLUSIONS

We have studied the strain effects on the low field magnetoresistance of the PSMO, LBMO, LSMO and LCMO films grown on different substrates and with various thickness. Very large LFMR was found on compressively-strained very thin PSMO, LCMO and LSMO samples, but it's not seen on LBMO films so far. Significant anisotropy and the thickness dependence of the LFMR were also observed. Based on the above experimental results and discussion, we conclude that the strain-induced magnetic anisotropy plays a crucial role in determining the LFMR properties in the ultrathin manganite films.

ACKNOWLEDGEMNET

The authors acknowledge T. E. Mallouk for helps in X-ray diffraction measurement. This work is supported in part by the NSF DMR-9876266 and 9972973 and Petroleum Research Fund.

REFERENCE

1. H. S. Wang and Qi Li, Appl. Phys. Lett. **73**, 2360 (1998).

2. H. S. Wang, Qi Li, Kai Liu, and C. L. Chien, Appl. Phys. Lett. **74**, 2212 (1999).

3. X. W. Xu, M. S. Rzchowski, H. S. Wang, and Qi Li, submitted to Phys. Rev. B.

4. H. S. Wang, E. Wertz, Y.F. Hu, Qi Li, and D. Schlom, submitted to J. Appl. Phys..

5. J. O'Donnell, M. S. Rzchowski, J. N. Eckstein and I. Bozovic, Appl. Phys. Lett. **72**, 1175 (1998).

6. J. N. Eckstein, I. Bozovic, J. O'Donnell, M. Onellion, and M. S. Rzchowski, Appl. Phys. Lett. **69**, 1312 (1996).

GROWTH AND CHARACTERIZATION OF EPITAXIAL THIN HETEROSTRUCTURES OF FERROMAGNETIC/ANTIFERROMAGNETIC SrRuO₃/Sr₂YRuO₆

R.A. PRICE, M.K. LEE, C.B. EOM
Department of Mechanical Engineering and Materials Science, Duke University, Durham, NC 27708
X.W. Wu, M.S. RZCHOWSKI
Department of Physics, University of Wisconsin-Madison, Madison, WI

ABSTRACT

We have grown epitaxial thin films of antiferromagnetic ruthenate Sr_2YRuO_6 on miscut (001) $SrTiO_3$ by 90° off-axis sputtering. Sr_2YRuO_6 is a unique material that allows us to grow epitaxial ferromagnetic/antiferromagnetic heterostructures. Antiferromagnetic Sr_2YRuO_6 has the same pseudo-cubic perovskite crystal structure as the ferromagnetic conductive oxide $SrRuO_3$. The Sr_2YRuO_6 perovskite crystal structure has Y and pentavalent Ru located on the octahedral sites and the pseudo-cubic lattice parameter of 4.08Å. The Neel temperature of bulk Sr_2YRuO_6 is known to be 26K. Four-circle X-ray diffraction analysis revealed the Sr_2YRuO_6 films are purely (110) normal to the substrate with two 90° domains in the plane. We have also grown epitaxial heterostructures of $SrRuO_3/Sr_2YRuO_6$. These bilayers permit detailed studies of the magnetic exchange bias phenomena at these interfaces, including the role of uncompensated spins thought to arise from interface roughness. Magnetization measurements on the $SrRuO_3/Sr_2YRuO_6$ heterostructures show a shifting of the hysteresis loop, indicating exchange bias. Such exchange-biased interfaces are important for electrode pinning in magnetic tunnel junctions.

INTRODUCTION

Sr_2YRuO_6 is an interesting material among oxides with ordered perovskite structure because of its antiferromagetic character as a bulk material. The successful growth of antiferromagnetic Sr_2YRuO_6 thin films makes possible the fabrication of a unique ferromagnetic/antiferromagnetic bilayer. This bilayer system allows the detailed study of interface interactions and the exchange bias effect in a high quality ferromagnetic/antiferromagnetic bilayer system.   Battle et al. have reported that bulk Sr_2YRuO_6 exhibits Type I antiferromagnetic ordering with a Neel temperature of 26K. In Type I antiferromagnetic ordering the nearest neighbor ions are antiferromagnetically coupled to the central ion and the next-nearest neighbors are ferromagnetically coupled. From their results, they concluded that bulk Sr_2YRuO_6 has a magnetic structure consisting of ferromagnetic (001) sheets that are coupled antiferromagnetically along the [001] direction with dipolar interactions causing the spins to be in the (001) planes, as seen in Figure 1, and a ordered magnetic moment of $1.85\mu_B$ per Ru^{5+} ion. [1] Pentavalent Ru

Mat. Res. Soc. Symp. Proc. Vol. 602 © 2000 Materials Research Society

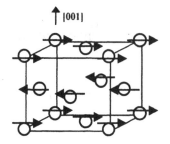

Fig. 1
Type I antiferromagnetic magnetic ordering in bulk Sr_2YRuO_6. (001) ferromagnetic sheets are coupled antiferromagnetically along [001].

↑ [001]

and Y are located on the octahedral sites of the perovskite Sr_2YRuO_6 crystal structure which has orthorhombic unit cell parameters a = 5.752Å, b = 5.773Å, and c = 8.158Å. [2] Using these parameters, the pseudocubic lattice parameter is calculated to be 4.08Å. Further studies show that bulk Sr_2YRuO_6 is not a metallic conductor and has an electrical resistivity of $5.9 \cdot 10^4$ Ω-cm at 300K. [3]

The reasons for choosing the conductive oxide $SrRuO_3$ for the ferromagnetic layer in the bilayer system are two-fold. $SrRuO_3$ thin films have a pseudocubic perovskite crystal structure and pseudocubic lattice parameter 3.905Å, similar to Sr_2YRuO_6. Also, structure and crystal quality studies have shown that single domain in-plane single crystal growth of $SrRuO_3$ is possible on the smooth surface of $SrTiO_3$. Uniform growth can occur because of the small unit cell step height, 4Å, of $SrTiO_3$. [4] Therefore, $SrRuO_3$ has the potential for providing a good growth surface for Sr_2YRuO_6 film growth with a very uniform interface between the Sr_2YRuO_6 and $SrRuO_3$ layers. A high quality bilayer allows unprecedented control and optimization of exchange bias, an important phenomena in modern magnetic technology.

EXPERIMENT

The Sr_2YRuO_6 films were deposited on (001) $SrRuO_3$ substrates, miscut 0°, 0.8°, and 2° towards [100], by 90° off-axis sputtering using a stoichiometric target. The films were deposited at an operating pressure of 200mTorr (60%Ar/40%O_2) and temperature of 600°C. The samples were cooled to room temperature at an oxygen pressure of 300 Torr. The thickness of the films is approximately 1000Å. Thicker films, approximately 3000Å, were deposited on the same substrate materials and under the same conditions except for an increased operating temperature of 700°C. The temperature was increased in an attempt to improve crystalline quality, but similar results for the two different films showed that this temperature change was not very significant. X-ray diffraction analysis, θ-2θ scans and φ scans, was used to characterize crystallographic orientation and domain structure and sensitive magnetic measurements were made with the SQUID magnetometer.

RESULTS

The Sr_2YRuO_6 films were deposited on (001) $SrTiO_3$ substrate samples that are miscut towards [100]. The rocking curve FWHM values ranging from 0.5507° to

0.6025° showed that the films had high crystalline quality. The out-of-plane lattice parameter of 4.05Å is determined from the x-ray diffraction θ-2θ scan of Sr₂YRuO₆ film on 2° miscut (001) SrTiO₃ substrate, as shown in Figure 2, which shows the film peaks at 21.97° for the (110) reflection and at 44.80° for the (220) reflection. This value is similar to the pseudocubic lattice parameter of bulk Sr₂YRuO₆ (4.08Å). Therefore, incoherent growth occurs. Azimuthal x-ray scans of the off-axis (110) and (221) reflection indicate an epitaxial arrangement of two 90° domains in the film plane, as shown in Figure 3.

Magnetic measurements using a SQUID magnetometer were performed as a function of temperature and magnetic field. After grinding the backside of the SrTiO₃ substrate, the magnetization measurement at 5K showed a diamagnetic signal from the SrTiO₃ substrate that deviated slightly around 0 Oe. This result was different from the result of the measurement made prior to grinding the back surface, which showed weak ferromagnetic behavior. This indicates that there is some weak ferromagnetic component in the Sr₂YRuO₆ film, which is most likely due to defects and impurities in the film. Figure 4 shows this magnetization curve after subtracting out the diamagnetic background signal; only a very weak signal is apparent.

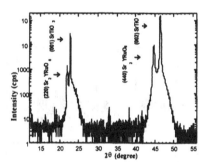

Fig. 2
X-ray diffraction θ–2θ scan of Sr₂YRuO₆ thin film on 2°miscut (001) SrTiO₃.

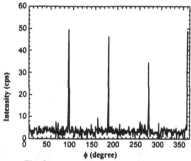

Fig. 3
X-ray diffraction off-axis φ scan of the Sr₂YRuO₆ (002)° reflection.

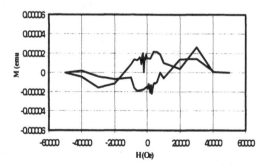

Fig. 4
Magnetization curve for Sr₂YRuO₆ film on miscut (001) SrTiO₃ substrate.

The magnetic measurements made on the $SrRuO_3/Sr_2YRuO_6$ bilayer samples also indicate the antiferromagnetic nature of Sr_2YRuO_6 films. $SrRuO_3/Sr_2YRuO_6$ bilayers were grown on miscut (001) $SrTiO3$ by 90° off-axis sputtering with similar deposition conditions used for Sr_2YRuO_6 thin film growth. The two layers were not deposited in situ. X-ray diffraction analysis confirmed the epitaxial growth of Sr_2YRuO_6 on the $SrRuO_3$ film, as shown in Figure 5. Magnetization measurements made on these samples indicate exchange bias phenomena. The magnetization results show a shifting of the hysteresis loop to the left for the $Sr_2YRuO_6/SrRuO_3$ bilayer. The amount of offset of the hysteresis loop is temperature dependent and is best detected at 5K. Figure 5 compares the hysteresis loops for the 100Å bilayer sample with the hysteresis loop of the 100Å $SrRuO_3$ thin film. The results show a shifting of 1.02kOe to the left for the bilayer sample. The large coercive field of the $SrRuO_3$ layer makes the shifting of the hysteresis loop difficult to detect. The thicker bilayer sample has a smaller coercive field, approximately 0.75kOe at 5K, but the shifting of the hysteresis loop is also reduced and even more difficult to detect. The hysteresis loop shifting suggests that the $SrRuO_3$ film layer is being "pinned" by the Sr_2YRuO_6 upper film layer, a characteristic of a ferromagnetic/antiferromagnetic bilayer.

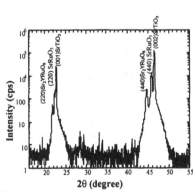

Fig. 5a
X-ray diffraction θ-2θ scan of $SrRuO_3/Sr_2YruO_6$ bilayer on 2° miscut (001) $SrTiO_3$ substrate.

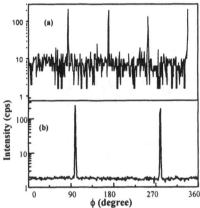

Fig. 5b
X-ray diffraction off axis φ scan of (a) Sr_2YRuO_6 (002) reflection and (b) $SrRuO_3$ with [110] normal orientation.

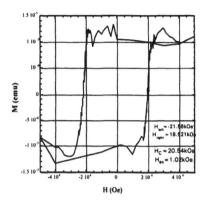

Fig. 6a
The shifting of the hysteresis loop for $SrRuO_3/SrYRuO_6$ bilayer sample at 5K can be detected in this magnetization curve.

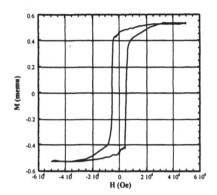

Fig. 6b
The symmetry of the hysteresis loop for a 100Å $SrRuO_3$ thin film on (001) $SrTiO_3$ substrate at 5K is seen in this magnetization curve. The field is applied parallel to the [110] in plane.

CONCLUSIONS

We have grown films and $SrRuO_3/Sr_2YRuO_6$ bilayers on miscut (001) $SrTiO_3$ by 90° off-axis sputtering and studied their epitaxial arrangement and magnetic properties. These studies have shown that antiferromagnetic Sr_2YRuO_6 films with two 90° domains can be grown. The successful growth of the $SrRuO_3/Sr_2YRuO_6$ bilayer provides a new system for studying exchange bias phenomena and spin ordering at the interfaces of oxide materials.

ACKNOWLEDGMENTS

This work was supported by the NSF-REU Program, the David and Lucile Packard Fellowship, the NSF Young Investigator Award, and the NSF-DMR.

REFERENCES

1. P.D. Battle and W.J. Macklin, J. of Solid State Chem. **52**, p. 138-145 (1984).
2. P.C. Donohue and E.L. McCann, Mat. Res. Bull. **12**, pp.519-524 (1977).
3. R. Greatrex, N. Norman, M. Lal, and I. Fernandez, J. of Solid State Chem. **30**, p. 137-148 (1979).
4. R.A. Rao, Q. Gan, C.B. Eom, Applied Physics Letters 71, p. 1171-1173 (1997).

GROWTH AND MAGNETIC PROPERTIES OF La$_{0.65}$Pb$_{0.35}$MnO$_3$ FILMS

Q. L. XU, M. T. LIU, Y. LIU, C. N. BORCA, H. DULLI, P. A. DOWBEN, S. H. LIOU
University of Nebraska, Behlen Laboratory of Physics and the Center for Materials Research and Analysis, Lincoln, NE, 68588-0111, sliou@unl.edu

ABSTRACT

We have successfully grown La$_{0.65}$Pb$_{0.35}$MnO$_3$ thin films by a RF magnetron sputtering method onto (100) LaAlO$_3$ single crystal substrates. X-ray diffraction measurements are consistent with a (100) cubic orientation of the films. The fourfold symmetry showed by LEED(Low Energy Electron Diffraction) patterns indicate that the films have surface order. STM(Scanning Tunneling Microscopy) measurements indicate that the surface of the films were smooth, with approximate 5 nm roughness. XPS (X-ray Photoemission Spectroscopy) shows that the surface defect density in the films is comparatively low. The bulk magnetization of the films at 6K in 1 T magnetic field reached 77 emu/g and a Curie temperature near 354 K, close to maximum resistivity. A negative magnetoresistance of 47% was observed at 320K in 5.5 T magnetic field.

INTRODUCTION

Since the discovery [1] of the large magnetoresistance (MR) in La$_{1-x}$A$_x$MnO$_3$ (A=dopant), the attention was focused on Sr and Ca doped manganites [2-4], because of both potential technological applications and basic condensed-matter physics studies. On the other hand, these type of materials exhibit a wide range of imperfectly understood, structure, magnetic and electronic properties.

The La$_{1-x}$Pb$_x$MnO$_3$ systems have received much less attention, although single crystal samples were successfully grown and studied [5-9]. More recently, several papers [10-13] reported that La$_{1-x}$Pb$_x$MnO$_3$ thin film exhibited large magnetoresistance. In this paper, we report results of fabrication and characterization of La$_{0.65}$Pb$_{0.35}$MnO$_3$ films made by RF magnetron sputtering process.

SAMPLE PREPARATION AND CHARACTERIZATION

La$_{0.65}$Pb$_{0.35}$MnO$_3$ films were deposited by RF magnetron sputtering method. The targets were made of from very homogeneous powders that were prepared by citric acid sol-gel process. The targets were pre-sintered in 900°C for 12 hours (form a stable phase) and then sintered at 1000°C for 24 hours (form a dense target). The deposition system was pumped down to a base pressure of 10^{-7} torr and then filled with a mixture of 20% oxygen and 80% argon up to a pressure of 23 mtorr during the sputtering. The target-substrate distance was around 2 inches. Films with a thickness of 1000 Å were deposited when on-axis sputtering was carried out at a RF power of 15 Watts (at 13.6 MHz) for 1.5 hours. The temperature of LaAlO$_3$ substrates was heated between 200 and 300°C during the deposition. The as-prepared films were subsequently annealed, first at 650°C for 10 hours (in Ar), then at 850°C for 2hours (in Ar), and last, at 650°C for 10 hours (in 21 psi O_2).

X-ray diffraction patterns were measured by using a Rigaku X-ray difractometer with Cu K_α radiation. DC magnetization and resistance in a four-probe configuration were carried out using a SQUID quantum magnetometer. X-ray photoemission spectroscopy was undertaken with the Mg-K_α line (1253.6 eV). Energy distribution curves of the elemental core levels were acquired with an hemispherical electron analyzer (PHI Model 10-360). The surface morphology and magnetic domain structure were characterized by a scanning probe microscope (Digital Instrument Dimension 3000).

RESULTS AND DISCUSSIONS

The best films have been deposited using an optimum substrate (single crystal LaAlO$_3$), kept at a temperature between 200 to 300°C. Higher deposition temperature resulted in a random orientation of the films. The as-prepared films were amorphous, and the perovskite phase was formed after annealing in oxygen. Figure 1 shows the magnetization versus temperature curves (in a field of 1T) for a 1μm thick $La_{0.65}Pb_{0.35}MnO_3$ film sintered at different temperatures for 10 hours. Magnetization and Curie temperature are very low for the film annealed at 700°C. Microstructure measurements indicate that the films have not crystallized well. Both magnetization and Curie temperature increase with increasing annealing temperature up to 800°C. The film annealed at 850°C for 10 hours has the highest magnetization (77 emu/g) and Curie temperature is as high as 354 K.

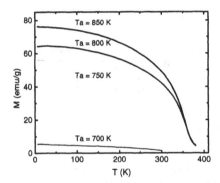

Figure 1. The curves of the magnetization versus temperature for 1μm La$_{0.65}$Pb$_{0.35}$MnO$_3$ film annealed at different temperatures for 10 hours.

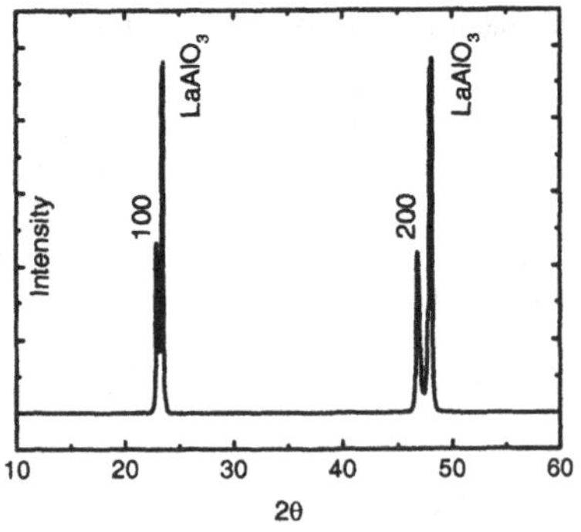

Figure 2. X-ray diffraction (θ-2θ scan) for 1000 Å thick film.

We prepared a thinner film of 1000 Å, which was subsequently annealed in argon at 650°C for 10 hours and at 850°C for another 2 hours. The film was additionally treated at 650°C for 10 hours in 21 psi O_2, in order to improve the oxygen content. Figure 2 shows the X-ray diffraction of the 1000 Å tick film. The two peaks, (100) and (200) are indexed with a cubic perovskite structure, and no extra phases were observed even on a logarithmic scale. The LEED pattern on the same sample was recorded at around 60 eV electron energy. The picture (data not shown) indicates that the surface has a high degree of order, with fourfold symmetry. The x-ray diffraction peaks of the (100) and (200) reflections are very sharp, comparable to that of the single crystal $LaAlO_3$ substrates, indicating an excellent order parallel to the surface normal. The average lattice constant, obtained from the peak position at the (200) x-ray reflection is 3.866 Å.

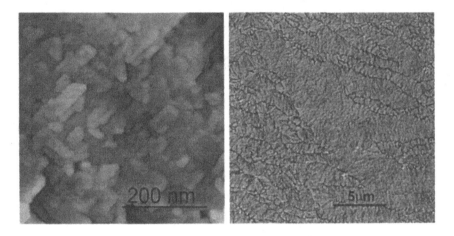

Figure 3. STM (left) and MFM (right) images of a 1000 Å thick $La_{0.65}Pb_{0.35}MnO_3$ film.

The surface morphology was examined using a scanning probe microscope. The scanning tunneling microscopy (STM) image (Figure 3, left panel) shows a 5 nm surface roughness for the 5 nm for the 1000 Å thick $La_{0.65}Pb_{0.35}MnO_3$ films. The crystallite size of the film is about 50-150nm and is aligned about 45° from the [100] direction of the $LaAlO_3$ substrate. The magnetic force microscopy (MFM) scan (Figure 3, right panel) shows the presence of complicated ferromagnetic domains structure. Many cross-tie walls were observed. The magnetic force microscopy tip was magnetized along the z-direction (perpendicular to the film surface). The domain size is much larger than the crystallite size.

Figure 4 shows the oxygen 1s core level as a function of emission angle obtained by X-ray Photoemission Spectroscopy (XPS). The peak of the lower binding energy, around 529.4 eV is the main O^{2-} lattice oxide peak and the shoulder at 531.3 eV is indicative of defects in the surface [14]. The intensity of 529.4 eV O²⁻ lattice peak decreases quickly and monotonically with increasing emission angle, similar to other manganese perovskites [14]. The satellite peak appears only at high emission angles, indicating that the defects are constraint in a very near surface region. Comparatively, this shoulder in the main O-1s core is much weaker than in the case of Sr, Ca, or Ba doped manganites [14], indicating that the defect density in our $La_{0.65}Pb_{0.35}MnO_3$ films is very low.

The temperature dependencies of resistivity for the 1000 Å thick $La_{0.65}Pb_{0.35}MnO_3$ film measured in 0 T, as well as 5.5 T field are shown in Figure 5. The zero field resistivity shows a maximum at 351 K very close to the Curie temperature (354 K). The value of resistivity at 351 K is 4.8 mΩcm close to the value of 5 mΩcm, obtained for a single crystal of $La_{1-x}Pb_xMnO_3$ with x=0.4 [8].

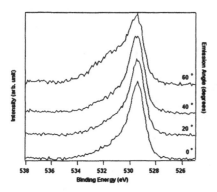

Figure 4. Oxygen 1s core level spectra of the $La_{0.65}Pb_{0.35}MnO_3$ oxide films as a function of emission angle, obtained by XPS.

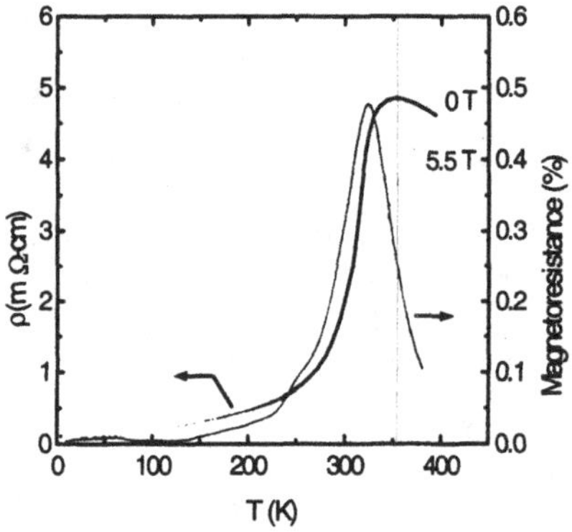

Figure 5. Temperature dependencies of resistivity (solid lines) measured at zero and 5.5T magnetic fields respectively. The dotted line is for MR.

The resistivity ρ can be extrapolated to about 0.2 mΩcm at T=0 K, which is an intrinsic property for single crystals or epitaxial films [15]. The peak of the ρ-T curve shifts to higher temperature in 5.5 T applied magnetic field. On the right axis of Figure 5 is plotted the magnetoresistance MR= Δρ/ρ(0,T)=-(ρ(H,T)-ρ(0,T))/ρ(0,T) versus temperature. The curve displays a sharp peak around 320 K and a negative magnetoresistance of 47% in a 5.5 T applied field. The value of the MR becomes small with decreasing temperature. With decreasing temperature, ρ drops rapidly as the neighboring spins become aligned and magnetization approaches its saturation value.

CONCLUSION

The $La_{0.65}Pb_{0.35}MnO_3$ films have fabricated by using RF magnetron sputtering and subsequent annealing steps. The LEED, STM and X-ray diffraction patterns show that the films have an ordered surface and (100) orientation. Microstructure analyses indicated that the films have a very flat surface, with an approximate 5 nm roughness. Compared to other manganese perovskites, the surface density of defects is very low, as indicated by the line shape of the O-1s core level. Magnetic properties confirmed that the films were fully ferromagnetic with the Curie temperature as high as 354 K, for samples thermally treated at 850 °C. The resistivity maximum appears at a temperature very close to the Curie point, and the magnetoresistance is as high as 47% at ambient temperatures.

ACKNOWLEDGMENTS

This work was supported by NSF grant DMR-98-02126, Department of Energy grant DE-FG02-90ER45427, and the Center for Materials Research and Analysis at the University of Nebraska.

REFERENCES

1. R. M. Kusters, J. Singleton, D. A. Keen, R. M. Greevy, and W. Hayes, Physica B **155**, p. 362 (1989).

2. S. Jin, T. H. Tiefel, M. McCormack, R. A. Fastnacht, R. Ramesh, and L. H. Chen, Science **264**, p. 413 (1994).

3. A. Asamitsu, Y. Moritomo, R. Kumi, Tomioka, and Y. Tokura, Phys. Rev. B **54**, p. 1716 (1996).

4. A. Urushibara, Y. Moritomo, T. Arima, A. Asamitsu, G. Kido, and Y. Tokura, Phys. Rev. B **51**, p. 14103 (1995).

5. A. H. Morrish, B. J. Evans, J. A. Eaton, and L. K. Leung, Canadian J. of Phys. **47**, p. 2691 (1969).

6. L. K. Leung, A. H. Morrish, and C. W. Searle, Canadian J. of Phys. **47**, p. 2697 (1969).

7. C. W. Searle, and S. T. Wang, Canadian J. of Phys. **47**, p. 2703 (1969).

8. C. W. Searle, and S. T. Wang, Canadian J. of Phys. **48**, p. 2023 (1970).

9. L. K. Leung, and A. H. Morrish, Phys. Rev. B **15**, p. 2485 (1977).

10. S. Sundar Manoharan, N. Y. Vasanthacharya, and M. S. Hegede, J. Appl. Phys. **76**, p. 3923 (1994).

11. K. M. Satyalakshmi, S. Sundar Manoharan, M. S. Hegede, V. Prasad, and S. V. Subramanyam, J. Appl. Phys. **78**, p. 6861 (1995).

12. G. Srinivasan, V. Suresh Babu, and M. S. Seehra, Appl. Phys. Lett. **67**, p. 2090 (1995).

13. J. Q. Wang, R. C. Barker, G. J. Cui, T. Tamagawa, and B. L. Halpern, Appl. Phys. Lett. **71**, p. 3418 (1997).

14. Jaewu Choi, Hani Dulli, S. H. Liou, P.A. Dowben, and M. A. Langell, Phys. Stat. Sol. (b) **214**, p.45 (1999).

15. Y. X. Jia, Li Lu, K. Khazeni, Vincent H. Crespi, A. Zettl, and Marvin L. Cohen, Phys. Rev. B **52**, p. 9147 (1995).

CATION ORDERING STRUCTURE IN $La_{0.8}Ca_{0.2}MnO_3$ THIN FILMS BY PULSED LASER DEPOSITION

Y. H. LI[+], M. RAJESWARI[*], A. BISWAS[*]. D. J. KANG[*], C. SEHMEN[*], L. SALAMANCA-RIBA[+], R. RAMESH[+*] and T. VENKATESAN[*]
NSF-MRSEC and [+]Dept. of Materials and Nuclear Engineering and [*]Center for Superconductivity Research, University of Maryland, College Park, MD 20742

ABSTRACT

A careful analysis of high-resolution transmission electron microscopy images from $La_{0.8}Ca_{0.2}MnO_3$ thin films indicates that the images can not explained based on either the classical Pnma structure or its monoclinic distortion. A cation ordered structure is proposed which could be responsible for the significantly higher Tc of the film (298K) compared with the bulk material of the same composition (190K).

INTRODUCTION

The physics of Colossal Magneto-Resistance (CMR) manganite systems has attracted a significant interest in the recent years [1-3]. These materials have very interesting physical properties which are still not well understood, in part because in these materials the spin, charge, orbital and lattice systems are intimately coupled. One important experimental finding is that many physical properties, most notably the value of the magnetic transition temperature, the magnetic moment observed in the low temperature magnetic phase, and the magnitude and temperature dependence of the electrical conductivity display an unusually high degree of sensitivity to apparently weak external perturbations. These include magnetic field, applied strain [4-6], surface preparation [7], and presence of grain boundaries [8-10] and changes in sample form (i.e. film vs. poly-crystal bulk vs. single-crystal bulk). This anomalous sensitivity is not at present understood. It has been argued [6] that it is related to a strong strain dependence of the materials' properties, stemming ultimately from a strong coupling of charge degrees of freedom to lattice distortions. One important finding has been that in thin films, particularly in the low-doping regime, the ferromagnetic transition temperature and the metal-insulating transition temperature are anomalously high compared to the bulk materials [11-13]. Part of this anomaly is related to possible cation vacancies and self-doping effects, but these alone can not explain the anomaly completely. In an attempt to understand possible structural contributions to such anomaly we have investigated the structure of these films by Transmission Electron Microscopy (TEM).

In this study, we present a pseudo-tetragonal structure due to cation-ordering in $La_{0.8}Ca_{0.2}MnO_3$ (LCMO) thin films which could be responsible for the dramatic increase of Tc for the film to 298K compared with bulk material of the same composition (190K).

EXPERIMENT

The epitaxial LCMO thin films used in this study were grown on LAO (100) single crystal substrates by pulsed laser deposition technique. A KrF excimer laser was employed and the ablation was performed at laser energy of 2 J/cm². The substrate temperature was 820 °C and oxygen pressure was 400 mTorr during the deposition and 400 Torr during cool down. The film thickness was about 100 nm. After deposition, the films were annealed in an oxygen atmosphere at 850 °C for 10 hours.

TEM studies were carried out using a JEOL 4000FX TEM operated at 300 kV. Plan-view TEM samples were made by mechanical thinning and ion milling from the backside of the substrate. Mactempas software was used for High-resolution image simulation.

RESULTS

It is known that the $La_{0.7}Ca_{0.3}MnO_3$ films normally have a face-centered pseudo-cubic structure with a lattice parameter of a ~ 2 a_p (a_p ~ 0.38 nm, the basic perovskite unit) [14]. Only when the film is annealed does a pseudo-tetragonal structure with lattice parameters of a ~ $\sqrt{2}$ a_p and c ~ 2 a_p develops locally [14]. An electron diffraction pattern from a plan-view sample of the $La_{0.8}Ca_{0.2}MnO_3$ film is shown in Fig.1a, which can be indexed as a simple cubic pattern along the [001] zone axis and with a lattice parameter of a ~ $2a_p$. However, when a smaller selected area aperture was used, the diffraction pattern changed as shown in Fig.1b. This pattern suggests a pseudo-tetragonal structure with a lattice parameter of a ~ $\sqrt{2}$ a_p and c ~ $2a_p$. The diffraction pattern in Fig.1a can then be explained as an overlap of the diffraction patterns of three possible epitaxial orientations of this pseudo-tetragonal structure [14]. This has also been confirmed by high-resolution pictures which show clearly regions with three possible expitaxial orientations as shown in Fig.2.

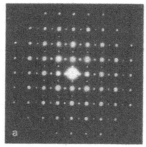 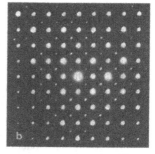

Fig.1. Electron diffraction pattern from the film (a) and from a single grain (b).

In addition, the doubling of the basic perovskite unit cell in the image is represented by the alternating brightness of the lattice fringes as shown in 3a. It has previously been shown that a classical Pnma structure could not give a similar high-resolution image with double periodicity and a monoclinic distortion of the Pnma structure was used to obtain simulated images [15,16]. A series of image simulations along 2 a_p direction based on this distorted structure were obtained in the present study in an attempt to match a wide range of experimental conditions. However, the simulated images do not give as good match as the cation ordered structure presented below. The image in Fig.4 shows clearly a $\sqrt{2}$ a_p square pattern formed by strong spots with a considerable weaker spot at the square center. But the image simulations based on the distorted structure over a wide

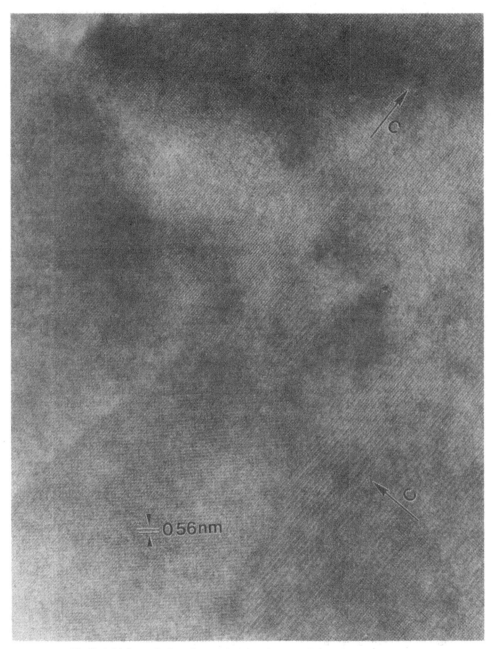

Fig.2. A high-resolution picture showing three possible oriented grains.

Fig.3a. A lattice image along [110] direction and its image simulation (inset).

Fig.3b. A lattice image with different defocus condition and its image simulation.

range of thickness and defocus values can only produce a $\sqrt{2}\ a_p$ square pattern with a spot at the square center with the same brightness as shown in Fig.5 (the image covers four unit cells). Therefore, we propose a new tetragonal structure with cation ordering which has lattice parameters of a = 0.56 nm and c = 0.77 nm. The unit cell of this structure is shown in Fig.6. In this structure, La ions take up A sites at the four corners and body center and bottom face center. The A sites at the four vertical edges are shared by both Ca with an occupancy of 0.8 and La with an occupancy of 0.2. This gives an over all Ca composition of 0.2. The image simulation along [001] direction with a defocus of 30 nm and a sample thickness of 9 nm based on this structure gives a quite good match to the experimental image as shown by the inset to Fig.4. The image simulation along the [110] direction assuming cation ordering with a defocus of 40nm and a sample thickness of 18nm also gives a good match to the experimental image as shown as an inset to Fig.3a. Fig.3b shows another image and the simulation with a defocus of 60nm and a sample thickness of 15nm.

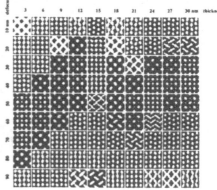

Fig.4. A lattice image along [001] direction and its image simulation (inset).

Fig.5. Image simulations based on the monoclinic distorted structure.

Resistivity data of this $La_{0.8}Ca_{0.2}MnO_3$ thin film is shown in Fig.7. Tc of the film is significantly higher than that reported for the bulk samples with the same composition. It is noteworthy that though the self-doping effect of cation vacancies has been known to cause higher Tc, this alone can not explain why the Tc of the film is higher than the maximum Tc in the bulk phase diagram. It has been reported that cation ordering in $LaBaMn_2O_6$ bulk material can increase Tc from 270K to 335K [17]. Therefore, we attribute the increase in Tc the thin films to cation ordering.

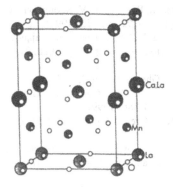

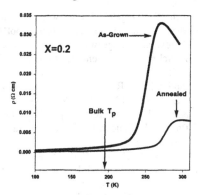

Fig.6. A schematic of cation ordered structure.

Fig.7. Resistivity data of the film compared with the bulk material.

CONCLUSION

Electron diffraction patterns, high-resolution images and image simulation obtained in this work are consistent with the formation of a pseudo-tetragonal structure due to cation ordering in the $La_{0.8}Ca_{0.2}MnO_3$ films. This cation ordering could be responsible for the significant increase of Tc in the films compared with the bulk material with the same composition.

ACKNOWLEDGMENT

This work has been supported by NSF-MRSEC under grant # DMR-9632521 and NSF-DMR-9705065.

REFERENCES

1. K. Chahara, Y. Ohno, M. Kasai and Y. Kosono, Appl. Phys. Lett. **63**, p.1,990 (1993).
2. S. Jin, T. H. Tiefel, M. McCromack, R. A. Fastnatch, R. Ramesh and L. H. Chen, Science **264**, p.413 (1994).
3. A. J. Millis, P. B. Littlewood, B. I. Shraiman, Phys. Rev. Lett. **74**, p.514 (1995).
4. H. Y. Hwang, T. T. M. Palstra, S. W. Cheong and B. Batlogg, Phys. Rev. B **52**, p.15,046 (1995).
5. A. J. Millis, A. Goyal, M. Rajeswari, K. Ghosh, R. Shreekala, R. L. Greene, R. Ramesh and T. Venkatesan, Appl. Phys. Lett. (unpublished).

6. A. J. Millis, T. Darling and A. Migliori, J. Appl. Phys. **83**, p.1,588 (1998).

7. K. Takenaka, K. Iida, Y. Sawaki, S. Sugai, Y. Moritomo and A. Nakamura (unpublished).

8. H. Y. Hwang, S. W. Cheong, N. P. Ong and B. Batlogg, Phys. Rev. Lett. **77**, p.2,041 (1996).

9. N. D. Mathur, G. Burnell, S. P. Isaac, T. J. Jackson, B. S. Teo, J. L. MacManus-Driscoll, L. F. Cohen, J. E. Evetts and M. G. Blamire, Nature (London) **387**, p.266 (1997).

10. J. Y. Gu, S. B. Ogale, M. Rajeswari, T. Venkatesan, R. Ramesh, V. Radmilovic, U. Dahmen, G. Thomas and T. W. Noh, Appl. Phys. Lett. **72**, p.1,113 (1998).

11. R. Shreekala, M. Rajeswari, K. Ghosh, A. Goyal, J. Y. Gu, C, Kwon, Z. Trajauovic, T. Boettcher, R. L. Greene, R. Ramesh and T. Venkatesan, Appl. Phys. Lett. **71**, p.282 (1997).

12. R. Shreekala, M. Rajeswari, R. C. Srivastava, K. Ghosh, A. Goyal, V. V. Srinivasu, S. E. Lofland, S. M. Ghagat, M. Downes, R. P. Sharma, S. B. Ogale, R. L. Greene, R. Ramesh and T. Venkatesan, Appl. Phys. Lett. **74**, p.1,886 (1999).

13. W. Prellier, M. Rajeswari, T. Venkatesan and R. L. Greene, Appl. Phys. Lett. **75**, p.1,446 (1999).

14. Y. H. Li, K. A. Thomas, P. de Silva, L. F. Cohen, A. Goyal, M. Rajeswari, N. D. Mathur, M. G. Blamire, J. E. Evetts, T. Venkatesan and J. L. MacManus-Driscoll, J. Mater. Res., **13**, p.2,161 (1998).

15. M. Hervieu, G. Van Tendeloo, V. Caignaert, A. Maignan and B. Raveau, Phys. Rev. B **53**, p.14,274 (1996).

16. O. I. Lebedev, G. Van Tendeloo, S. Amelinckx, B. Leibold and H. U. Habermeier, Phys. Rev. B **58**, 8,065 (1998).

17. F. Millange, V. Caignaert, B. Domenges and B. Raveau, Chem. Mater. **10**, p.1,974 (1998).

LARGE MAGNETORESISTANCE ANISOTROPY IN STRAINED PR$_{0.67}$SR0$_{.33}$MNO$_3$ THIN FILMS

H. S. Wang, Y. F. Hu, E. Wertz, and Qi Li
Department of Physics, Pennsylvania State University, University Park, PA 16802

ABSTRACT

We have studied the anisotropic magnetoresistance (AMR) of strained Pr$_{0.67}$Sr$_{0.33}$MnO$_3$ thin films by measuring the MR as a function of the angle between the magnetic field direction and the substrate normal (out-of-plane). The results show that the compressive- and tensile-strained ultrathin films (5-15 nm) grown on LaAlO$_3$ (001) (LAO) and SrTiO$_3$ (001) (STO) substrates show unusually large out-of-plane AMR, but with opposite signs. In contrast, the almost strain-free films on the NdGaO$_3$ (110) substrates show much smaller AMR over all the temperature and field ranges studied. Thick films on LAO and STO substrates also show much smaller AMR.

I. INTRODUCTION

The effect of the anisotropic magnetoresistance (AMR) is an important property of ferromagnetic (FM) materials. It has a wide range of practical applications in magnetic devices such as magnetic reading heads,[1] and magnetic sensors. By measuring the AMR effect, it is also possible to obtain information about the magneto-crystal anisotropy and spin-orbit coupling.[2,3] For thin film samples, one can measure the AMR in the film plane by changing the magnetic field direction relative to the current flow direction. In this case, the measured AMR includes both the Lorentz MR[4] and the magneto-crystalline AMR. One can also measure the out-of-plane AMR, in which the magnetic field and the current can always be kept perpendicular to each other as the magnetic field rotates from in-plane to out-of-plane. The Lorentz MR can then be eliminated, and the AMR from magneto-crystalline anisotropy can be exclusively obtained.

The AMR effect in single crystal samples of the colossal magnetoresistance (CMR) manganites has been found negligible. However, in epitaxial thin film samples the magnetic anisotropy can be much stronger and the AMR can be much larger than in bulk materials due to uniaxial strain effect induced by lattice mismatch between the film and substrate.[5-10] Recently, Eckstein et al.[6] have reported relatively large in-plane AMR (~8 %) in the La$_{0.66}$Ca$_{0.33}$MnO$_3$ thin films. Wolfman et al.[8] have studied the out-of-plane AMR of relatively thick Pr$_{0.7}$Sr$_{0.3}$MnO$_3$ films at different magnetic fields and found a correlation between the magnetization and the AMR. They reported a small (< 1 %) magneto-crystalline AMR. Very recently,[11,12] we have shown that the uniaxial strain can strongly affect the low-filed MR anisotropy in Pr$_{0.67}$Sr$_{0.33}$MnO$_3$ (PSMO) thin films. We attributed the effect primarily to spin dependent scattering and magnetic domain rotation process. In this work, we report on the out-of-plane AMR effect of differently strained PSMO thin films measured at high magnetic field where the magnetization has reached its saturation and the intrinsic AMR can be measured.

II. EXPERIMENTAL

High quality PSMO thin films with thickness of 4-400 nm have been epitaxially grown on LaAlO$_3$ (001) (LAO), SrTiO$_3$ (001) (STO), and NdGaO$_3$ (110) (NGO) substrates by pulsed laser deposition. Details of the film preparation and characterization have been reported previously.[11,12] The MR measurement was carried out using a Quantum Design PPMS 6000

Mat. Res. Soc. Symp. Proc. Vol. 602 © 2000 Materials Research Society

system with maximum magnetic field of 9 Tesla and equipped with a motorized sample rotator. Standard four-terminal method was used for the resistance measurement. At a given magnetic field H and temperature T, the out-of plane angular dependent resistance $R(\vartheta)$ is measured by rotating the sample around the axis parallel to the current direction so that the angle between the film plane and the magnetic field is varied. The AMR is calculated from the $R(\vartheta)$ curve as AMR=$(R_{\perp}-R_{//})/R_{//}$, where $R_{\perp}$ and $R_{//}$ is the resistance at the field perpendicular ($\vartheta=0°$) and parallel ($\vartheta=90°$) to the film plane, respectively. For comparison, films with compressive- (PSMO/LAO), tensile-(PSMP/STO), and very little- (PSMO/NGO) strains were measured. Field and temperature dependence of the AMR for each type of samples were also examined.

III. RESULTS

Before discussing the AMR effect, we first show in Fig. 1 the thickness dependence of the peak resistance temperature T_P (see inset) of a group of relatively thin PSMO films grown on different substrates. It is seen that with decreasing thickness, the T_P decreases rapidly for all the three types of the films. However, the T_P values of the strained PSMO/LAO and PSMO/NGO samples are much lower than those of the almost strain-free PSMO/NGO samples at the same thickness. The anomalous AMR effects which will be discussed later were observed only in the strained ultrathin (<20 nm) films.

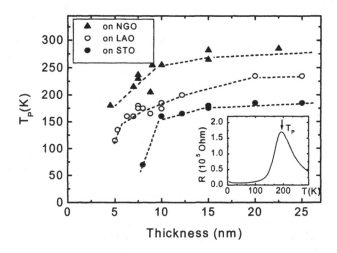

Fig. 1 Thickness dependence of the T_P of thin PSMO films grown on different substrates. The dashed lines are guide to the eyes. Inset is a typical R(T) curve of the PSMO film, in which the T_P is defined.

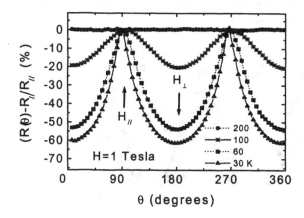

Fig. 2 The $(R(\vartheta)-R_{//})/R_{//}$ vs. ϑ curves of a 5 nm thick PSMO/LAO film measured at different temperatures and at H=1 T.

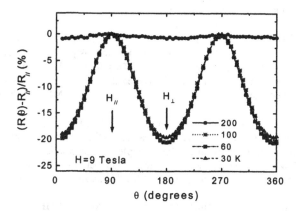

Fig. 3 The $(R(\vartheta)-R_{//})/R_{//}$ vs. ϑ curves of the same sample of Fig. 2, but measured at H= 9 T.

Fig. 2 and Fig. 3 show the $(R(\vartheta)-R_{//})/R_{//}$ vs. ϑ curves of a 5 nm thick PSMO/LAO film measured at H=1 and H= 9 T, respectively. One of the striking features of Fig. 2 is the very strong angular dependence of the resistance. The AMR reduces as the temperature increases and is very small at 200 K, which is well above the T_p. Another interesting feature of Fig. 2 is the appearance of sharp peaks at $\vartheta = 90°$ for 30 K and 60 K curves.

Fig. 3 shows that the AMR at high field is much smaller than at low field, although it is still much larger than what is reported for conventional FM materials.[1-3] The sharp peaks as seen in Fig. 2 also become rounded in Fig. 3. Now, the curves at T=30, 60 and 100 K almost coincide with each other, indicating that the AMR is almost independent of temperature at low temperatures and very high magnetic field.

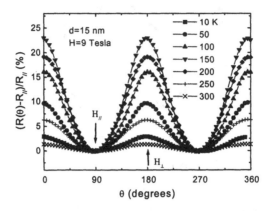

Fig. 4 The $(R(\vartheta)-R_{//})/R_{//}$ vs. ϑ curves of a 15 nm thick PSMO/STO sample measured at H=9 T and at different temperatures.

Fig. 4 presents the $(R(\vartheta)-R_{//})/R_{//}$ vs. ϑ curves of a 15 nm thick PSMO/STO sample measured at H=9 T and at different temperatures. It is seen that the peak positions in Fig. 4 are shifted by 90° compared to those shown in Fig. 2 and Fig. 3. This behavior has been observed in the PSMO/STO samples at all the temperatures and fields measured. The maximum AMR is strongly temperature dependent, and reaches 23 % at 150 K, at which the CMR also shows maximum value.

The $(R(\vartheta)-R_{//})/R_{//}$ vs. ϑ curves of the thick (400 nm) PSMO/STO film is shown in Fig . 5. Compared to the thin sample shown in Fig. 4, the basic features of the curve are essentially the same. However, the magnitude of the AMR of the thick sample is much smaller than the thin one. Also, the maximum AMR now appears at 250 K, higher than that of the thin sample (150 K). This is because the resistive transition temperatures of the thick sample (260 K) is higher than the thin sample (165 K).

The AMR exhibits interesting temperature dependence. A maximum AMR was observed at temperatures lower than the T_p. At very low (far below T_p) and very high (well above T_p) temperatures, the AMR is very small. The AMR also shows different field dependences at temperatures above and below T_p. For the compressive-strain films, the maximum AMR appears at the temperatures where the low-field MR is the largest. For the tensile-strain films, however, the maximum AMR appears at maximum CMR temperatures.

Compared to the highly strained PSMO/LAO and PSMO/STO samples, the almost strain-free PSMO/NGO samples show much less (a few percent) AMR. The AMR of the PSMO/NGO samples also shows very weak dependence on the film thickness.

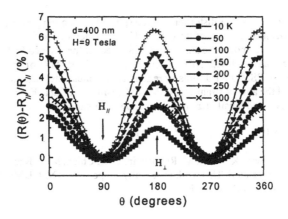

Fig. 5 The $(R(\vartheta)-R_{//})/R_{//}$ vs. ϑ curves of a 400 nm thick PSMO/STO sample measured at H=9 T and at different temperatures. Note the Y-scale is different from Fig. 4.

It should be mentioned that the data shown above are not corrected for demagnetization field. However, our direct measurement and estimation[13] have shown that the demagnetization correction does not change the basic features of the curves show from Fig. 2 to Fig. 5. It modifies the AMR by less than a few percent and makes the absolute AMR even larger for the PSMO/LAO samples and slightly smaller for the PSMO/STO samples. This is because the demagnetization always favors an in-plane magnetization and the internal field in the perpendicular field geometry ($\vartheta=180°$) for the compressive-strain film is smaller than the external field.

IV. DISCUSSION

The AMR results shown above are interesting in several aspects. First, the magnitude of the AMR is unusually large compared to the reported AMR in conventional ferromagnetic (FM) materials[2-3] and in thick manganite films.[8] The AMR of the FM is typically a few percent and is no more than 25% even at very low temperatures.[2] The maximum AMR in our PSMO films is more than 30 times larger than that of the Ni-Fe films (1-2 %) used the in the magnetic recording devices,[1] although the temperature where the maximum AMR appears in the PSMO films is lower than the latter. Second, the AMR of the compressive- and tensile-strained films show opposite signs. This is interesting both practically and physically. From a practical point of view, this phenomenon indicates that the AMR can be effectively tuned by applying different types of strains. This may provide one more parameter for film and device design. Third, the AMR shows a relatively strong field dependence. This is also interesting in consideration of the fact that all the results shown in this paper were obtained at the field well above the saturation field.[12]

The physical origin of the observed anomalous AMR in our PSMO films is not well understood yet. However, although the peak (valley) positions of the $R(\vartheta)$ curves is shifted by 90° between the compressive- and tensile-strained samples, the lower resistive state is always achieved when the magnetization is parallel to the long axis of the MnO_6 octahedral, i.e., along the long Mn-O-Mn bond direction. This may imply that the atomic orbital ordering caused by the strain-induced static Jahn-Teller distortion and the spin-orbit coupling play crucial roles in the AMR properties. Detailed analysis of the data will be presented in a forthcoming paper.[13]

In conclusion, we have studied the anisotropic MR effect of the strained PSMO thin films grown on different substrates and with different film thickness. The AMRs of the highly strained ultrathin PSMO/LAO and PSMO/STO films were found to be unusually larger than the less strained thick films and the almost strain-free PSMO/NGO ultrathin films. Furthermore, the AMR of the compressive- and tensile-strained films show opposite signs.

ACKNOWLEDGMENT:

Valuable discussions with X. X. Xi, M. Rubinstein, C. L. Chien, M. S. Rzchowski, and D. G. Schlom are greatly appreciated. This work is partially supported by NSF DMR-9876266, DMR-9972973, and Petroleum Research Fund.

REFERENCE:

1. H. Neal Bertran, "The physics of magnetic recording", *Applied Magnetism*, ed. R. Gerber, C. D. Wright, and G. Asti, (Kluwer Acadmic Publishers, 1992).

2. Th. G. S. M. Rijks, R. Coehoorn, M. J. M. de Jong, and W. J. M. de Jonge, Phys. Rev. **B51**, 283 (1995); Th. G. S. M. Rijks, S. K. J. Lenczowski, R. Coehoorn, and W. J. M. de Jonge, *ibid.* **B56**, 362 (1997).

3. E. Dan Dahlberg, K. Riggs, and G. A. Printz, J. Appl. Phys. **63**, 4270 (1988).

4. U. Ruediger, J. Yu, S. Zhang, A. D. Kent, and S. S. P. Parkin, Phys. Rev. Lett. **80**, 5639 (1998).

5. J. O'Donnell, M. S. Rzchowski, J. N. Eckstein, and I. Bozovic, Appli. Phys. Lett. **74**, 1775 (1998).

6. J. N. Eckstein, I. Bozovic, J. O'Donnell, M. Onellion, and M. S. Rzchowski, Appl. Phys. Lett. **69**, 1312 (1996).

7. Y. Susuki, and H. Y. Hwang, J. Appl. Phys. **85**, 4797 (1999).

8. J. Wolfman, W. Prellier, Ch. Simon, and B. Mercey, J. Appl. Phys. **83**, 7186 (1998).

9. M. Ziese, and S. P. Sena, J. Phys.: Condens. Matter **10**, 2727 (1998).

10. X. T. Zeng, and H. K. Wong, Appl. Phys. Lett. **72**, 740 (1998).

11. H. S. Wang, and Qi Li., Appl. Phys. Lett. **73**, 2360 (1998).

12. H. S. Wang, Qi Li, Kai Liu, and C. L. Chien, Appl. Phys. Lett. **74**, 2212 (1999).

13. H. S. Wang *et al.*, unpublished.

MICROSTRUCTURE AND ELECTRICAL CHARACTERISTICS OF $La_{1-x}Sr_xMnO_3$ (0.19≤x≤0.31) THIN FILMS PREPARED BY SPUTTER TECHNIQUES

H. Heo*, S.J. Lim*, G.Y. Sung**, and N.-H. Cho*
*Department of Materials Science and Engineering, Inha University, Inchon, Korea 402-751,
nhcho@dragon.inha.ac.kr.
**Electronics and Telecommunications Research Institute, Daejeon, Korea 305-600

ABSTRACT

$La_{1-x}Sr_xMnO_3$(0.19≤x≤0.31) thin films were prepared on silicon wafers by sputter techniques. The effect of substrate temperature, chemical composition and post-deposition heat-treatment on the crystalline structure and electrical characteristics of the films was investigated. The films grown at a substrate temperature of 500°C were found to be of the pseudo-tetragonal system (0.97≤a/c≤1) and exhibited a strong tendency of {001} planes to lie parallel to substrate surface. With the increase of x, the electrical resistivity of the films decreased and the transition temperature between the metallic and semiconducting electrical transport behaviors shifted to high temperature. With a magnetic field of 0.18 Tesla, the maximum magneto-resistance ratio (MR%) of $La_{0.69}Sr_{0.31}MnO_3$ polycrystalline thin films was about 390%.

INTRODUCTION

$La_{1-x}A_xMnO_3$(A=Ca, Sr, Ba) oxides have known to exhibit prominent magneto-resistance characteristics at metal-semiconductor transition temperature. $La_{1-x}A_xMnO_3$ epitaxial thin films and single crystals have been investigated intensively for the last ten years mainly because of their MR ratios ranging from a few hundred to several million % at the transition temperature. Such large MR effect has attracted much attention from relevant industries as well as research fields for the understanding and application of the unique and useful electrical transport behaviors.[1-4]

$La_{1-x}A_xMnO_3$ oxides are antiferromagnetic insulator if x is zero and, as x increases, the oxides begin to exhibit ferromagnetic metallic transport behaviors. Such a change in electro-magnetic characteristics with x has been understood by double exchange model suggested by Zener[2], and the electrical transport characteristics are expected to be affected by lattice distortion, stress[5-7], structural defects[8], and resultantly electron spin arrangement as well as energy band structure of the materials[9]. However, most of the researches regarding the colossal magneto-resistance (CMR) materials has been focused on epitaxial thin films. In particular, details of the correlation of the structure with the MR effects of the films have not been revealed enough to make clear the relation with physical characteristics[3,4].

In this paper, we report the crystalline structure and MR characteristics of $La_{1-x}Sr_xMnO_3$ polycrystalline thin films prepared by sputter techniques. Some of the results were compared with those of the films prepared by PLD methods in this study. In addition, the variation of the unit cell and microstructural characteristics of the films with deposition parameters as well as post-deposition heat-treatment conditions were analyzed, and correlated with MR effects.

EXPERIMENT

$La_{1-x}Sr_xMnO_3$ polycrystalline thin films were prepared by rf magnetron sputter

Mat. Res. Soc. Symp. Proc. Vol. 602 © 2000 Materials Research Society

techniques. A chamber base pressure of 1×10^{-6} Torr was obtained using a diffusion pump, and then argon gas with a purity of 5N was introduced into the chamber such that the gas pressure was 20 mTorr. Substrate temperature ranged from room temperature to 500°C, and a sputter power density of 9.84 watt/cm^2 was used. La$_{0.67}$Sr$_{0.33}$MnO$_3$ ceramics(Seattle Specialty Ceramics, Inc.) as well as La$_{0.3}$Sr$_{0.7}$MnO$_3$ ceramics were used as a main as well as auxiliary targets, respectively. X-ray diffraction(Philips, PW3719) and TEM(Philips, CM200) analysis were performed to analyze the crystalline structure and microstructural features of the films. For X-ray diffraction, a Cu Kα ray was used with diffraction angle ranging from 20 to 65°. Conventional TEM specimen preparation method was applied to produce plane-view as well as cross-sectional TEM specimens. Chemical analysis of the films was performed by Rutherford backscattering spectroscopy, in which He^{2+} ions with a kinetic energy of 3.06 MeV were used as an incident beam and a backscattered angle of 170° was applied.

The oxidation state of the Mn ions in the films was investigated by X-ray photoelectron spectroscopy(Surface Sci. Instr., 2803-S). Quantitative information on the Mn^{3+} and Mn^{4+} ions was obtained by fitting the experimental spectra with relevant Gaussian curves. A four point probe method was applied to measure the electrical resistivity of the films at temperatures ranging from liquid nitrogen temperature to room temperature

RESULTS AND DISCUSSION

Fig. 1 shows the XRD patterns of the La$_{1-x}$Sr$_x$MnO$_3$ thin films deposited at various substrate temperatures ranging from room temperature to 500°C. The films prepared at room temperature are amorphous, and as the substrate temperature increases the crystallinity as well as orientation preference increases. In particular, the films grown at a substrate temperature of 500°C exhibit a strong tendency of {001} planes to lie parallel to substrate surface. Such a {001} preferential orientation seems to be produced due to the stress in the films. The peak at 36° indicates the presence of a superlattice with a period of 3a/5. Such a superlattice is expected to be present due to the periodic arrangement of Mn^{4+} ions, which were produced by the substitution of Sr^{2+} ions for La sites.

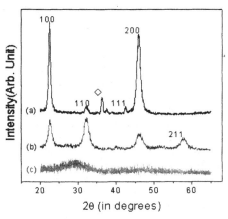

Fig. 1 XRD patterns of the thin films. The films were deposited on silicon wafers at (a) 500°C, (b) 300°C, and (c) room temperature, respectively.

The polycrystalline films, as shown in Fig. 2, consist of grains with a diameter of approximately 30 nm. The grains are of the cubic structure with a lattice parameter of 0.3879 nm. On the other hand, the epitaxial films prepared on LaAlO$_3$ substrate by PLD techniques in this study exhibit the cubic structure with a lattice parameter of 0.3823 nm. Considering the lattice parameters of 0.386 ~ 0.389 nm for the La$_{1-x}$Sr$_x$MnO$_3$ ceramics [10], the films prepared by PLD techniques are believed to be under considerably higher strain than the films prepared by sputter techniques.

As the content of Sr increases, the 001 peak in Fig. 3 shifts to low diffraction angle, indicating that the planar distance of {001} increases. The lattice parameter of c-axis of the

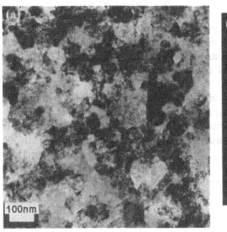
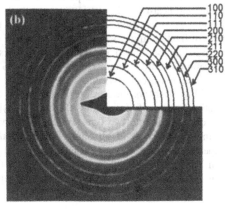

Fig. 2 (a) Bright-field TEM image. (b) Selected-area diffraction pattern.

films ranges from 3.95 to 3.98, resulting in the pseudo-tetragonal structure ($0.97 \leq a/c \leq 1$). Such increase is regarded to be due to the substitution of Sr (r=0.126 nm) for La (r=0.116 nm), and particular compressive strain is formed along the surface of substrates. In addition, the coefficients of thermal expansion (CTE) of the $La_{1-x}Sr_xMnO_3$ oxides increases with the content of Sr, varying from 11.2 to 12.8×10^{-6} K^{-1} [11]. A substantial difference between the CTE of the oxide films and of the Si wafers (14.94×10^{-6} K^{-1}) might also cause such strong compressive stress to the films, and as a result structural variation from the cubic to the pseudo-tetragonal system occurred.

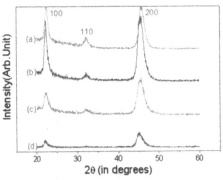

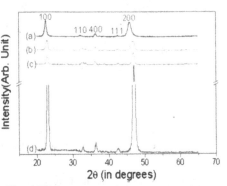

Fig. 3 Variation of 001 XRD peak position with the content(x) of Sr; (a) x=0.19, (b) x=0.22, (c) x=0.26, (d) x=0.31.

Fig. 4 XRD patterns of the post-deposition heat-treated films. The films were prepared at 500°C(a), and then annealed at 700°C(b), 800°C(c), and 900°C(d) in O_2 atmosphere for 6hrs, respectively.

Fig. 4 shows the XRD patterns of the $La_{0.69}Sr_{0.31}MnO_3$ thin films, which were post-deposition heat-treated at various temperatures for 6 hrs. All the films exhibit strong {001} preferential orientation. As the heat-treatment temperature increases, the 001 peak shifts to high diffraction angle, indicating the reduction of lattice parameters along the c-axis. This is regarded to be caused by the relaxation of strain. During the heat-treatment, there was reduction of lattice parameters by about 2-3% and the structure changed from the pseudo-tetragonal to the cubic system.

Fig. 5 shows the Mn 2p orbital region of the XPS for the $La_{1-x}Sr_xMnO_3$ films. The relative fraction of Mn^{4+}, $[Mn^{4+}/(Mn^{3+}+Mn^{4+})]$, was obtained by fitting the experimental Mn 2p-related peak into two Gaussian curves relevant to the Mn^{3+} as well as Mn^{4+} ions. In this fitting, the presence of Mn^{2+} ions is ignored, mainly because very little fraction of Mn^{2+} is expected from the XPS peak shape in Fig. 5(a) [12]. In Fig. 5(b) and (c), the ratios are 31.46 and 15.92% for films with x=0.31 and 0.19, respectively. Consequently, the hole concentration of the films with x=31.46% is regarded larger, compared with that of the films with x=0.19.

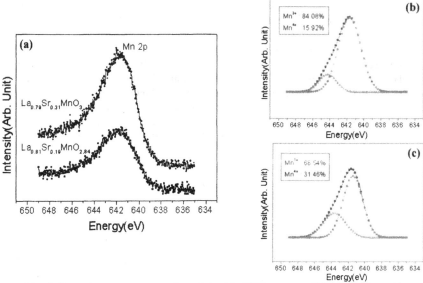

Fig. 5 Core electron energy region of the Mn XPS spectra of the thin films. (a) raw spectra, (b) and (c) fits for the spectra shown in (a).

In Fig. 6, the variation of electrical resistivity with temperature under a magnetic field of 0.18 Tesla is shown for thin films with x=0.31. The maximum MR ratio of the $La_{1-x}Sr_xMnO_3$ films increases from 1.5 to 390% as x increases from 0.19 to 0.31. Such a variation of MR characteristics with x is regarded to be caused by the variation of spin configuration as well as mutual interaction of t_{2g} and e_g of Mn ions. Such interactions are regarded to be affected by the distortion of lattice.

The variation of electrical resistivity with temperature of the films which were post-deposition heat-treated at 900°C for 6 hrs is shown in Fig. 7; the maximum MR ratio is about

10%, small compared with that of the as-prepared films. As mentioned above, the post-deposition heat-treatment changed the structure of the films from the pseudo-tetragonal to the cubic system. This structural change is believed to reduce the influence of magnetic field on the movement of d-shell electrons of the Mn^{4+} ions.

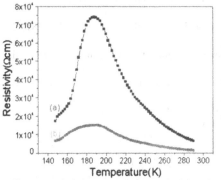

Fig. 6 Resistivity-temperature relation of the $La_{0.69}Sr_{0.31}MnO_3$ thin films. The relations were obtained (a) without, and (b) with a magnetic field of 0.18 T, respectively.

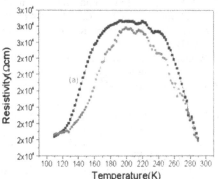

Fig. 7 Resistivity-temperature relation of the post-deposition heat-treated $La_{0.69}Sr_{0.31}MnO_3$ thin films. The relation was obtained (a) without, and (b) with a magnetic field of 0.18 T, respectively.

CONCLUSIONS

1. The $La_{1-x}Sr_xMnO_3$ ($0.19 \leq x \leq 0.31$) thin films prepared by rf magnetron sputter techniques exhibit strong orientation preference as the substrate temperature increases. As the content of Sr in the $La_{1-x}Sr_xMnO_3$ films increases, the lattice becomes the pseudo-tetragonal system ($0.97 \leq a/c \leq 1$). The electrical resistivity decreases with raising the content of Sr because of the increase of Mn^{4+} fraction.

2. For the $La_{0.69}Sr_{0.31}MnO_3$ thin films, the maximum MR ratio is about 390% under a magnetic field of 0.18 T. As the content of Sr decreases, the MR ratio is also decreases. Such variation of MR ratio with the content of Sr is regarded to be affected by the lattice distortion, depending on x values. In order to maximize the influence of the magnetic field on the mobility of the d-orbital hole of Mn^{4+}, it is considered to be very important to control the type of structure as well as the a/c ratio.

3. When the films were post-deposition heat-treated in oxygen atmosphere, the electrical resistivity decreases and a wide range of transition temperature is present. This phenomena are believed to be due to the boundary effect as well as the inhomogeneity of the film composition.

ACKNOWLEDGEMENT

This research was supported by the academic research fund of Ministry of Education, Republic of Korea 1998.

REFERENCES

1. C.N.R. Rao, R. Mahesh, A.K. Raychaudhuri, and R. Mahendiran, J. Phys. Chem. Solids, **59**, 487 (1998).
2. C. Zener, Phys. Rev., **82**, 403 (1951).
3. A.J. Millis, P.B. Littlewood, and B.I. Shraiman, Phys. Rev. Lett., **74**, 5144 (1995).
4. H.Y. Hwang, Phys. Rev. Lett., **75**, 914 (1995).
5. G.-M. Zhao, M.B. Hunt, K. Conder, H. Keller, and K.A. Muller, Physica C **282-287**, 202 (1997).
6. B. Garcia-Landa, M.R. Ibarra, J.M. DeTeresa, Guo-meng Zhao, K. Conder and H. Keller, Solid State Commun., **105**, 567 (1998).
7. J.-S. Zhou, J.B. Goodenough, A. Asamitsu, and Y. Tokura, Phys. Rev. Lett., **79**, 3234 (1997).
8. H.L. Ju, and Jyunchul Sohn, J. Solid State Commun., **102**, 463 (1997).
9. Warren E. Pickett, and David J. Singh, Phys. Rev. B **53**, 1146 (1996).
10. J.Z. Sun, L.K. Elbaum, A. Gupta, G. Xiao, P.R. Duncombe, S.S.P. Parkin, IBM J. Res. Develop., **42**, 89 (1998).
11. S.C. Singhal, and H. Iwahara, Solid Oxide Fuel Cells, Vol. 93-4, pp. 205, The High Temperature Materials and Battery Divisions, Edited by S.C. Singhal, The Electrochemical Society Inc., Pennington (1993).
12. T. Saitoh, A.E. Bocquet, T. Mizokawa, H. Namatame, and A. Fujimori, Phys. Rev. B **51**, 13942 (1995).

MAGNETOTRANSPORT IN EPITAXIAL TRILAYER JUNCTIONS FABRICATED FROM 90° OFF-AXIS SPUTTERED MANGANITE FILMS

J. S. Noh, T. K. Nath, J. Z. Sun* and C. B. Eom
Department of Mechanical Engineering and Materials Science, Duke University, Durham, NC 27708
* IBM T.J. Watson Research Center, PO Box 218, Yorktown Heights, NY 10598

ABSTRACT

We report studies on $La_{1-x}Sr_xMnO_3$ /SrTiO$_3$/ $La_{1-x}Sr_xMnO_3$ (x=0.33) trilayer junctions made using 90° off-axis sputtering. Both (110) NdGaO$_3$ and (001) (LaAlO$_3$)$_{0.3}$-(Sr$_2$AlTaO$_6$)$_{0.7}$ (LSAT) are used for substrates. Optical lithography is used for junction formation. These sputtered trilayers show improved junction resistance uniformity over trilayers made using laser ablation. A magnetoresistance of ~100% is observed for junctions on LSAT with 30 Å barrier at 13 K and around 100 Oe. The shape of junction magnetoresistance *vs.* field depends both on barrier thickness and on substrate type, suggesting that both inter-layer coupling and substrate-induced-strain play a role in determining the junction's micromagnetic state. These results indicate better junction interfaces can be obtained for manganite trilayer junctions by 90° off-axis sputtering.

INTRODUCTION

$La_{0.67}Sr_{0.33}MnO_3$ (LSMO) is a doped perovskite manganite exhibiting colossal magnetoresistance (CMR) behavior[1-7]. The materials physics of spin-polarized transport in this class of materials has been of recent interest. One good model system for investigating the interface-related spin-dependent transport is a trilayer current-perpendicular junction with a configuration of ferromagnet (top)/insulating barrier/ferromagnet (bottom)[8]. Such trilayer junctions are useful for the investigation of spin-dependent transport, and it may also have potential technological importance as field-sensors with large junction magnetoresistance (MR)[9].

So far, most manganite trilayer films have been fabricated with films deposited using pulsed-laser ablation[10,11]. Laser ablation is known to generate surface particulate during film deposition, although it is advantageous for stoichiometric transfer of materials with complex composition, and it has a relatively high deposition rate. For manganite trilayers, it has been recently reported that particulate of ~200Å were formed during the process[12,13], and it is undesirable because it can cause a junction short across the barrier. We therefore tried to make the trilayer films by 90° off-axis sputtering which does not generate particulates during deposition. With improved film quality, other factors controlling junction MR could also be examined systematically. This led to our observation of a systematic dependence of junction magnetoresistance behavior on barrier thickness as well as on the different types of substrates used – namely, (110) NdGaO$_3$ and (001) (LaAlO$_3$)$_{0.3}$-(Sr$_2$AlTaO$_6$)$_{0.7}$ (LSAT).

EXPERIMENT

The La$_{0.67}$Sr$_{0.33}$MnO$_3$ (top) /SrTiO$_3$/ La$_{0.67}$Sr$_{0.33}$MnO$_3$ (bottom) trilayer films were grown on both (110) NdGaO$_3$ and (001) LSAT substrates by a 90° off-axis sputtering technique. While bottom and top electrodes were deposited on the substrates held at 750 °C in the atmosphere of 120 mTorr Ar and 80 mTorr O$_2$, the barrier was grown in the environment of 180 mTorr Ar and 20 mTorr O$_2$ while keeping the substrates at 720 °C. Ag overlayer was deposited on the top layer for protection during photolithography. The film thicknesses are: Ag/top layer/barrier/bottom layer=700/400/variable/600Å. Barrier thicknesses were 50, 40, 30, and 20Å. On separate dummy films atomic force microscope (AFM) study was taken before and after the barrier layer deposition to investigate surface morphologies.

Optical lithography was used to form junction devices from these trilayer films. The process involved three major steps. They were: (1) a base electrode patterning step, (2) a junction pillar definition step, and (3) contact metallization. For each step, photolithography and ion-milling were used for pattern definition and pattern transfer, respectively. The junction patterns are rectangles, with aspect ratios of 1:1 or 1:2. The junction size ranged from 1×1 µm^2 to 16×16 µm^2 for 1:1 junctions, and 1×2 µm^2 through 40×80 µm^2 for 1:2 junctions. Magnetotransport measurements were made for the junctions as a function of temperature between 300 K and 13K.

RESULTS

After fabrication, each 1×1 cm^2 chip contains 330 junctions. We measured for around 10 junctions among them. Junctions chosen were inspected under microscope to ensure a damage-free surface. Figure 1 shows resistance (R) *vs.* magnetic field (H) at 13 K for the best measured junction in each chip with different substrates and barrier thickness. Magnetic field was applied parallel to the surface of the junction and perpendicular to base electrode stripe. The R(H) scans show distinctive magnetoresistive behavior. The MR is calculated as (R$_{high}$- R$_{low}$)/ R$_{low}$, where R$_{high}$ and R$_{low}$ represent the junction resistance at its resistive-high and -low state, respectively. In our work, the largest MR is ~100% from a junction on (001) LSAT substrate.

There are several interesting trends for data shown in Figure 1. Junction magnetic switching fields from MR measurements agree in general with coercive fields obtained from magnetic hysteresis measurements on unpatterned LSMO films[8,14]. For junctions, the switching fields also depend somewhat on barrier thickness, becoming smaller for thinner barriers. This suggests inter-layer magnetic coupling plays a role in determining the junction magnetic switching characteristics. For junctions of the same barrier thickness, those on LSAT substrate in general show smaller switching fields than those on NdGaO$_3$. This difference suggests the involvement of substrate in determining the junction's micromagnetic state. It is not yet clear whether lattice-matching-related strain is the underlying cause, as the lattice parameters of both substrates are actually quite close ($(a^2+b^2)^{1/2}/2$=3.862Å for NdGaO$_3$, a=3.868Å for LSAT)[15]. Interface chemistry between LSMO and the substrate may also be important.

Junctions made using sputtered films show improved electrical uniformity. This is most clearly shown in Figure 2(a). Here, the RA product *vs.* junction size at both room temperature and 13 K are plotted against junction size. Ideally, the RA product should be independent of junction size. In reality it often deviates from constant, due to pin-hole short or interface inhomogeneity [18], or departure of junction shape from a designed (mask) size, etc. It was observed earlier in junctions made with laser-ablated films that junction RA product decreases with increasing junction size – a trend that could signify pin-hole shorts distributed over the lengthscale of several microns[8,18]. In our junctions made with sputtered films as shown in Figure 2(a), the RA product stayed constant as should be with junction size at both temperatures.

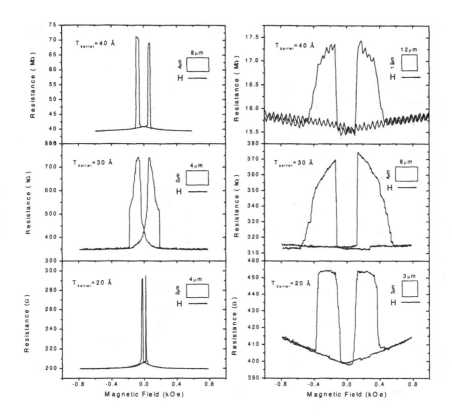

Fig. 1. Resistance vs. magnetic field at 13 K. Left: trilayer junctions on (001) LSAT substrate, right: trilayer junctions on (110) NdGaO₃ substrate. The inset shows each junction geometry and the applied field direction.

One exception was for the chip on NdGaO₃ with 40Å barrier, which also showed worst MR behaviors, and may indicate a process-related anomaly. These results indicate that our trilayer films have a more uniform junction interface. Actually, AFM examination on both base electrode and barrier layer showed that both surfaces were morphologically smooth with root mean square (RMS) surface roughness less than 10Å, and were free of surface particles.

For most chips, the best junction has a rectangular shape with an aspect ratio of 1:2, which suggests shape anisotropy tends to assist the formation of a magnetically anti-parallel state (which gives a larger R_{high}, hence larger MR)[16,17].

Data in Figure 1, also show that junction resistance decreases with decreasing barrier thickness. To investigate this relationship more closely, we plot the resistance-area (RA) product as a function of barrier thickness. This is shown in Figure 2(b), where both ambient and 13 K data are plotted. Indeed a nearly exponential increase of junction RA is observed as a function of barrier thickness. This is suggestive that a tunneling process is involved for electrical transport across the barrier.

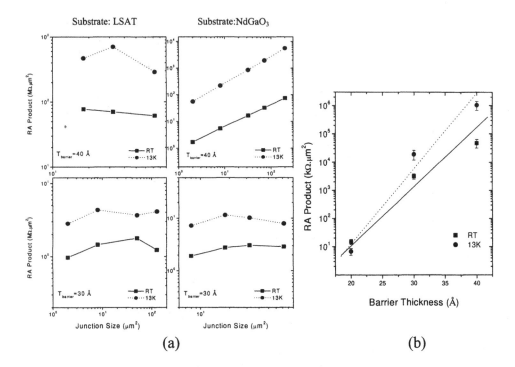

(a)   (b)

Fig. 2. Junction resistance scaling behavior: (a) RA product vs. junction size for junctions with 40Å, 30Å barrier respectively on both types of substrates; left: on LSAT substrate, right: on NdGaO$_3$ substrate, (b) RA product vs. barrier thickness,.

In Figure 3 we show the temperature evolution of a junction's R_{high} and R_{low}. They were extracted from a sequence of R(H) loops taken at 15 K temperature intervals for a sample junction with 30Å barrier on LSAT. This junction at low temperatures show the largest MR (~100%) in this batch of samples. MR disappears at room temperature, but several % of MR was still visible above 200 K. The highest temperature, at which we could observe distinguishable MR spikes, was around 270 K for this junction.

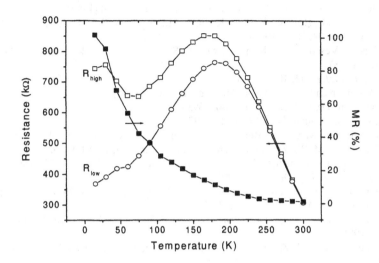

Fig. 3. Temperature dependence of R_{high}, R_{low}, and MR for a 2×4 μm^2 sized junction with 30Å barrier on LSAT.

CONCLUSION

We have grown $La_{0.67}Sr_{0.33}MnO_3$ /$SrTiO_3$/ $La_{0.67}Sr_{0.33}MnO_3$ trilayer films epitaxially on both (110) $NdGaO_3$ and (001) LSAT substrates by 90° off-axis sputtering. The magnetotransport measurement for junctions defined from these films showed distinctive MR behaviors for most chips, with a maximum MR of ~100% at 13 K. We found that RA products of the junctions decrease exponentially with the decreasing barrier thickness and remain almost constant regardless of junction size, reflecting good area-scaling and hence junction uniformity. We also saw some improvement in the temperature dependence of junction MR.

ACKNOWLEDGEMENT

This work was partly supported under NSF Grant DMR-9802444 and NSF Young Investigator Award (CBE), and by IBM. The authors would like to thank Roger Koch, Bill Gallagher, Steve Brown, John Connolly at IBM Research Yorktown Heights for assistance at various stages of the experiment.

REFERENCES

1. J. Inoue and S. Maekawa, *Phys. Rev. Lett.* **74**, 3407 (1995).
2. P. Lyu, D. Y. Xing, and J. Dong, *Phys. Rev. B.* **58**, 54 (1998).

3. H. Asano, J. Hayakawa, and M. Matsui, *Appl. Phys. Lett.* **71**, 844 (1997).
4. B. Wiedenhorst, C. Hofener, Y. Lu, J. Klein, L. Alff, R. Gross, B. H. Freitag, and W. Mader, *Appl. Phys. Lett.* **74**, 3636 (1999).
5. J.-H. Park, E. Vescovo, H.-J. Kim, C. Kwon, R. Ramesh, and T. Venkatesan, *Phys. Rev. Lett.* **81**, 1953 (1998).
6. C. Zener, *Phys. Rev.* **82**, 403 (1951).
7. P. W. Anderson and H. Hasegawa, *Phys. Rev.* **100**, 675 (1955).
8. J. Z. Sun, *Phil. Trans. R. Soc. Lond.* **356**, 1693 (1998).
9. J. M. Daughton. *J. Appl. Phys.* **81**, 3758 (1997).
10. Z. Tajanovic, C. Kwon, M. C. Robson, K.-C. Kim, M. Rajeswari, R. Ramesh, T. Venkatesan, S. E. Lofland, S. M. Bhagat, and D. Fork, *Appl. Phys. Lett.* **69**, 1005 (1996).
11. X. W. Li, Yu Lu, G. Q. Gong, G, Xiao, A. Gupta, P. Lecoeur, J. Z. Sun, Y. Y. Wang, and V. P. Dravid, *J. Appl. Phys.* **81**, 5509 (1997).
12. J. Z. Sun, D. W. Abraham, K. Roche, S. S. P. Parkin, *Appl. Phys. Lett.* **73**, 1008 (1998).
13. J. Z. Sun, *J. Magn. Magn. Mater.* **202**, 157 (1999).
14. J. Z. Sun, In *Colossal Magnetoresistance Oxides*, ed. Y. Tokura. Singapore: Gordon & Breach (1998).
15. R. A. Rao, D. Lavric, T. K. Nath, C. B. Eom, L. Wu, and F. Tsui, *Appl. Phys. Lett.* **73**, 3294 (1998).
16. W. J. Gallagher, S. S. P. Parkin, Yu Lu, X. P. Bian, A. Marley, and K. P. Roche, R. A. Altman, S. A. Rishton, C. Jahnes, T. M. Shaw, and G. Xiao, *J. Appl. Phys.* **81**, 3741 (1997).
17. Yu Lu, R. A. Altman, A. Marley, S. A. Rishton, P. L. Trouilloud, G. Xiao, W. J. Gallagher, and S. S. P. Parkin, *Appl. Phys. Lett.* **70**, 2610 (1997).
18. A. Gupta and J. Z. Sun, *J. Magn. Magn. Mater.* **200**, 24 (1999).

Transport and Optical Properties

SPECTROSCOPIC STUDIES OF MANGANESE DOPED LaGaO$_3$ CRYSTALS

M. A. NOGINOV*, N. NOGINOVA*, G. B. LOUTTS*, K. BABALOLA**,
R. R. RAKHIMOV*
*Center for Materials Research, Norfolk State University, Norfolk, VA, mnoginov@vger.nsu.edu
** CMR NSU summer school student

ABSTRACT

We have grown Mn:LaGaO$_3$ crystals (Mn=0.5%, 2%, 10%, and 50%) and studied their properties with the methods of optical and EPR spectroscopy. We have found Mn^{2+}, Mn^{4+}, and Mn^{5+} ions in low doped crystals (Mn=0.5% and 2%) and Mn^{3+} ions in crystals with higher Mn concentrations (10% and 50%). The formation of Mn^{3+} valence state in Mn:LaGaO$_3$ is discussed.

INTRODUCTION

In order to better understand the properties of Colossal Magneto-Resistance (CMR) materials (Ca:LaMnO$_3$, Sr:LaMnO$_3$, etc.), we have grown and optically characterized bulk single crystals of manganese doped LaGaO$_3$ (Mn=0%, 0.5%, 2%,10%, and 50% in the melt). The crystals were grown by the Czochralski technique in slightly oxidizing atmosphere.

We recently have synthesized and spectroscopically studied Mn doped YAlO$_3$, a potential material for holographic data storage [1-4]. The spectroscopic properties of different Mn valence states in YAlO$_3$ were compared to those in Gd$_{1-x}$La$_x$AlO$_3$ (x=0, 0.27, 0.33, and 0.5), YbAlO$_3$, and Y$_3$Al$_5$O$_{12}$. We have found that the variance in optical spectra of Mn ions in different hosts is not very dramatic (even comparing orthoaluminates to garnets). That is why we expect that the spectra of Mn ions in LaGaO$_3$ can be interpreted using the analogy with those in Mn:YAlO$_3$.

In Mn:YAlO$_3$ crystals grown in slightly oxidizing atmosphere without charge compensators, manganese ions exist not in 3+ valence state (as it can be expected since both Al and Y in YAlO$_3$ are three valent ions), but in valence states 2+ and 4+. The charge balance is maintained partially due to compensation of 2+ and 4+ Mn valence states and partially due to defects. The absorption and emission spectra of Mn^{4+} ions in the crystal are shown in Figures 1a and 2a [2]. Mn^{2+} ions are almost optically inactive, since all ground state transitions are spin forbidden [3].

Mn^{3+} ions were observed only in Mn:YAlO$_3$ crystals co-doped with Ce ions (probably in result of the reaction Ce^{4+}+Mn^{4+}→Ce^{4+}+Mn^{3+}). The absorption spectrum of Mn^{3+} in YAlO$_3$ is shown in Figure 1a. As-grown single doped Mn:YAlO$_3$ crystals have a yellowish color and crystals co-doped with Ce ions have a pinkish color [3].

In Mn:YAlO$_3$ crystals co-doped with Ca ions and in photoexposed single Mn doped YAlO$_3$ samples we have found Mn^{5+} ions in octahedral coordination [4]. (In illuminated crystals the change of the Mn valence state was due to photoionization Mn^{4+}→Mn^{5+}+e.) The absorption of Mn^{5+} ions is intense and covers practically the whole visible range of the spectrum, Figure 1b. That is why YAlO$_3$ crystals containing even relatively small concentration of Mn^{5+} ions have a dark grayish-bluish color. Strong Mn^{5+} absorption was also observed in the Mn:YAlO$_3$ crystal grown in slightly reducing atmosphere and annealed in air [4]. The emission spectrum of Mn^{5+} ions in YAlO$_3$ (recorded under excitation of the crystal with short pulses of light at 532 nm) is shown in Figure 2b.

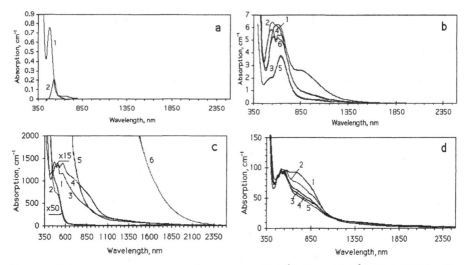

Figure 1. (a) Room temperature absorption spectra of Mn^{4+} (1) and Mn^{3+} (2) in $YAlO_3$. The estimated Mn^{3+} and Mn^{4+} concentrations (in the crystal) are $\approx 0.0125\%$ (each).

(b) Room temperature absorption spectra of strongly photoexposed (at $\lambda=514.5$ nm) $Mn:YAlO_3$ at different directions of light propagation, polarization of the electric field and polarization of the magnetic field. The estimated Mn^{5+} concentration (in the crystal) is $\approx 0.02\%$.

(c) Room temperature absorption spectra in $LaGaO_3$. 1 - undoped, 2 - Mn=0.5%, 3 - Mn=2% (inner slice), 4 - Mn=2% (outer slice), 5 - Mn=10%, and 6 - Mn=50%.

(d) Room temperature absorption spectra in 2%Mn:$LaGaO_3$ (outer slice) at different temperatures. 1 - 15°C, 2 - 70°C, 3 - 135°C, 4 - 145°C, and 5 - 190°C.

ELECTRIC CONDUCTIVITY STUDIES

The electric resistance of the crystals studied strongly depended on temperature. The temperature dependence of resistivity, ρ, could be described with the hopping model as

$$\rho=BT \exp(E/kT), \tag{1}$$

where B is the resistivity coefficient, T is the temperature, k is the Boltzman constant, and E is the activation energy. In our samples the coefficient B and the activation energy E increased significantly with decrease of Mn concentration. Thus, E changed from 336 meV in 50%Mn doped sample to 570 meV in 2% doped sample. The results of electric studies in Mn:$LaGaO_3$ will be discussed in detail elsewhere.

ELECTRON PARAMAGNETIC RESONANCE STUDIES

Manganese valence states 2+ and 4+ have an odd spin and can be observed by electron paramagneric resonance (EPR) technique. The EPR spectra of 0.5% and 2% manganese doped $LaGaO_3$ crystals contain multiple narrow lines corresponding to paramagnetic Mn^{2+} and Mn^{4+} ions, Figure 3. At Mn concentration equal to 10%, individual Mn^{2+} and Mn^{4+} lines disappear and

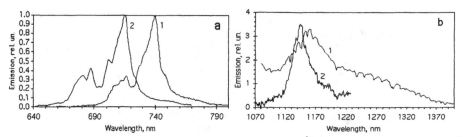

Figure 2. (a) Room temperature emission spectrum of Mn^{4+} in 0.5%Mn:LaGaO$_3$ (1) and 0.5%Mn:YAlO$_3$ (2); (b) Emission spectrum of Mn^{5+} in 2%Mn:LaGaO$_3$, T=300K, (1) and photoexposed 0.5%Mn:YAlO$_3$, T=77K (2).

a single broad line forms instead, Figure 3. This broad line is due to coupling between neighboring Mn ions. It becomes significant at high Mn concentration.

At Mn concentration equal to 10%, another new line with the g-factor equal to 2 appears in the EPR spectrum, Figure 3. It is getting more prominent in the crystal with 50% Mn concentration. This line resembles EPR signals typically observed in CMR materials in paramagnetic state [5]. In CMR materials, this signal is assigned to mixed Mn^{3+}-Mn^{4+} centers. In Mn:LaGaO$_3$ crystals, this line may point to the presence of Mn^{3+} ions at high Mn doping concentrations.

OPTICAL SPECTROSCOPIC STUDIES

Undoped LaGaO$_3$ crystals had a brownish color. The brownish cast was, most likely, due to color centers which are typical to perovskites. Small concentration of Mn ions helped to reduce the amount of coloration, slightly shifting the strong absorption band extending from ultraviolet (UV) toward shorter wavelengths, Figure 1c. The same effect is known in transition metal doped YAlO$_3$ crystals [6].

The 0.48 mm absorption band belonging Mn^{4+} ions, almost identical to that in Mn:YAlO$_3$ [1,2], can be seen in the optical spectrum of 0.5%Mn:LaGaO$_3$, Figure 1c. Assuming that the Mn^{4+} absorption cross section in Mn:LaGaO$_3$ is approximately equal to that in Mn:YAlO$_3$, we can estimate the Mn^{4+} concentration in nominally 0.5% doped Mn:LaGaO$_3$ to be equal to 0.12%. (The total concentration of Mn ions in Mn:LaGaO$_3$ crystals was not determined. In Mn:YAlO$_3$, the concentration of Mn ions in the crystal was approximately ten times smaller than that in the melt). The emission of Mn^{4+} ions with the maximum at $\approx$740 nm, very similar to that in Mn^{4+}:YAlO$_3$, Mn^{4+}:Y$_3$Al$_5$O$_{12}$, etc., was observed in 0.5%Mn:YAlO$_3$ at 532 nm short-pulsed ($t_{pulse} \approx$10 ns) excitation, Figure 2a. Since (1) Mn^{4+} absorption band is not well resolved from the charge transfer (CT) band extending from UV and (2) Mn^{3+} absorption band is expected to be overlapped with Mn^{4+} absorption band, no conclusion concerning the presence of Mn^{3+} ions in 0.5%Mn:LaGaO$_3$ crystal can be made.

At Mn concentration (in the melt) equal to 2% the crystal was almost black. An intense visible and near IR absorption of the crystal, covering the spectral range from $\approx$0.45 µm to $\approx$1.05 µm (k_{abs}^{max}=90 cm^{-1} at λ=0.53 µm), has a relatively complex structure and apparently is multi-component, Figure 1c. This absorption spectrum appears to be somewhat similar to that of Mn^{5+} ions in YAlO$_3$ and is very different from the absorption spectra of Mn^{3+}:YAlO$_3$ and

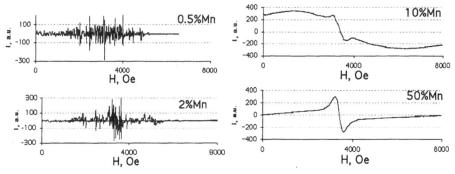

Figure 3. EPR spectra of Mn:LaGaO$_3$.

Mn^{4+}:YAlO$_3$. An emission peaking at $\approx$1.16 μm was found in 2%Mn:LaGaO$_3$ under 532 nm short-pulsed pumping, Figure 2b. The shape of this emission spectrum is similar to the shape of the emission spectrum in Mn^{5+}:YAlO$_3$ ($\lambda_{max}\approx$1.14 μm). Assuming that Mn^{5+} absorption cross section in LaGaO$_3$ is approximately equal to that in YAlO$_3$, Mn^{5+} concentration in nominally 2% Mn doped LaGaO$_3$ can be estimated to be equal to 0.3%.

As follows from Figures 1a, 1b, the absorption intensity of Mn^{3+} ions and Mn^{4+} ions is much smaller than that of Mn^{5+} ions. That is why no conclusion concerning the presence (or absence) of Mn^{3+} and Mn^{4+} ions in 2%Mn:LaGaO$_3$ crystal can be drawn from its absorption spectrum. No Mn^{4+} emission was found in 2% Mn doped LaGaO$_3$ crystal as well as no Mn^{5+} emission was found in 0.5%Mn:LaGaO$_3$. The absence of Mn^{4+} emission does not mean that Mn^{4+} ions are not present in the crystal. Reasonably high concentration of Mn^{5+} ions and very intense Mn^{5+} absorption band ($k_{abs}\approx$90 cm^{-1}) overlapping with Mn^{4+} emission band can cause a dramatic energy transfer quenching of Mn^{4+} luminescence. A similar quenching of luminescence of Mn^{4+} ions, although much less efficient, was observed in photoexposed Mn:YAlO$_3$ crystals [2]. Note that the presence of Mn^{4+} ions in 2%Mn:LaGaO$_3$ crystal was evidenced by EPR measurements.

The outer layer of the 2%Mn:LaGaO$_3$ crystal ($\approx$1.5 mm thick) was noticeably darker than the inside volume. Accordingly, the absorption spectra of the inner and outer crystal slices were different: the outer layer had an additional absorption shoulder centered at approximately 0.8 μm, Figure 1c. A slight change of the color (from brownish to brownish-violet) and an increase of the absorption intensity were also observed in 0.5%Mn:LaGaO$_3$ sample annealed in the air. This suggests that, at least partially, the additional coloration in the material is caused by its oxidation.

The long-wavelength part of the absorption band in the outer slice of the 2%Mn:LaGaO$_3$ crystal (the additional absorption which is relevant to oxidation) significantly reduces at increased temperatures, Figure 1d. At t=190°C, the absorption spectrum of the outer slice resembled the room temperature absorption spectrum of the inner slice, compare Figures 1c and 1d. This temperature behavior is normally expected from CT bands [7]. A similar behavior, a decrease of the intensity of the absorption band with the increase of temperature, was also observed at 0.6$\div$1.2 μm in CMR materials (La$_{0.6}$Sr$_{0.4}$MnO$_3$) [8].

In Ref. [9], an absorption band at $\approx$1 μm was attributed to metal-oxygen CT. However, the long-wavelength position of this band is rather typical of metal-metal CT than metal-oxygen CT. On the other hand, if following Refs. [10, 11], we will assume that the 0.8 μm absorption band in our crystals is due to $Mn^{3+} \rightarrow Mn^{4+}$ CT, we will have to conclude that Mn^{3+} ions, which were not seen in $0.5\%Mn:LaGaO_3$ and inner volume of $2\%Mn:LaGaO_3$, first emerged in annealed outer layer of $2\%Mn:LaGaO_3$. This result is in line with the results of the EPR studies. It can be understood if we will make the following two assumptions: (1) similarly to $Mn:YAlO_3$, in low doped $LaGaO_3$ crystals Mn ions present in 2+ and 4+ valence states and not 3+ valence state; and (2) Mn^{3+} ions appear in $Mn:LaGaO_3$ samples annealed in air in result of oxidation of Mn^{2+}.

In 10% Mn doped sample the absorption was much stronger than that in $2\%Mn:LaGaO_3$ and the shape of the absorption spectrum was different form that in $2\%Mn:LaGaO_3$, Figure 1c. The very intense absorption band in $Mn(10\%):LaGaO_3$ extended from UV and visible range of the spectrum to near-infrared range of the spectrum. A very strong absorption intensity may be an evidence of a CT transition. The fact that the CT absorption band in $10\%Mn:LaGaO_3$ has a maximum at much shorter wavelength than in $2\%Mn:LaGaO_3$ suggests that this is possibly a different, metal-oxygen, CT. The absorption in $50\%Mn:LaGaO_3$ was much stronger than that in $10\%Mn:LaGaO_3$, Figure 1c.

Note that even at relatively high Mn concentrations, the absorption in our samples was ten-to-hundred times weaker than that in the 600 nm thick CMR film of $La_{0.75}Sr_{0.25}MnO_3$ synthesized by the MOCVD technique, Figure 4. This large difference could be expected. In fact, according to our resistance measurements, the conductivity in $50\%Mn:LaGaO_3$ crystal is approximately thousand times lower than the conductivity of the CMR film in paramagnetic state.

SUMMARY AND CONCLUSIONS

We have grown and spectroscopically studied manganese doped $LaGaO_3$ bulk single crystals. At low manganese concentrations, characteristic signatures of several Mn valence states were found. Mn^{2+} and Mn^{4+} ions were detected by EPR spectroscopy in $0.5\%Mn:LaGaO_3$ and $2\%Mn:LaGaO_3$ crystals. The presence of Mn^{4+} ions in $0.5\%Mn:LaGaO_3$ and the presence of Mn^{5+} ions in $2\%:LaGaO_3$ was evidenced by optical spectroscopy measurements. At high Mn concentrations ($\geq 10\%$), strong interactions between Mn ions strongly changed both optical and EPR spectra. Several bands and lines belonging to complexes of interacting ions and CT transitions were found in highly Mn doped crystals instead of spectra of individual ions. Some of them could be assigned to Mn^{3+} containing complexes.

We suggest that similar to $Mn:YAlO_3$, at low doping levels (0.5%) manganese ions enter $LaGaO_3$ crystal in form of Mn^{2+} and Mn^{4+}. Mn^{3+} ions can be obtained in $2\%Mn:LaGaO_3$ crystal

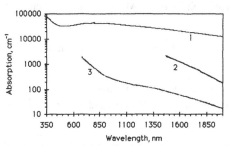

Figure 4. Absorption in the $La_{0.75}Sr_{0.25}MnO_3$ CMR film (1), $10\%Mn:LaGaO_3$ (2), and $50\%:LaGaO_3$ (3).

111

annealed in air, probably in result of oxidation of Mn^{2+} ions. The mechanism of formation of Mn^{3+} ions at high Mn concentrations is not clear. One can assume that at high concentrations Mn ions also prefer to enter the crystal in 2+, 4+ and 5+ valence states. In this case, to produce Mn^{3+}-Mn^{4+} pairs, two-valent charge compensator or an excess of oxygen in the growth atmosphere is needed. The role of this charge compensator would be to oxidize Mn^{2+} ions to Mn^{3+}. This scenario is different from the commonly accepted one, where Mn ions are assumed to enter the crystal in form of Mn^{3+} and the charge compensator is needed to oxidize Mn^{3+} to Mn^{4+}.

ACKNOWLEDGMENTS

The work was supported by the AFOSR/BMDO grant #F49620-98-1-0101, DOE grant #DE-FG01-94EW11493, NSF CREST #HRD-9805059, and NASA Grant NRA-99-OEOP-4

REFERENCES

1 G. B. Loutts, M. Warren, L. Taylor, R. R. Rakhimov, H. R. Ries, G. Miller III, M. A. Noginov, M. Curley, N. Noginova, N. Kukhtarev, J. C. Caulfield, P. Venkateswarlu, Phys. Rev. B, **57**, p. 3706-3709 (1998).

2. M. A. Noginov , G. B. Loutts, JOSA B, **16**, p. 3-11 (1999).

3. M. A. Noginov, G. B. Loutts, M. Warren, JOSA B, **16**, p. 475-483 (1999).

4. M. A. Noginov, M. A. Noginov, G. B. Loutts, N. Noginova, S. Hurling, S. Kück, accepted for publication in Physical Review B.

5. M. T. Causa, M. Tovar, A. Caniero, F. Prado, G. Ibanez, C. A, Ramos, A. Butera, and B. Alascio, Phys. Rev. B, **58**, p. 3233-3239 (1998).

6. G. B. Loutts, M. Warren, Eastern Regional Conference on Crystal Growth and Epitaxy, Atlantic City, NJ, September 28-October 1, 1997.

7. R. G. Burns, *Mineralogical applications of crystal field theory*, Second Edition, Cambridge University Press, Cambridge, 1993, 551 p.

8. Y. Morimoto, A. Matchida, K. Matsuda, M. Ichida, and A. Nakamura. Phys. Rev. B, **56**, p. 5088-5091 (1997).

9. K. Matsuda, A. Matchida, Y. Morimoto, and A. Nakamura. Phys. Rev. B, **58**, p. R4303-R4306 (1998).

10. Y. G. Zhao, J. J. Li, R. Shreekala, H. D. Drew, C. L. Chen, W. L. Cao, C. H. Lee, M. Rajeswari, S. B. Ogale, R. Ramesh, G. Baskaran, and T. Venkatesan, Phys. Rev. Lett., **81**, p. 1310-1313 (1998).

11. J. F. Lawler, J. G. Lunney, and J. M. D. Coey, Appl. Phys. Lett., **65**, p. 3017-3018 (1994).

GIANT 1/f NOISE IN LOW-T_c CMR MANGANITES: EVIDENCE OF THE PERCOLATION THRESHOLD

V. PODZOROV[1], M. UEHARA[1], M. E. GERSHENSON[1], T. Y. KOO[2], and
S-W. CHEONG[1,2]
1) Serin Physics Laboratory, Rutgers University, Piscataway, NJ 08854-8019
2) Bell Labs, Lucent Technologies, Murray Hill, NJ 07974

ABSTRACT

We observed a dramatic peak in the $1/f$ noise at the metal-insulator transition (MIT) in low-T_c manganites. This many-orders-of-magnitude noise enhancement is observed for both polycrystalline and single-crystal samples of $La_{5/8-y}Pr_yCa_{3/8}MnO_3$ ($y = 0.35 - 0.4$) and $Pr_{1-x}Ca_xMnO_3$ ($x = 0.35 - 0.5$). This observation strongly suggests that the microscopic phase separation in the low-T_c manganites causes formation of a percolation network, and that the observed MIT is a percolation threshold. It is shown that the scale of phase separation in polycrystalline samples is much smaller than that in single crystals.

INTRODUCTION

It is well known that the electronic phase diagrams of perovskite manganites are very complex; they exhibit numerous ground states and phase transitions as the carrier concentration is varied [1]. The phase diagram for the system $La_{1-x}Ca_xMnO_3$ in the plane of the doping concentration x and temperature T is shown in Fig. 1. The concentration of charge carriers in this system is proportional to the Ca doping level. At high temperatures, the system is in a paramagnetic insulating phase. Recent resistivity and X-ray studies revealed that this paramagnetic phase comprises three different lattice types [2]. In the temperature range $T \simeq 200 - 800K$, the lattice is orthorhombic (Jahn-Teller distorted at low x, and octahedron rotated at higher doping levels); at higher temperatures, it is rhombohedral. In the range $x \simeq 0.3 - 0.5$, the system undergoes a transition into the ferromagnetically-ordered state at $T \simeq 200 - 250K$; this magnetic transition is accompanied by the metal-insulator transition (MIT) [3, 4, 5]. A very high sensitivity of this MIT to the external magnetic field results in the so-called colossal magnetoresistance (CMR) [6]. The change of the resistivity across the transition and, correspondingly, the CMR are more dramatic in compounds with a lower transition temperature T_c. The MIT in low-T_c manganites bears many features which are intrinsic to the first-order transitions, including a strong thermal hysteresis of the resistivity ρ and magnetization M [7, 8].

There is a growing theoretical and experimental evidence that transport properties of the insulating state above T_c are dominated by small polarons or magnetic polarons, and that the band-like carriers become important below T_c [9, 10, 11]. However, the details of the MIT remain unclear. The experimental data suggest that the phase separation occurs *gradually* with decreasing temperature [7, 8]. Appearance of small ferromagnetic (FM) regions in the charge-ordered (CO) phase has been reported at $T >> T_c$, well beyond the conventional fluctuation regime [12]. A very large and *temperature-independent* resistivity in the FM state, much greater than the Ioffe-Regel limit for a uniform metallic system, also indicates

113

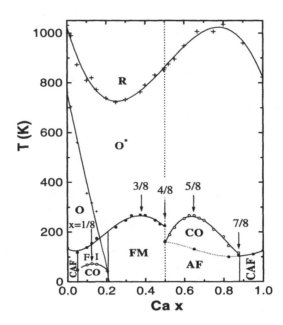

Figure 1: The phase diagram for $La_{1-x}Ca_xMnO_3$ [2]. The electronic and structural phases: O* - orthorhombic (octahedron rotated), O - orthorhombic (Jahn-Teller distorted), R - rhombohedral, CO - charge-ordered, AF - antiferromagnetic, CAF - canted antiferromagnetic, FM - ferromagnetic metallic, FI - ferromagnetic insulating.

that the insulating CO and metallic FM phases coexist on the "metallic" side of the MIT [7, 8]. The transition into the FM state is accompanied by an increase of the magnetization M, which saturates below T_c (see Fig. 2). In contrast to the resistance, M varies smoothly across the transition. The magnetization is proportional to the volume fraction of the FM phase; this volume fraction is $\sim 15 - 20\%$ at T_c, which is close to the percolation threshold in a three-dimensional percolated system. All these observations suggest that the low-T_c manganites can be viewed as macroscopically inhomogeneous systems, where the metallic FM domains are imbedded into the insulating CO matrix. Percolation phenomena might be important in this situation near the MIT, provided the scale of the phase separation is much smaller than the sample's dimensions.

The noise measurements can open a new window on the MIT in the CMR manganites. Indeed, it is well known that in classical percolation systems, the $1/f$ noise diverges at the percolation threshold [13, 14]. The $1/f$ noise reflects fluctuations of the resistance, and its spectral density S_V is proportional to the fourth power of the bias current density j [14]. Close to the percolation threshold, the current density becomes strongly non-uniform, and, with approaching the threshold, the contribution of the regions with a large j to the $1/f$ noise increases more rapidly than their contribution to the resistance [14]. We report on the

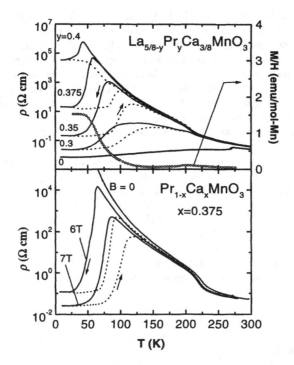

Figure 2: Temperature dependence of the resistivity in $La_{5/8-y}Pr_yCa_{3/8}MnO_3$ (at $B = 0$) and in $Pr_{1-x}Ca_xMnO_3$ (at $B = 0 - 7T$). The solid lines correspond to cooling, the dashed ones - to heating. The temperature dependence of the magnetization for $La_{5/8-y}Pr_yCa_{3/8}MnO_3$ with $y = 0.35$ is shown for cooling as open circles in the upper panel.

$1/f$ noise measurements in polycrystalline and single crystal samples of low-T_c manganites. The combined transport and noise measurements strongly suggest that the so-called Curie temperature in the low-T_c materials is, in fact, a percolation threshold temperature rather than the temperature of the long-range ferromagnetic phase transition. The scaling analysis of the $1/f$ noise is consistent with the percolation model of conducting domains randomly distributed in an insulating matrix [15].

EXPERIMENT

The resistance drop at the MIT is more dramatic in the CMR manganites with a lower transition temperature T_c. One can decrease T_c, for example, by partial substitution of La with Pr, which changes the chemical pressure in the system [6]. Alternately, it is possible to start with a compound that does not demonstrate the MIT even at the lowest T (e.g.,

$Pr_{1-x}Ca_xMnO_3$), and to induce the MIT by the magnetic field. In this work, we have used both methods. The transport and noise measurements have been carried out on poly- and single crystal bulk samples of $La_{5/8-y}Pr_yCa_{3/8}MnO_3$ ($y = 0.35 - 0.4$) and $Pr_{1-x}Ca_xMnO_3$ ($x = 0.25 - 0.5$). The former compound demonstrates the MIT at zero magnetic field, whereas in the latter compound, the MIT can be induced by the magnetic field (Fig. 2). The sample preparation is described elsewhere [6]. Typically, the polycrystalline samples were $4 \times 1 \times 1$ mm^3, single crystals - $3 \times 1 \times 0.5$ mm^3. The spectral density of the $1/f$ noise and the resistivity ρ have been measured in the four-probe configuration over the temperature range $T = 4.2 - 300K$ in the magnetic field $B = 0 - 8T$ for both cooling and heating.

The bias current I was driven through the sample by a low-noise current source with the output resistance much greater than the sample's resistance. Typical values of the bias current were $I \simeq 10^{-6} - 10^{-4}A$. All the data discussed below were obtained in the linear regime, where the rms noise was linear in current. The noise signal was amplified by a preamplifier PAR 113 and measured by a lock-in amplifier SR 830 in the mean average deviation mode. This regime allows to measure continuously the spectral noise density $S_V = \langle (V - \bar{V})^2 \rangle$ ($\bar{V}$ is the average value of the voltage across the sample, $\langle .. \rangle$ stands for the time averaging), while the temperature (or the magnetic field) is slowly varying. Because of a very strong temperature dependence of the resistance in manganites, especially in the vicinity of the MIT, the rate of the T (or B) sweep must be very slow (otherwise, the change of $\bar{V}$ during the measurement time will contribute to S_V). This also requires a good temperature stabilization (the long-term temperature stability in our measuring set-up was better than $1mK$ at $T = 10 - 300K$). Typically, the noise was measured at $f = 10 - 30Hz$, with the equivalent noise bandwidth $\Delta f = 1 - 5Hz$ and the time constant $\tau = 0.1 - 1s$. It has been verified that the noise spectrum has a power-law form $1/f^{\gamma}$ in the frequency range $f = 1 - 10^3 Hz$ with γ close to unity for all temperatures.

RESULTS

Typical temperature dependences of the resistivity for the $LaPrCaMnO$ system at $B = 0$ and for the $PrCaMnO$ system at different magnetic fields are shown in Fig. 2. Qualitatively, the dependences $\rho(T)$ for these systems are similar: with cooling, the resistivity grows exponentially below the CO transition ($T_{CO} \sim 200 - 220K$), decreases rapidly when the system undergoes the MIT, and remains almost temperature-independent in the FM state. The temperature of the MIT is strongly y-dependent: T_c increases from $35K$ for $y = 0.4$ to $75K$ for $y = 0.35$. For the $LaPrCaMnO$ system, the change of the resistivity is the most dramatic at $y = 0.35 - 0.375$, the "sharpness" of the transition is also the largest for these concentrations of Pr. Both systems exhibit a very strong hysteresis upon cooling and heating. The transition into the FM state is accompanied by the increase of the magnetization M, which saturates at $T < T_c$ (see Fig. 2). In contrast to ρ, M changes gradually across T_c for both poly- and single crystal samples. According to the magnetization data, the fractional volume of the FM phase, on the one hand, always exceeds $\sim 20\%$ on the "metallic" side of the MIT (we recall that the percolation threshold for both discrete and continuum percolation models is $\sim 17\%$ in three dimensions), on the other hand, is significantly smaller than 100%. This observation, as well as an anomalously large and temperature-independent resistivity in the FM state (observed even for single crystal samples), is consistent with the idea of coexistence of the FM and CO phases on the ferromagnetic side of the MIT.

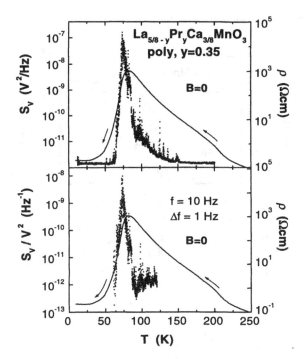

Figure 3: Temperature dependences of the resistivity (solid lines), the spectral noise density (dots, the upper panel), and the normalized spectral noise density corrected on the background (dots, the lower panel) for a polycrystalline sample $La_{5/8-y}Pr_yCa_{3/8}MnO_3$ with $y = 0.35$. All the dependences have been measured for the zero-field cooling.

The upper panel of Fig. 3 shows a typical temperature dependence of the spectral noise density, measured for a polycrystalline sample $La_{5/8-y}Pr_yCa_{3/8}MnO_3$ with $y = 0.35$. The noise exceeds the background level at $T \simeq 150K$ (the preamplifier was not used in this measurement, which explains a relatively large magnitude of the background), increases by many orders of magnitude with approaching the MIT, and drops sharply below the background level on the FM side of the transition. The frequency dependence of this noise is close to $1/f$. To compare the $1/f$ noise for different samples, it is convenient to use the normalized spectral noise density, $S_V/V^2 = \langle(\delta\rho/\rho)^2\rangle$: in the linear regime, this quantity does not depend on the bias current and on the sample's geometry. The lower panel of Fig. 3 shows that S_V/V^2 for this sample is weakly T-dependent in the CO state far from the MIT: it varies by a factor of 2-3 over the range $T = 90 - 150K$, though ρ changes by 2-3 orders of magnitude over the same T interval (see also the lower panel of Fig. 4). With approaching T_c, the $1/f$ noise exhibits a giant peak. The peak value of $S_V/V^2 = 10^{-10} - 10^{-8}\ Hz^{-1}$ exceeds by many orders of magnitude S_V/V^2 observed for macroscopically uniform metallic or semiconducting samples ($10^{-19} - 10^{-23}\ Hz^{-1}$ and $10^{-16} - 10^{-19}\ Hz^{-1}$, correspondingly,

117

for the same frequency and sample's volume) [14].

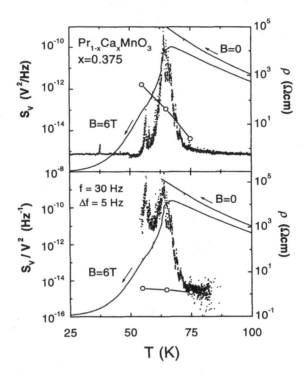

Figure 4: Temperature dependences of the resistivity (solid lines), the spectral noise density (dots and open circles, the upper panel), and the normalized spectral noise density corrected on the background (dots and open circles, the lower panel) for a polycrystalline sample $Pr_{1-x}Ca_xMnO_3$ with $x = 0.375$. All the dependences have been measured for cooling. At $B = 0$, this compound does not exhibit the MIT. The normalized $1/f$ noise measured for several fixed T at $B = 0$ coincides with the data at $B = 6T$, measured with a slow temperature sweep far from the transition. This indicates that the sweep rate $(4mK/s)$ was sufficiently low to exclude contribution of the time-dependent $\bar{V}$ to $S_V = \langle (V - \bar{V})^2 \rangle$.

The sharp peak of the 1/f noise allows to determine T_c with a high accuracy; below we identify T_c with the temperature of the S_V/V^2 maximum. There is a correlation between the magnitude of the noise peak and the ratio $\rho(T_c)/\rho(300K)$. For example, for the sample in Fig. 3, ρ increases by 4 orders of magnitude with cooling from room temperature down to $80K$; the normalized noise magnitude at the transition also increases by a factor of $\sim 10^4$. Since the high-temperature portion of the $\rho(T)$ dependences is approximately the same for all compounds $La_{5/8-y}Pr_yCa_{3/8}MnO_3$ with $y > 0.35$ [8], the noise peak is more pronounced for materials with a higher value of $\rho(T_c)$, e. g. with a lower T_c.

Qualitatively similar behavior is observed for the $Pr_{1-x}Ca_xMnO_3$ system, where the MIT can be induced by the magnetic field. The upper panel of Fig. 4 shows the $1/f$

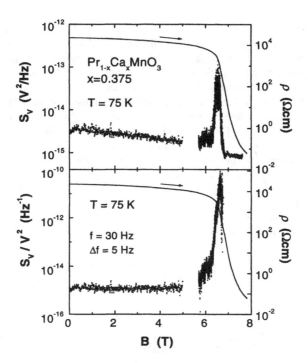

Figure 5: Magnetic field dependences of the resistivity (solid lines), the spectral noise density (dots, the upper panel), and the normalized spectral noise density corrected on the background (dots, the lower panel) for a polycrystalline sample $Pr_{1-x}Ca_xMnO_3$ with $x = 0.375$. All the dependences have been measured at $T = 75K$ after zero-field cooling from room temperature. The magnetic field sweep rate was $2 \cdot 10^{-4}T/s$.

noise versus T for the $Pr_{0.625}Ca_{0.375}MnO_3$ sample at zero magnetic field (where the MIT is absent), and for the field cooling at $B = 6T$ (the MIT is observed at $T = 63K$). At $B = 0$, the normalized noise varies weakly with temperature over a range $T = 50 - 80K$. If the sample is cooled down at a fixed B, the $1/f$ noise increases sharply at the transition. A "shoulder" on the FM side of the $\rho(T)$ dependence coincides with the second peak of the $1/f$ noise. This shoulder, more or less pronounced, has been observed for all the samples $Pr_{1-x}Ca_xMnO_3$. A very sharp peak of noise is also observed if the sample is initially zero-field-cooled down to a certain T, and then the MIT is induced by increasing magnetic field (see Fig. 5). Comparison of Figs. 4 and 5 shows that the field cooling results in the peak magnitude of the $1/f$ noise approximately one order of magnitude greater than that for the isothermic field-induced MIT if the sample was initially zero-field-cooled. In both cases, the magnitude of the noise peak correlates with sharpness of the transition. The noise peak is much smaller for compounds with $x < 0.35$ and $x > 0.4$, where the resistance drop at the transition is less dramatic, and the $R(T)$ and $R(B)$ curves are much smoother.

Our 1/f noise measurements provide strong evidence of the percolation nature of the CO-FM transition in the polycrystalline bulk samples of low-T_c manganites. Indeed, a diverging behavior of the 1/f noise is typical for the percolation metal-insulator transition [13, 15]. On the insulating side of the MIT, the $1/f$ noise is relatively small (it is still greater than in macroscopically homogeneous metals and semiconductors), and almost temperature-independent. The concentration of disconnected FM domains increases with approaching T_c on cooling. Switching of these domains between FM and CO states, which can be especially clearly seen for single crystals ([16], see also Fig. 7), results in a rapid growth of the 1/f noise with $T \rightarrow T_c$. The noise intensity exhibits a giant maximum at T_c, where the concentration of the FM phase reaches the percolation threshold, and a cluster of connected FM domains extends through the whole sample. Further strengthening of the backbone of the infinite cluster on the "metallic" side of the MIT causes a rapid fall of the 1/f noise. Note that T_c is lower than the temperature of the maximum resistivity; instead, T_c coincides with the maximum of $d\rho(T)/dT$. This is expected for a percolating mixture of two phases where the "insulating" phase has a finite ρ which increases rapidly with cooling.

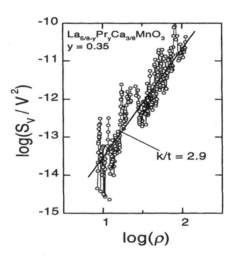

Figure 6: The scaling dependence of the normalized spectral density of the 1/f noise on ρ for the polycrystalline sample $La_{5/8-y}Pr_yCa_{3/8}MnO_3$ with $y = 0.35$ ($f = 10Hz$, $\Delta f = 1Hz$). Solid line is a power-law fit with the exponent $k/t = 2.9$.

A clearly diverging behavior of the 1/f noise allows to determine T_c with a high accuracy and to perform the scaling analysis of S_V and ρ on the "metallic" side of the MIT. In the vicinity of a percolation metal-insulator transition, the scaling behavior of ρ and S_V/V^2 is expected [13, 14, 15]:

$$S_V/V^2 \propto (p - p_c)^{-k} ,$$ (1)

$$\rho \propto (p - p_c)^{-t} . \tag{2}$$

Here p is the concentration of the metallic phase, p_c is the critical concentration, k and t are the critical exponents of the noise and resistivity, correspondingly. It is convenient to represent S_V/V^2 as a function of ρ (in this case, no assumption on the value of p_c is necessary):

$$S_V/V^2 \propto \rho^{k/t} . \tag{3}$$

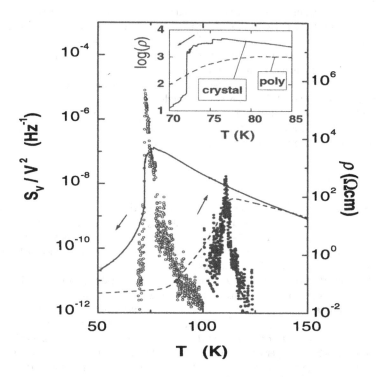

Figure 7: Temperature dependence of the resistivity (solid and dashed lines) and the normalized spectral density of the 1/f noise for cooling (open dots) and warming (solid dots) for the single crystal $La_{5/8-y}Pr_yCa_{3/8}MnO_3$ with $y \sim 0.35$ (at $f = 10Hz$, $\Delta f = 1Hz$). The inset shows the temperature dependence of the resistyvity for polycrystalline and single crystal samples in the vicinity of the MIT.

The normalized magnitude of the 1/f noise versus ρ for the polycrystalline sample $La_{5/8-y}Pr_yCa_{3/8}MnO_3$ with $y = 0.35$ is shown in the double-log scale in Fig. 6. Within the experimental accuracy, this dependence can be fitted by the power law (3) with $k/t =$

2.9 ± 0.5. For the other samples, the values of k/t fall into the range $1.2 - 3$. These values of k/t are consistent with the result $k/t = 2.4$ obtained for the continuum percolation model of conducting regions, randomly placed in an insulating matrix (the so-called inverted random-void model)[15]. Previously, similar values of $k/t = 2 - 3$ have been observed experimentally for the mixed powders of conducting and insulating materials [17, 18].

For high-quality single crystals of $La_{5/8-y}Pr_yCa_{3/8}MnO_3$ $(y \sim 0.35)$ [19], we have also observed a dramatic increase of the $1/f$ noise at the transition: S_V/V^2 reaches 10^{-5} Hz^{-1} for cooling and 10^{-8} Hz^{-1} for heating (Fig. 7). An evidence of the phase inhomogeneity of the single crystal is provided by the smooth dependence $M(T)$, which is similar to that for the polycrystalline samples. However, there are several important distinctions between the temperature dependences of ρ and S_V/V^2 for poly- and single crystals. Although the magnitude of ρ is similar for both types of samples, the temperature dependence of ρ for single crystals exhibits reproducible steps in the vicinity of T_c (see the inset of Fig. 7). The sharp drop of ρ by more than an order of magnitude, observed for this single crystal at $T \sim 72K$, can be interpreted as formation of a chain of a few connected FM domains between the voltage leads. The step-like behavior of ρ at the transition indicates that the size of the FM regions in the single crystal is significantly bigger than that in the polycrystalline samples. When the voltage leads become "shortened" by a chain of the metallic FM domains, an abrupt drop of the $1/f$ noise magnitude occurs. The percolation approach is not applicable in this case, since we probe the inhomogeneous system at the scale smaller than the percolation correlation length.

CONCLUSIONS

To summarize, we observed the dramatic peak of the $1/f$ noise in low-T_c CMR manganites $La_{5/8-y}Pr_yCa_{3/8}MnO_3$ and $Pr_{1-x}Ca_xMnO_3$ at the transition between the charge-ordered insulating and ferromagnetic metallic states. The peak value of the normalized noise density is by several orders of magnitude greater than that in macroscopically homogeneous metals and semiconductors. The combination of these data with the temperature dependence of ρ and M provides a strong evidence that the metal-insulator transition in these compounds is of a percolation nature, it corresponds to formation of a cluster of connected metallic FM domains which extends through the whole sample. The percolation theory describes adequately the $1/f$ noise in the vicinity of the MIT in polycrystalline samples, where the size of the FM domains is much smaller than the sample's dimensions. A well-pronounced step-like temperature dependence of the resistivity, observed for high-quality single crystals, suggests that the scale of the phase separation in single crystals is much greater than in polycrystalline samples.

ACKNOWLEDGEMENTS

We thank Sh. Kogan, M. Weismann, and V. Kiryukhin for helpful discussions. This work was supported in part by the NSF grant No. DMR-9802513.

References

[1] M. Imada, A. Fujimori, and Y. Tokura, Rev. Mod. Phys. **70**, 1039 (1998).

[2] M. Uehara, K. H. Kim, and S-W. Cheong, unpublished.

[3] R. M. von Helmholt et. al., Phys. Rev. Lett. **71**, 2331 (1993).

[4] S. Jin et. al., Science **264**, 413 (1994).

[5] E. L. Nagaev, Sov. Phys.-Uspekhi, **39**, 781 (1996).

[6] S-W. Cheong and H. Y. Hwang, in *Colossal Magnetoresistance Oxides*, edited by Y. Tokura (Gordon & Breach, London, 1999), ch. 7.

[7] N. A. Babushkina et. al., Phys. Rev. B **59**, 6994 (1999).

[8] M. Uehara, S. Mori, C. H. Chen, and S-W. Cheong, Nature (London) **399**, 560 (1999).

[9] A. J. Millis, P. B. Littlewood, and B. I. Shraiman, Phys. Rev. Lett. **74**, 5144 (1995).

[10] H. Röder, Jun Zang, and A. R. Bishop, Phys. Rev. Lett. **76**, 1356 (1996).

[11] J.-S. Zhou and J. B. Goodenough, Phys. Rev. Lett. **80**, 2665 (1998).

[12] J. M. De Teresa et al., Nature **386**, 256 (1997).

[13] R. Rammal et. al., Phys. Rev. Lett. **54**, 1718 (1985).

[14] Sh. Kogan, *Electronic Noise and Fluctuations in Solids*, Cambridge University Press 1998.

[15] A.-M. S. Tremblay, S. Feng, and P. Breton, Phys. Rev. B. **33**, 2077 (1986).

[16] R. D. Merithew, M. B. Weissman, J. O'Donnel, and J. Eckstein "Mesoscopic Fluctuations in Collosal Magnetoresistance", preprint, 1999.

[17] D. A. Rudman, J. J. Calabrese, and J. J. Garland, Phys. Rev. B **33**, 1456 (1986).

[18] C. C. Chen and Y. C. Chou, Phys. Rev. Lett. **54**, 2529 (1985).

[19] The Pr concentration for the single crystal, estimated from the $T_c(y)$ dependence, was close to 0.35 (the nominal concentration was $y = 0.42$).

EVIDENCE OF ANISOTROPIC THERMOELECTRIC PROPERTIES IN La₂/₃Ca₁/₃MnO₃ THIN FILMS STUDIED BY LASER-INDUCED TRANSIENT VOLTAGES

H.-U. Habermeier*,**, Xiaohang Li*, Pengxiang Zhang*,**
*Max-Planck-Institut - FKF, Heisenbergstr.1 D-70569, Stuttgart, Germany.
**Institute of Lasers, Yunnan Polytechnic University, Kunming 650051, P. R. China.

ABSTRACT

Pulsed UV laser illumination of doped $LaMnO_3$ thin films deposited on vicinal cut $SrTiO_3$ [STO]substrates causes unexpected transient voltages at room temperature in remarkable analogy to those observed for intrinsically anisotropic $YBa_3Cu_3O_{7-x}$ [YBCO] thin films. The experimental data for $La_{2/3}Ca_{1/3}MnO_3$ [LCMO] thin films are consistent with the description of thermoelectric fields caused by off-diagonal elements of the Seebeck tensor. Systematic measurements of the dependence of the laser-induced signal on the geometry of the sample, the temperature and doping dependence suggest that the anisotropy of the Seebeck tensor is due to combination of long range co-operative Jahn-Teller distortions and substrate-induced biaxial strain effects.

INTRODUCTION

The observation of a colossal magnetoresistance (CMR) close to the spin ordering temperature, T_c, in doped rare earth manganites of the type $Ln_{1-x}B_xMnO_3$ [with Ln = La, Pr, Nd and B = Ca, Sr, Ba, Pb] has generated considerable research activities due to their peculiar electronic, magnetic and structural properties [1-4]. The CMR materials have a perovskite-type crystal structure in which the corner shared oxygen octahedra surrounding the Mn ions can co-operatively rotate around the cubic [110] or [111] direction, causing a reduction of the original cubic to an orthorhombic or rhombohedral symmetry. It is now accepted that both, double exchange [5] and Jahn-Teller [JT] lattice distortions [6], play an important role in determining their electronic and magnetic properties. If the JT distortions of the oxygen octahedra surrounding the Mn^{3+} ions are not distributed randomly in the material, an anisotropy of the transport properties can be expected in addition to the anisotropy introduced by spin ordering at temperatures below the spin ordering temperature..

In principle, experimental studies of anisotropy effects of transport properties can directly be performed using sufficiently large single crystals or - in the case of unavailability of the appropriate crystals - with single-crystalline thin films. In reality, using the thin film approach, the technological problems for the measurement of the out-of-plane properties can only be overcome if, e.g., the electrical conductivity in plane and out-of-plane differ by orders of magnitude. In this paper, we use an unconventional thin film approach and describe experimental evidence of a pronounced anisotropy in the transport properties of doped $LaMnO_3$ that is not compatible with a cubic symmetry or faintly distorted cubic symmetry.

Our experiment is designed in analogy to the well established so called thermopile effect in $YBa_2Cu_3O_7$ single-crystal thin films grown on vicinal cut STO single-crystal substrates [7-11]. This approach probes the anisotropic thermoelectric properties of a material rather than the dc conductivity. The origin of the thermopile effect lies in the generation of thermoelectric fields transverse to a temperature gradient, ∇T. The thermoelectric fields are caused by the nonzero off-diagonal elements of the Seebeck tensor S_{ij}. Off-diagonal Seebeck coefficients occur in tetragonal, trigonal, hexagonal and orthorhombic crystals only, if ∇T has an orientation not along

125

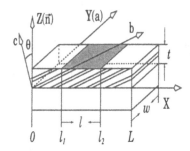

Fig. 1 Schematic sketch of an YBCO film deposited on a vicinal cut STO substrate

the low-indexed crystallographic axes [12]. In this case, the gradient of the electrical potential $\nabla\Phi$ is related to the gradient of the lattice temperature ∇T by

$$\nabla\Phi_i = S_{ij} \nabla_j T \qquad (1)$$

In the thermopile experiment, the non-crystallographic system orientation is realised by depositing YBCO thin films onto STO (001) substrates intentionally miscut by an angle α towards the [010] direction as shown schematically in Fig.1. This causes the CuO_2 planes of the YBCO to grow with a tilt angle α with respect to the substrate surface plane. An illumination of the film with a short laser pulse generates a transient tem-perature gradient $\nabla_\perp T$, which gives rise to a voltage, U. Quantitatively, assuming $\nabla_\perp T$ being, the integration of Equ. (1) yields

$$U_{[010]} = 1/2t\Delta\ S\Delta T\ sin2\alpha$$

$$U_{[100]} = 0 \qquad (2)$$

[l = length of illuminated film strip, t = film thickness, ΔT = temperature difference of top and bottom of the film, $\Delta S = S_{ab} - S_c$, the difference of the Seebeck coefficients in the ab-plane and along the c-axis]. In YBCO the signals detected at room temperature even for moderate laser fluence of 50 mJ/cm^2 can be as large as 18 V for a sample with t = 250 nm and an illuminated length of 2 mm [11].

The experiments reported here, demonstrate the existence of an unexpected laser-induced voltage [LIV] in $La_{2/3}Ca_{1/3}MnO_3$ thin films in the order of 1/10 of those measured in YBCO. The existence of these laser-induced signals suggests a microscopic mechanism comparable to that in YBCO implying a substantial anisotropy of the thermoelectric properties of $La_{2/3}Ca_{1/3}MnO_3$.

II. EXPERIMENTAL TECHNIQUES

Expitaxial $La_{2/3}Ca_{1/3}MnO_3$ films were prepared on STO (001)-oriented substrates with intentional miscut angles α using the standard pulsed laser deposition technique[13], subsequently patterned and contacted by the usual lithographic and deposition procedures. The films, characterised by x-ray diffraction, magnetotransport and AFM, consist of single - phase material with properties typical of epitaxial CMR thin films [14,15]. The thermoelectric experiment is performed using an excimer laser [KrF, l = 248 nm, pulse duration 28 ns] as energy source in conjunction with an Oxford optical cryostat allowing measurements in the range 10 K < T < 300 K. For the measurement of the transient voltage signal, a HP 500 MHz oscilloscope was used allowing a time resolution of 2 nsec. The digitised data are collected by a computer using a standard IEEE 488 interface. Figure 2a shows the original raw data as recorded for a specimen deposited on a perfectly oriented (001) substrate together with the signals for a specimen deposited on a substrate with 10^0 tilt vs. the [010] direction.

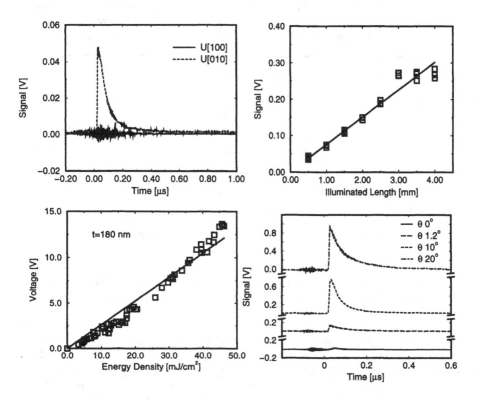

Fig 2   Laser-induced voltage signals   measured at room temperature for $La_{2/3}Ca_{1/3}MnO_3$ thin films of thickness 180 nm in [100] and [010] direction (a), as a function of illuminated length (b), as a function of illuminating laser energy density (c) and the vicinal angle of the substrate (d).

III. RESULTS AND DISCUSSION

As shown in Fig.2a, the existence of a laser-induced voltage in $La_{2/3}Ca_{1/3}MnO_3$ thin films oriented not along the principal crystal axes, suggests its thermoelectric origin. Consequently, the predictions of Equ..(2) such as its dependence on the crystallographic orientation, length, thickness and laser fluence, i.e. ΔT must hold. The results of the most crucial experiment are given in Fig. 2a-d for a $La_{2/3}Ca_{1/3}MnO_3$ films deposited on vicinal cut substrates and patterned in a L- shape with branches oriented along the [010] and [100] directions, respectively. According to Equ (2), U_s must vanish. for the strip oriented along [100]. Along the tilt direction [010], however, U_s should be nonzero and scale linearly with the illuminated length, laser fluence and the tilt angle α. The data given in Fig 2a-d are in accordance with the predictions of Equ.(2). Furthermore, the temperature dependence of the laser-induced signal, U_s, should reflect the temperature dependence of ΔS. The analysis of the temperature dependence of the signal, however, requires the approximation that a laser pulse with a given energy density causes a constant value for ΔT independent of the sample temperature. Fig. 3 represents the temperature dependence of U_s together with R(T). The data show a maximum of U_s close to the metal-insulator transition and a gradual decrease with decreasing temperature in the metallic state.

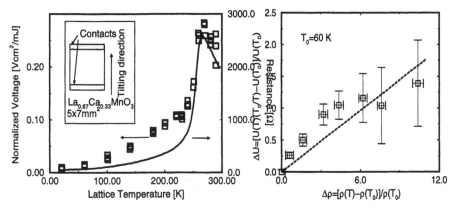

Fig 3. Temperature dependence of the LIV signal and ρ for a 180 nm LCMO film.

Fig.4. Relative changes of the measured ΔS vs. relative changes of ρ.

These phenomenological findings are interpreted in terms of an interplay of JT- induced anisotropy of the Seebeck tensor and the magnitude and temperature dependence of ΔS. The temperature dependence of $\Delta S(T)$ and $\rho(T)$ given in Fig. 3 suggests a close relationship between these quantities. Assuming that the components $S_{ij}(T)$ have the same - or at least similar - temperature dependence, the close relation of $\Delta S(T)$ and $\rho(T)$ can qualitatively be understood in terms of Mott's formula for the charge contribution to the Seebeck coefficient in the metallic state [16]

$$ S \sim T\{ \sigma'(E_F)/ \sigma(E_F)\} \qquad (3) $$

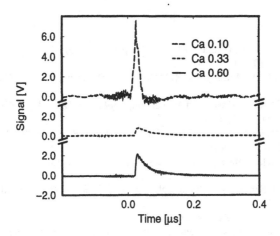

Fig 5. Laser-induced voltage for LCMO films of different composition
at room temperature

σ (E_F) is the conductivity at the Fermilevel and σ' stands for $\partial/\partial E\{\sigma(E_F)\}$. The decrease in S for $T < T_{MI}$ is therefore attributed to the large increase in conductivity due to the reduction of spin scattering by ferromagnetic ordering. In the approximation σ' to be constant and σ to be isotropic [i.e. $\sigma^{-1} = \rho$] a linear dependence

$$[S(T)(T_0 /T) - S (T_0)] / S (T_0)] \sim [\rho(T) - \rho (T_0)] / \rho (T_0) \qquad (5)$$

is expected. Measurements performed on La-Sr-Mn-O single crystals strongly support this view [17,18]. Figure 4 represents the data of Fig. 3 [taken from the metallic region below T = 250 K] plotted according to Equ.(5). Here, the values for the normalisation are arbitrarily taken for a low temperature [T_0 = 60 K]. Within a certain scatter the data scale linearly as expected, thus supporting the concept of the LIV being of thermoelectric nature. Finally, with the hypothesis that the thermoelectric effect is responsible for the LIV signals and their magnitude arises from the Jahn-Teller distortions causing intrinsic anisotropy of the system, the LIV signals are expected to depend on the doping level. We have studied the LIV signals for specimens with 10% and 60% Ca and found for the 10% doped sample a strong, for the 60% doped sample a weak increase of the room temperature signal as shown in Fig. 5. Even if the strong increase of the LIV signal for the sample with 10% Ca doping supports the model of JT induced anisotropy, the effect is superimposed by the doping depend-ence of r and S according to Equ. (4). For a complete understanding a detailed investigation including the dependence of the Seebeck-coefficients on doping is required.

In conclusion, we have demonstrated the existence of a laser-induced voltage in La-Ca-Mn-O thin films deposited on vicinal cut substrates which is compatible with the predictions given by a model based on the existence of off-diagonal elements of the Seebeck tensor. This implies the presence of an anisotropy of the material. Tentatively, we ascribe the origin of this anisotropy to the interplay of the Jahn-Teller distortion caused by the Mn^{3+} ions and biaxial interfacial strain of the films.

REFERENCES

1. G. H. Jonker and J. H. Van Santen, *Physica* **16**, p.337 (1950)

2. J. Volger, *Physics* **20**, p.49 (1954)

3. S. Jin, T. H. Tiefel, M. McCormack, R. Fastnacht, R. Ramesh, and L. H. Chen, *Science,* **264**, p. 413 (1994)

4. K. Chahara, T. Ohno, M. Kasai, and Y. Kozono, *Appl Phys. Lett.* **63**, p. 1990 (1993)

5. C. Zener, *Phys. Rev.* **82**, p. 403 (1951)

6. A. J. Millis, P. B. Littlewood, and B. I. Shraiman, *Phys. Rev. Lett.* **74**, p. 5144 (1995), and *Phys. Rev. Lett.* **77**, p.175 (1996)

7. C..L. Chang, A. Kleinhannes, W.G. Moulton, and R. L. Testardi, *Phys. Rev. B* **41**, p.11,564 (1990)

8. H. Lengfellner, G. Kremb, A. Schnellbögl, J. Betz, K. F. Renk, and W. Prettl, *Appl. Phys. Lett.* **60**, p.501 (1992)

9. L. R. Testardi, *Appl. Phys. Lett.* **64,** p. 2,347 (1994)

10. H.-U. Habermeier, N. Jisrawi, and G. Jäger-Waldau, *Inst. Phys. Conf. Ser.* **148,** p.1,023 (1995)

11.H.-U. Habermeier, N. Jisrawi, G. Jäger-Waldau, U. Sticher, and B. Leibold, *Mat. Res. Soc. Symp. Proc.* **379**, p.229 (1996)

12. J. F. Nye, *Physical Properties of Crystals* (Clarendon, Oxford 1985), Chap. 7

13. H.-U. Habermeier, *Eur. J. Solid State Inorg. Chem* **28**, p. 201 (1991)

14. M. McCormack, S. Jin, T. H. Tiefel, R. M. Fleming, J. M. Philips, and R. Ramesh, *Appl Phys.Lett.***64**, p.3,045 (1994)

15. R. B. Praus, B. Leibold, G. M. Gross, and H.-U. Habermeier, *Appl. Surf. Sci.* **138-139,** p. 40 (1999)

16. N. F. Mott and E. A. Davis, *Electronic Processes in Non-Crystalline Materials* (Oxford University Press, Oxford 1979) p.52

17. A. Asamitsu, Y. Moritomo and Y. Tokura, *Phys. Rev. B* **53**, p.R2952 (1996)

18. S. Uhlenbeck, B. Büchner, R. Gross, A. Freimuth, A. Maria de Leon Guevara, and A. Revco levschi, *Phys. Rev. B* **57**, . R5,571 (1998)

INVESTIGATION OF THE TRANSPORT MECHANISM IN DOPED LA-BASED MANGANITE THIN FILMS BY TRAVELING WAVE METHOD

L. WANG, S. HUANG, J. YIN, X. HUANG, J. XU, Z. LIU, K. CHEN
National Laboratory of Solid State Microstructures and Department of Physics,
Nanjing University, Nanjing 210093, P. R. China

ABSTRACT

The traveling wave (TW) method has been utilized to investigate the transport mechanism in paramagnetic-insulator state of $La_{0.75}Sr_{0.11}Ca_{0.14}MnO_3$ films. The drift mobility of the films increased from 2.5×10^{-2} cm^2/Vs at 310 K to about 9.2×10^{-2} cm^2/Vs at 400 K. The Arrhenius behaviors of the conductivity and drift mobility indicate that the transport process in manganites above the Curie temperature is dominated by the thermally assisted hopping of small polarons.

INTRODUCTION

Lanthanum-based transition-metal oxide perovskites have received wide attention from the theoretic and experimental researchers [1,2,3] since these materials exhibit a large negative magnetoresistance called colossal magnetoresistance (CMR) effect near the point of ferromagnetic-metal to paramagnetic-insulator phase transition. Due to this character, it can be applied in magnetic record and switching devices[4]. And these materials in paramagnetic-insulator phase are suitable to act as the semiconductor layer of the ferroelectric field effect transistor (FET) [5] since they are closely lattice-matched to the common perovskite-type ferroelectrics. Although CMR has been discovered for a few decades, the mechanism of it is still not clear. Initially, the mechanism of CMR is ascribed as the double exchange model [6]. But now many theoretical [7,8] and experimental papers [9-13] have shown that the magnetic interactions alone are insufficient to explain CMR and a coupling between charge and lattice is significant. Under this situation, it is necessary and important to well understand the transport mechanism of the materials in ferromagnetic and paramagnetic states. Recently, Hall measurements [12,14,15] have been used to investigate the transport properties in Lanthanum-based manganites. Since the anomalous part of Hall coefficient is not well understood, it is impossible to get information on the behavior of the drift mobility or the carrier concentration in these materials using Hall measurements [14]. However, the information on the drift mobility is indispensable to clearly understand the mechanism.

In order to get insight into the transport process, we adopt the traveling wave (TW) technique to investigate the transport mechanism in manganite films. Since it was introduced by Adler et al [16], TW method has been utilized to investigate the transport in amorphous semiconductor for several years [17-19]. One of the characteristics of this method is the possibility to get information on the transport mechanism in the materials that have high resistance and low mobility.

In this paper, we focus on the transport properties of $La_{0.75}Sr_{0.11}Ca_{0.14}MnO_3$ thin films in the paramagnetic-insulator phase. The conductivity and the drift mobility in the films at different temperatures were measured and the values at room temperature are 1.6×10^{-2} S/cm and 2.5×10^{-2} cm^2/Vs, respectively. Both conductivity and drift mobility increase with increasing temperature. The Arrhenius behaviors of the conductivity and drift mobility are consistence with a small polaron model, which indicates that the transport mechanism above the Curie temperature is dominated by the hopping of small polarons.

EXPERIMENT

The samples were fabricated by pulsed laser deposition method using $La_{0.75}Sr_{0.11}Ca_{0.14}MnO_3$ ceramic target. A 248 nm KrF excimer laser source with 5Hz in repetition rate and 30ns in pulse width. The average laser energy density was 2.5 J/cm^2. The target and the substrate holder were rotated during the deposition. The films were deposited on fused silica substrates, with a thickness of 500 nm, at 800 $^{\circ}$C in 30 Pa of oxygen. Immediately after the deposition the films were *in situ* annealed at 850 $^{\circ}$C in pure oxygen of 0.5 atm. for one hour. A more detailed description on sample preparation procedure has been published in our previous work [20]. The chemical composition of the films was analyzed by inductively coupled plasma quantemeter(ICP). The result of ICP measurement shows that the films are stoichiometry. The samples are single-phase, as checked by X-ray diffraction. The Curie temperature T_c of the target used in this paper is 280 K. To meet the need of the TW measurements, co-planar electrodes with a distance of 1 mm were deposited by evaporation. The measurements were carried out in vacuum.

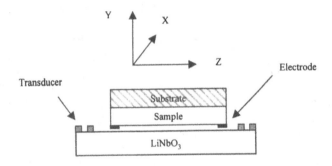

FIG.1 A schematic diagram of the traveling wave experiment.

A schematic diagram of the traveling wave experiment is shown in Fig.1. The sample is placed above the surface of LiNbO$_3$ single crystal plate with a gap of 12 μ m. The transducers on the LiNbO$_3$ generate a surface acoustic wave (SAW) that propagates along the surface of the LiNbO$_3$ plate in the z direction. Due to the piezoelectric properties of LiNbO$_3$, an electric field associated with the SAW couples into the sample to bunch the carriers into charge packets and drifts these carriers through the material, consequently an acoustic-electronic direct current is produced and can be measured from two electrodes. Fritzsche and Chen et al. [17-19] have derived the relationship between drift mobility and the acoustic-electric current or the acoustic-electric voltage as the following [19]:

$$\mu = \frac{2I_{ae}}{w\Phi_0 \sinh(kd)(k\sigma/v_s)\left[\left(|A|^2 + |B|^2\right)\cosh(kd) - 2(A \cdot B)\sinh(kd)\right]}$$

(1)

$$\sigma = \frac{I_{ae}L}{V_{ae}dw}$$

(2)

where I_{ae} is the acoustic-electronic short current, V_{ae} the acoustic-electronic open voltage, k the wave vector of the SAW, v_s the velocity of the SAW, d the thickness of the sample, L the distance between the electrodes, w the width in the x direction of the sample. The symbol Φ_0 is the potential at the surface of the piezoelectric plate and is given by the Rayleigh wave power [21]. It increases as the square of the power carried by the SAW, P_{SAW} and depends on the properties of the piezoelectric plate and propagation axis. The proportionality constant depends on the properties of the piezoelectric plate and propagation axis of the SAW.

The coefficient A and B are determined by the boundary conditions imposed on the electric field. They are generally complex, which means that the charge waves and the field have nontrivial phase relations. And they depend on the conductivity and the dielectric constant of the sample and the dielectric constant of the substrate. They can be described as

$$A = \left[\cosh(kh) + \varepsilon^*(B/A)\sinh(kh)\right]^{-1}$$

(3)

$$B/A = \left(\varepsilon_s + \varepsilon^* kd\right)/\left(\varepsilon^* + \varepsilon_s kd\right)$$

(4)

$$\varepsilon^* = \varepsilon - i4\pi\sigma/\omega$$

(5)

where h is the gap between the sample and the piezoelectric plate, ε_s the dielectric constant of the substrate, ε the dielectric constant of the sample, ω the frequency of the SAW.

TW method can be used to investigate the transport process in the materials with high resistivity and low mobility, for which Hall measurement meets some difficulties. The transport process in doped La-based manganites in the paramagnetic state is semiconductor-like, and these materials usually have high resistivity and the mobility is rather low. At this case, TW method is suitable and advantageous to be used for investigating these materials. Therefore we can obtain conductivity and drift mobility in these samples with I_{ae} and V_{ae} from Equation (1) and (2) as all the other quantities are either known or measurable. The sign of I_{ae} depends on the type of the majority carriers and the direction of the SAW. Since the direction of the SAW is controlled by the experiments, the type of majority carriers can be directly determined from the sign of I_{ae}.

RESULTS

The conductivity of the films was measured by TW method over the temperature range from 300K to 400 K. The conductivity obtained by the TW method is good agreement with that derived from normal I-V measurement. The value from the TW method is 1.6×10^{-2} S/cm and the value from I-V measurement is 1.8×10^{-2} S/cm at room temperature. The temperature dependence of the conductivity is shown in Fig.2. The conductivity increases with the increase of temperature and extends from 1.6×10^{-2} S/cm at 300 K to 9.6×10^{-2} S/cm at 400 K, which shows that the conduction property of the samples above T_c is universally semiconductor-like. This transport behavior is consistent with the previous reported.

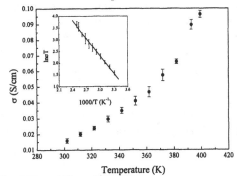

FIG.2   The resistivity vs temperature. In the inset, the resistivity plotted in the adiabatic limit. And the fine line is a linear fit indicating an activation energy of 219 meV.

The drift mobility of the films at temperature range from 300 K to 410 K was also measured by TW method. The drift mobility at 300 K is about 2.2×10^{-2} cm^2/Vs, which approaches to the mobility (0.03 cm^2/Vs) given in ref.12, 22. The temperature dependence of the drift mobility was shown in Fig.3. The drift mobility increases while the temperature is increasing and extends to 9.2×10^{-2} cm^2/Vs at 410 K. And the sign of I_{ae} in our work shows the behavior of the carriers in the films is electronlike.

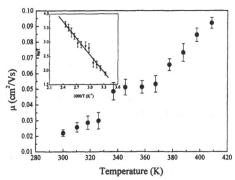

FIG.3   The drift mobility vs temperature. In the inset, the drift mobility plotted in the adiabatic limit. And the fine line is a linear fit indicating an activation energy of 171 meV.

As shown in Fig.2 and Fig.3, the Arrhenius behavior of the conductivity and drift mobility in our samples is observed, which is the distinctive character of small polaron model [23]. We plot $\ln(\sigma T)$ vs $1000/T$ in the insect of Fig.2. It is obvious that the conductivity is thermally activated with the energy of 219 meV. We also plot the $\ln(\mu T)$ vs $1000/T$ in the insect of Fig.3. From the insect, the thermal activation of the mobility is clearly observed with the activation energy of 171 meV. The activation energy of the drift mobility is much larger than the activation energy of the Hall coefficient (91 meV) from Hall measurements [12]. In a small polaron model, the activation energy of the drift mobility is always more than that for the Hall mobility [12]. And the value $(2.2 \times 10^{-2}$ cm^2/Vs) obtained for the drift mobility in our samples is much smaller than 1 cm^2/Vs, which is a distinct character of small polarons. Based on the above analysis, we can get a conclusion that the transport behavior above T_c in the films is dominated not by spin disorder scattering but by the hopping process of small polarons.

CONCLUSION

It is by traveling wave method that the drift mobility in manganite films has been firstly measured. The drift mobility extends from 2.5×10^{-2} cm^2/Vs at 310 K to about 9.2×10^{-2} cm^2/Vs at 400 K. The temperature dependence of the drift mobility reveals that transport process in doped manganites is dominated by the thermally assisted hopping of small polarons. Our results show that traveling wave method has more advantages compared to Hall measurement for that the more information on the transport process can be got by this method. This method also can be used to investigate transport process of the interface carriers in the perovskite heterostructure and in all-perovskite FET and will attract wide attentions in the future.

ACKNOWLEDGEMENTS

The authors thank Dr. Di Wu, Zhiming Zheng and Yunong Qi for fruitful discussions. This work is supported by The Climbing Program-National Key Project for Fundamental Research. One of the authors (L. Wang) would like to thank the financial support by the Motorola fellowship.

REFERENCES

1. J. H. Jonker and J. H. Van Santen, Physica (Amsterdam) **16**, p. 599 (1950).
2. S. Jin, M. McCormack, R. A. Fastnach, R. Ranesh and L. H. Chen, Science **264**, p. 413 (1994).
3. R. Von Helmholt, J. Wecker, B. Holzaptel, L. Schultz and K. Samwer, Phys. Rev. Lett. **71**, p. 2331 (1993).
4. P. Majumdar and P. B. Littlewood, Nature (London) **395**, p. 479 (1998)
5. S. Mathews, R. Ramesh, T. Venkantesan, J. Benedetto, Science **276**, p. 238(1997).
6. Zener, Phys. Rev. **82**, p. 403 (1951).
7. J. Millis, P. B. Littlewood and B. I. Shraiman, Phys. Rev. Lett. **74**, p. 5144 (1995).

8. H. Röder, J. Zhang and A. R. Bishop, Phys. Rev. Lett. **76**, p. 1356 (1996).

9. S. G. Kaplan et al., Phys. Rev. Lett. **77**, p. 2081 (1996).

10. J. M. De Teresa et al., Nature (London) **386**, p. 256 (1997)

11. M.Jaime, M.B. Salamon, M. Rubinstein, R. E. Treece, J. S. Horwitz and D. B. Chrisey, Phys. Rev. B **54**, p. 11914 (1996)

12. M. Jaime, H. T. Hardrer, M. B. Salamon, M. Rubinstein, P. Dorsey and D. Emin, Phys. Rev. Lett. **78**, p. 951 (1997).

13. T. T. M. Palstra, A.P. Ramirez, S-W. Cheong, B. R. Zegarski, P. Schifer and J. Zannen, Phys. Rev. B **56**, p. 5104 (1997).

14. P. Matl, N. P. Ong, Y. F. Yan, Y. Q.Li, D. Studebaker, J.Baum and G. Doubinina, Phys. Rev. B **57**, p. 10248 (1998).

15. G. J. Snyder, M. R. Beasley and T. H. Geballe, Appl. Phys. Lett. **69**, p. 4254 (1996).

16. R, Adler, D. James, B. J. Hunsinger and S. Datta, Appl. Phys. Lett. **38**, p. 102 (1981).

17. H. Fritzsche and K. J. Chen, Phys. Rev. B **28**, p. 4900 (1983).

18. K. J. Chen and H. Fritzsche, J. Non-cryst. Solid, **59&60**, p. 441 (1983).

19. Robert E. Jahanson, Phys. Rev. B **45**, p. 4089 (1992).

20. J. Yin, X. S. Gao, Z. G. Liu, Y. X. Zhang, X. Y. Liu, Appl. Surf. Sci. **141**, p. 21 (1999)

21. A. Auld, *Acoustic Fields and Waves in Solids* (Wiley, New York, 1973), Vol.2

22. A. S. Alexandrov and A. M. Bratkovsky, Phys. Rev. Lett. **82**, p. 141 (1999)

23. D. Emin, Ann. Phys. (N. Y.) **53**, p. 439(1969)

rf SUSCEPTIBILITY OF $La_{1-x}Sr_xMnO_3$ SINGLE CRYSTALS: MAGNETIC SIGNATURES OF STRUCTURAL CHANGES

P.V. PARIMI, H. SRIKANTH, M. BAILLEUL, and S. SRIDHAR,
Department of Physics, Northeastern University, Boston, MA 02115.

R. SURYANARAYANAN, L. PINSARD and A. REVCOLEVSCHI.
Laboratoire de Chimie des Solides, UA 446, Universitě Paris-Sud, Orsay 91405, France.

ABSTRACT

A sensitive tunnel diode oscillator (TDO) technique operating at $4MHz$ is used to probe the dynamic response of $La_{1-x}Sr_xMnO_3$ single crystals for $x = 0.125, 0.175, 0.28$ and 0.33 doping. Systematics of the measured change in reactance (ΔX) as a function of temperature ($30K < T < 320K$) and DC magnetic field ($0 < H < 6kOe$) reveal distinct temperature and field scales associated with the dynamic response of spin. It is notable that these features are far more striking than the corresponding features in static measurements. The results are discussed in the context of structural changes leading to polaron ordering.

INTRODUCTION

The perovskite oxides of the form $Re_{1-x}A_xMnO_3$ (where Re is a rare earth such as La and A is a divalent element such as Sr or Ca) have generated considerable interest in recent times because of the discovery of the colossal magnetoresistance (CMR) effect [1]. The CMR is a direct consequence of an unusual paramagnetic insulator (PMI) to ferromagnetic metal (FMM) transition driven mainly by the double exchange mechanism [2]. However, double exchange alone cannot describe the complete phase diagram of the manganites and it has been pointed out that the interplay of strong electron-phonon coupling and double exchange is required to understand the existence of the high temperature insulating phase, the CMR effect and its sensitivity to magnetic field [3][4].

The deficiency of the double exchange model is the fact that it does not consider spin-lattice or charge-lattice interactions, namely, Jahn-Teller interactions and polarons [5]. Experimental results clearly suggest that lattice contributions are important for a thorough understanding of manganites. Besides MI transition, charge ordering (CO) is one of the characteristic phenomena observed in these materials especially in the low doping regime. CO and stripe correlations of concentrated holes and spins have attracted much attention in recent times, particularly due to their possible role in high T_c superconductivity[6, 7, 8].

A variety of experiments including structural [9], transport [10, 11] and thermal [12] measurements have revealed novel features in the $Re_{1-x}A_xMnO_3$ directly associated with the interplay between structural, electronic and magnetic properties. Most of the experiments on manganites have been *static* and there have been relatively few experiments which probe the *dynamic* response of these systems. Dynamic experiments are likely to provide significant information about the collective response of spin and charge to the oscillating electric and magnetic fields impressed on the materials. In the present work rf dynamic response of $La_{1-x}Sr_xMnO_3$ for concentrations x=0.125, 0.175,0.28 and 0.33 are reported. We focus on the interplay of between holes and lattice distortions to understand the relation between the magnetic and structural properties.

Mat. Res. Soc. Symp. Proc. Vol. 602 © 2000 Materials Research Society

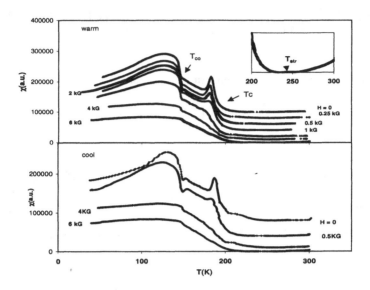

Figure 1: Top panel shows differential susceptibility for x=0.125 doping during warming. Bottom panel shows the same during cooling. The first order nature of the CO transition can be seen as a clear difference in the response between warming and cooling. Inset shows a magnified view of the structual transition.

EXPERIMENT

Single crystals of $La_{1-x}Sr_xMnO_3$ were grown using an image furnace technique [13]. Samples used in these measurements had cylindrical disk like shapes with diameter $5mm$ and thickness $2mm$ with polished surfaces and edges. The rf experiments were performed using a tunnel diode oscillator (TDO) which has very high sensitivity in measuring the electro- and magneto-dynamic properties of materials. The crystal is placed inside a copper coil which forms part of an LC-tank circuit driven by a stable tunnel diode oscillator. The inductive coil with the sample is mounted at the end of a rigid co-axial cable can be inserted into a continuous flow Helium cryostat. The temperature of this system can be regulated between $4.2K$ and $320K$ and an electromagnet is used to apply a dc magnetic field up to $6kOe$. The resonant frequency (f_0) is typically in the range of $2 - 4MHz$ depending on the geometric characteristics of the inductive coil and sample dimensions. The quantity that is measured, the change in frequency $\Delta f = f(T, H) - f_0$ as a function of T and H, is proportional to the change in reactance ΔX. For magnetic metals, from elementary considerations and applying Maxwell's equations, it can be shown that: $\Delta X \propto \sqrt{\chi}$, where χ is the differential susceptibility, dM/dH of the material.

RESULTS

Temperature dependence: $La_{0.875}Sr_{0.125}MnO_3$

The high sensitivity of the rf technique enables us to clearly detect a paramagnetic to ferromagnetic transition at T_c=180K as well as two additional transitions at T_s=270K and

$T_{co} \doteq 150K$, as shown in Fig. 1. Interestingly, this composition is observed to undergo structural transitions which are manifested in the change of lattice parameters at 150K, 180K and 270K [15]. At T_s the susceptibility shows a dip which is due to a structural phase transition from orthorhombic (pseudo cubic) to a cooperative Jahn-Teller distorted phase at lower temperature. In the presence of a magnetic field two characteristic changes are observed to take place at T_c. First, the peak disappears and secondly, the transition is broadened. In the absence of dc magnetic field the susceptibility is zero above T_c and raises rapidly at T_c. In the presence of dc field the susceptibility is finite at all temperatures and increases with applied field. Therefore, the sharp transition at T_c becomes broadened when field is applied. The hump observed at T_{co} is very clear and strong unlike the CO transition observed in resistivity and magnetization measurements[12]. This fact emphasizes the importance of high frequency measurements to detect CO transitions. The hump at T_{co} is caused by a magnetic transition accompanied by a change in structure[12][15, 9].We also observed hysteretic behavior in the susceptibility around T_{co} with decreasing and increasing temperature which indicates the first order nature of this transition. As can be seen from the Fig. 1 the rf reactance shows a dip at T_{co} during cooling which is not observed while warming.

The hump associated with T_{co} appears to be a purely ac response of the charge ordering as the dc response[12] does not show any hump. It is worth mentioning that the CO observed in $Nd_{0.45}Ca_{0.55}MnO_3$ at 260K also shows a hump in the ac susceptibility measurement[14]. The reason for the hump only in ac measurement is that in ac measurement the differential susceptibility is measured. The reversible response of the ferromagnetic domains to the rf field just below Tc gives rise to an increase in the differential susceptibility. With further decrease in T the onset of saturation magnetization locks the individual domain and hence the $\chi(T)$ starts decreasing.

The key to understanding the contribution of the structural transitions to the electronic and magnetic properties lies in the Mn-O interionic distance of the octahedra. The interionic distances $m(T), s(T)$, and $l(T)$, which are along a, c and b axes, respectively, are calculated from the representation $m^2 = 0.031(a^2 + b^2 + c^2)$, $s^2 = 0.125c^2 - m^2$ and $l^2 = a^2s^2/(16s^2 - a^2)$. In these calculations the rotation of the octahedra with respect to the axes is neglected. Fig 2 shows the temperature dependence of these parameters. As can be seen from the Fig. $m(T)$ is constant over the entire temperature range, while $s(T)$ and $l(T)$ show clear anomalies at 140K and 270K. These results imply that for temperatures below 140K or above 270K there is no contribution of the rhombic J-T Q2 mode to the formation of crystal lattice. The turning on of the Q2 mode as the sample is warmed above 140K results in structural phase transition from low temperature. The response of a ferromagnet in a magnetic field is also important to describe the first order transitions observed at T_{str} and T_{co}. Below T_{str} the system shows a spontaneous cooperative JT distorted phase[12]. The strong dependence of $\chi(T)$ on magnetic field suggests magnetoelastic coupling for CO besides coulomb repulsion.

An isolated hole in $LaMnO_3$ can be considered a small polaron which is given by a localized hole in the $3d_{x^2-y^2}$ orbital surrounded by inverse Jahn-Teller distortion. The polaron phase is an ordered arrangement of Mn^{3+} and Mn^{4+} ions for which one of the two alternating atomic layers in the (001) plane contains both Mn^{3+} ions, as in pure LaMnO3, while the other layer contains both Mn^{3+} and Mn^{4+}ions, i.e. holes. The local distortion is because the hole site Mn^{4+} is JT inactive whence the electron-phonon energy is lowered by restoring higher symmetry around the hole[16]. In this picture, at high temperatures

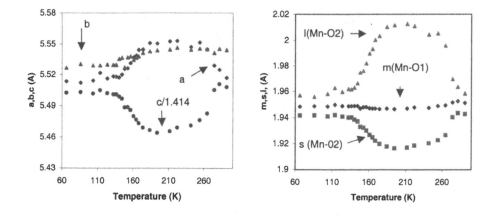

Figure 2: The change in the lattice parameters is shown on the left panel. Right panel shows the Mn-O interionic distances, m(T), s(T) and l(T) which are along a,c and b axes.

the $La_{1-x}Sr_xMnO_3$ may be viewed as a polaron liquid which will eventually transform into polaron lattice as the temperature is lowered. We, therefore, identify T_{co} as the onset point for polaron lattice formation, where holes start to freeze on lattice points. From this point of view $La_{1-x}Sr_xMnO_3$ is considered to undergo successive transitions from polaron liquid (insulator) to Fermi liquid (metal) to polaron lattice (insulator).

Field Dependence: $La_{1-x}Sr_xMnO_3$

The field dependence of the differential susceptibility, $\frac{dM}{dh_{ac}}|_{H_{dc}}$, at various temperatures below 300K is shown in Fig. 3. As can be seen from the figure for all the doping levels of Sr the $\chi(H)$ response shows an overall decrease with the increase in magnetic field. The magnetization M(H) shows a monotonic increase with field with an eventual saturation at high fields, for all the compositions studied. Therefore, the decrease in the differential susceptibility is not surprising. There are, however, many subtle changes in the $\chi(H)$ response at low magnetic fields, H < 2000G. In the case of x=0.125 composition $\chi(H)$ initially increases, reaches a maximum and starts decreasing forming a peak. For the remaining three compositions the peak is not prominent as can be seen from the figure. The behavior of $\chi(H)$ can be understood by a simple picture of domain response to weak and strong magnetic fields. When a weak field is applied the magnetization process is reversible. In the presence of strong fields the domains are locked and tend to form a single domain. Therefore, the response of the domains to the ac field at low dc fields is greater than that at high dc fields, thus contributing to the initial increase.

CONCLUSIONS

Dynamic rf susceptibility of $La_{1-x}Sr_xMnO_3$ revealed several magnetic signatures in both temperature and field dependent measurements. These magnetic signatures have direct correlation with structural changes in terms of Mn-O interionic distances of the octahedra,

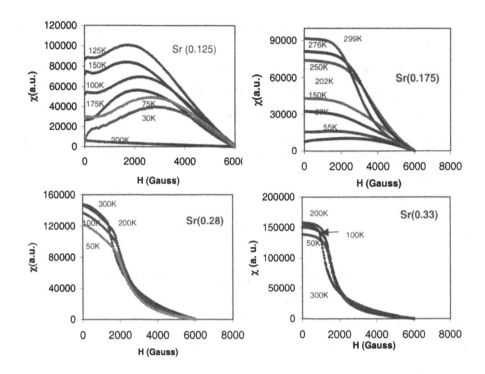

Figure 3: Field dependence of differential susceptibility for x=0.125, 0.175, 0.28 and 0.33 doping.

at the corresponding temperatures. Field dependent differential susceptibility is found to decrease monotonically with field with a rich structure at low fields.

ACKNOWLEDGMENT

This work was supported by a US-NSF- 9711910 and NSF-CNRS grant NSF-INT-9726801.

References

[1] K. Chahara, T. Ohno, M. Kasai and Y. Kozono, Appl. Phys. Lett. 63, 1990 (1993).

[2] C. Zener, Phys. Rev. 2, 403 (1951).

[3] A.J.Millis, P.B.Littlewood, and B.I.Shraiman, Phys. Rev. Lett. 74, 5144 (1995)

[4] H. Roder et al., Phys. Rev. Lett. 76, 1356 (1996).

[5] J.-S. Zhou, J. B. Goodenough, A. A samitsu, and Y. Tokura, Phys. Rev. Lett. 79, 3234 (1997).

[6] J. M. Tranquada et al., Nature 375, 561 (1995).

[7] M. I. Salkola et al., Phys. Rev. Lett. 77, 155 (1996),

[8] J. Q. Li et al., Phys. Rev. Lett. 82, 2386 (1999).

[9] H. Kawano, R. Kajimoto, M. Kubota, and H. Yoshizawa, Phys. Rev. B 53, R14709 (1996), H. Kawano, R. Kajimoto, M. Kubota, and H. Yoshizawa, Phys. Rev. B 53, 2202 (1996).

[10] A. Asamitsu, Y. Moritomo, R. Kumai, Y. Tomioka and Y. Tokura, Phys. Rev. B 54, 1716 (1996),

[11] A. Anane, et al., J. Mag. Mag. Mater, 165, 377 (1997).

[12] S. Uhlenbruck, R. Teipen, R. Klinger, B. Buchner, O. Friedt, M. Hucker, H. Kierspel, T. Niemoller, L. Pinsard, A. Revcolevschi, and R. Gross, Phys. Rev. Lett. 82, 185 (1999).

[13] A. Revcolevschi and D. Dhallene, Adv. Mater., 5, 657 (1993).

[14] A. A. Mukhin et al., JETP Letters, 68, 356 (1998).

[15] L. Pinsard et al. J. Alloys Compd. 262-263, 152 (1997).

[16] Y. Yamada, O. Hino, S. Nohdo, and R. Kanao, Phys. Rev. Lett. 77, 904 (1996).

[17] A. Anane, C. Dupas, K. Le Lang, J.P. Renard, P. Veillet, A. M. de Leon Guevara, F. Millot, L. Pinsard and A. Revcolevschi, J. Phy: Condensed Matter 7, 7015 (1995).

Charge and Orbital
Ordering Effects

MAGNETISM AND CMR IN ELECTRON DOPED PEROVSKITE AND LAYERED MANGANITES

A. MAIGNAN, C. MARTIN, M. HERVIEU AND B. RAVEAU
Laboratoire CRISMAT –UMR6508 – ISMRA – CNRS , Boulevard du Maréchal Juin, 14050 CAEN, France, antoine.maignan @ismra.fr

ABSTRACT

From the magnetic phase diagrams established for $Ln_{1-x}AE_xMnO_3$ manganites with $x>0.5$, it is shown that the magnetoresistance is only obtained for compositions of smallest average A-site cations $<r_A>$. The magnetic structure study reveals that the small $<r_A>$ favors the coexistence of ferromagnetism with G-type antiferromagnetism. It is also shown that ferromagnetism can be induced in the $Ln_{2-x}AE_xMnO_4$ type electron doped manganites by using compositions with small Ln and AE cations and x values close to 1.90. Finally, manganse site doping by Cr, Co, Ni can be used to weaken the charge-ordering of the Mn^{4+} rich perovskite manganites ($0.5<x\leq0.8$) and thus to induce CMR properties.

INTRODUCTION

One striking feature of the $Ln_{1-x}^{3+} AE_x^{2+} MnO_3$ manganites is the lack of symmetry of their structural and physical properties for the hole doped ($x<0.5$) and electron doped ($x>0.5$) counterparts [1-3]. For the formers, the majority Jahn-Teller Mn^{3+} species responsible for the polaron formation and the Mn^{3+}-O-Mn^{4+} double-exchange mechanism are the basic ingredients to describe the properties [4-7]. From the chemical point of view three important parameters – manganese valency [1-3], A-site average cationic size ($<r_A>$) [8-10] and A-site cationic size mismatch [11-12] – have been shown to control both the Curie temperature (T_C) value and the ratio of the coexisting ferromagnetic metallic (FMM) and antiferromagnetic insulating (AFMI) phases which govern the CMR magnitude [13]. In this respect, the Mn^{4+} rich compounds of small $<r_A>$ such as $Ln_{1-x}Ca_xMnO_3$ ($x>0.5$) which exhibit FMM behaviors for $x\approx0.85-0.90$ behave in an opposite manner to the hole-doped compositions [14-17]. Their resistivities are metallic in the paramagnetic state in contrast to the semi-conducting like behavior of Mn^{3+} rich manganites characterized by similar small $<r_A>$. The CMR of the Mn^{4+} rich compositions occurs for compositions at the boundary between the FMM compositions and the AFMI charge ordered ones, the latter corresponding to higher Mn^{3+} contents. The CMR compound $Sm_{0.15}Ca_{0.85}MnO_3$ illustrates this situation [16-17]. In order to get the essential informations concerning the origin of the CMR properties of Mn^{4+} riche side, the magnetic and nuclear structures of $Sm_{0.15}Ca_{0.85}MnO_3$ [18] and $Sm_{0.10}Ca_{0.90}MnO_3$ [19] have been established and correlated to the transport properties and to the macroscopic magnetization measurements. For these electron doped manganites, the CMR properties disappear as $<r_A>$ increases as shown for $Pr_{1-x}Sr_xMnO_3$ with $x>0.5$ [3]. In order to better understand this phenomenon, a comparative study of the $Sm_{1-x}Ca_xMnO_3$ and $Pr_{1-x}Sr_xMnO_3$ compositions has been performed.

As in the case of the Mn^{3+} rich manganites, the CMR properties of the Mn^{4+} compositions are not restricted to the 3D $Ln_{1-x}AE_xMnO_3$ perovskite manganites. Significant magnetoresistance

Mat. Res. Soc. Symp. Proc. Vol. 602 © 2000 Materials Research Society

has also been found in the layered $Ln_{2-x}Ca_xMnO_4$ manganites for $x \approx 1.9$ [20]. This magnetoresistance is again linked to the appearance of a ferromagnetic like behavior for these compositions. Similarly to the perovskite manganites, this ferromagnetic behavior is enhanced as $<r_A>$ decreases, i.e. inversely to the Mn^{3+} rich compounds.

Finally, the B-site substitution of chromium for Mn has been previously shown to be a powerful way to control the collapse of the charge-ordering (CO) in $Ln_{0.5}Ca_{0.5}MnO_3$ manganites [21-24]. We will show that this method can also be applied to weaken the CO of Mn^{4+} rich $Ln_{1-x}Ca_xMnO_3$ composition ($0.5 < x \leq 0.8$) and to restore CMR properties [25-26]. The possible coexistence of two phases with different electronic properties at low temperature (phase separation) will also be discussed.

EXPERIMENT

The preparation of the $Ln_{1-x}AE_xMnO_3$ samples under concern has been previously described [3, 12, 16-26]. The purity of the as-prepared ceramics was systematically checked, using X-ray powder diffraction. Neutron powder diffraction (NPD) experiments were performed on selected samples at the LLB (Saclay France) ; for Ln=Sm, the Sm^{152} isotope has been used to diminish the neutron absorption and to increase the contrast of diffraction. The structures of the obtained ceramics have also been studied by electron microscopy (EM). The electron microscopes are equipped with energy dispersive spectroscopy (EDS) analysers so that the cationic compositions and electron diffraction (ED) could be performed on the same crystallites. One of the microscope was also equipped with a liquid-nitrogen cryostat allowing to study the T dependence of the ED patterns from 92K to 300K.

T-dependent (5K-400K) or H-dependent (0-7T) resistance data were collected by means of a Quantum Design physical properties measurements system with a four probe technique. Magnetization as a function of temperature or of magnetic field was measured with either a Quantum Design SQUID magnetometer (0-5T) or with a vibrating sample magnetometer(0-1.45T).

RESULTS

CMR in the Mn^{4+} rich perovskites.

The difference between the Mn^{4+} rich compositions of large and small $<r_A>$ values are summarized in Fig.1a and 1b for $Pr_{1-x}Sr_xMnO_3$ and $Sm_{1-x}Ca_xMnO_3$, respectively. For the first series, $<r_A>$ increases as x increases from $<r_A>=1.245$ Å for x=0.5 to $<r_A>=1.297$ Å for x=0.9. Concomitantly, an A-type AFM structure, with $T_N < T_C$, is first observed for narrow composition range, $0.50 \leq x \leq 0.56$, involving two magnetic transitions successively, PMI (paramagnetic insulator) to FMM to AFMI. Then all the compositions corresponding to $0.56 < x \leq 0.90$, are characterized by a PMI-AFMI transition. Note that all the structural investigations performed at low temperature reveal the absence of CO for these compositions. The absence of underlying FMM state in these compounds for x>0.56 provides an explanation for the lack of magnetoresistance for the Mn^{4+} rich $Pr_{1-x}Sr_xMnO_3$ compositions. Note also that, with the same synthesis conditions, the compositions with x>0.9 cannot be prepared with the cubic structural form due to the formation of hexagonal-type phases. In contrast, for the second series Sm_{1-}

$_x$Ca$_x$MnO$_3$, a complete solid solution up to x=1 is obtained as shown in Fig. 1b. In this series, only a slight <r$_A$> increase is obtained as x increases, in comparison with the Pr$_{1-x}$Sr$_x$MnO$_3$ series [<r$_A$>=1.156 Å and <r$_A$>=1.18 Å for Sm$_{0.5}$Ca$_{0.5}$MnO$_3$ and CaMnO$_3$, respectively]. The smaller A-site average cationic size of the Sm$_{1-x}$Ca$_x$MnO$_3$ series favors the Mn^{3+}/Mn^{4+} CO for the Mn^{4+} rich compositions (see the T$_{CO}$ curve from x=0.5 to x=0.8 in Fig. 1b). These CO phenomena studied by ED vs T have been previously described and discussed [27].

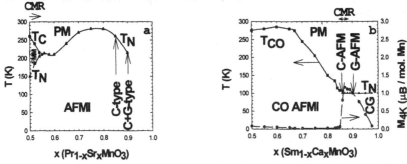

Fig. 1 : Magnetic phase diagrams of the Pr$_{1-x}$Sr$_x$MnO$_3$ (a) and Sm$_{1-x}$Ca$_x$MnO$_3$ (b) series for x≥0.5. The CMR compositions are delimited by horizontal arrows in the upper part.

For x>0.8 in Sm$_{1-x}$Ca$_x$MnO$_3$, the CO signature, consisting of super lattice reflexions in the ED patterns registered at 92K, is no more observed, and for 0.86≤x≤0.95 an unconventional ferromagnetic ('cluster glass' : CG in Fig.1b) metallic phase is obtained [28]. In the intermediate region, 0.82≤x≤0.86, CMR properties are observed (see the ρ(T)$_{H=0}$ and ρ(T)$_{H=7T}$ curves of Sm$_{0.15}$Ca$_{0.85}$MnO$_3$ in Fig.2 showing the CMR properties ; for sake of comparison, the Pr$_{0.15}$Sr$_{0.85}$MnO$_3$ curves are also given).

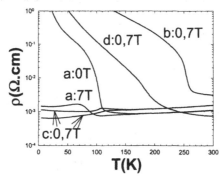

Fig. 2 : T dependent resistivity curves registered upon cooling in 0 an 7T.(a) Sm$_{0.15}$Ca$_{0.85}$MnO$_3$, (b) Pr$_{0.15}$Sr$_{0.85}$MnO$_3$, (c) Sm$_{0.10}$Ca$_{0.90}$MnO$_3$, (d) Pr$_{0.10}$Sr$_{0.90}$MnO$_3$.

The NPD studies of four samples, Sm$_{0.15}$Ca$_{0.85}$MnO$_3$, Sm$_{0.10}$Ca$_{0.90}$MnO$_3$, Pr$_{0.15}$Sr$_{0.85}$MnO$_3$ and Pr$_{0.10}$Sr$_{0.90}$MnO$_3$ have been carried on [18-19]. Both x=0.85 samples are C-type AFM, i.e. consist of antiferromagnetically coupled FM chains (Fig.3). Nevertheless these two compounds exhibit a different nuclear structure at room temperature (Fig.4a-b) though for both of them the structural and magnetic transitions are coupled. One indeed observes a Pm3m $(a_p \times a_p \times a_p)$ to I4/mcm $(a_p\sqrt{2} \times a_p\sqrt{2} \times 2a_p)$ transition and a Pnma $(a_p\sqrt{2} \times 2a_p \times a_p\sqrt{2})$ to $P2_1/m$ $(a_p\sqrt{2} \times 2a_p \times a_p\sqrt{2}, \beta \approx 91°)$ transition coupled with the PM to C-type AFM transitions at $T_N \approx 260K$ and 125K for "$Pr_{0.15}Sr_{0.85}$" and "$Sm_{0.15}Ca_{0.85}$", respectively. Note that the strong interplay between the nuclear and magnetic structures has also been revealed by magnetostriction measurements of $Sm_{0.15}Ca_{0.85}$ which exhibit a large volume increase as the sample enters in the high volume $P2_1/m$ C-type phase [29].

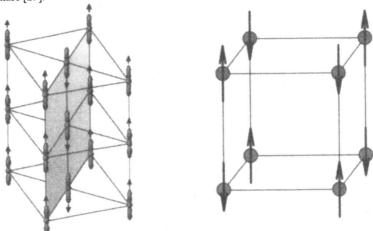

Fig. 3 : Orbital polarization and the C-type arrangement (left). G-type AFM structure (right).

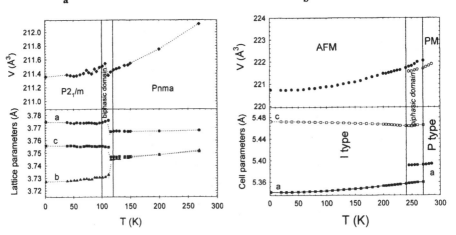

Fig. 4 : Cell volume and lattice parameters *vs* temperature for (a) $Sm_{0.15}Ca_{0.85}MnO_3$ and (b) $Pr_{0.15}Sr_{0.85}MnO_3$.

More importantly, the NPD study of $Sm_{0.15}Ca_{0.85}MnO_3$ indicates that, besides the majority C-type monoclinic phase (95%), there exists 5 % of a G-type orthorhombic phase with $T_N \approx 115K$, i.e. 10K lower than the T_N of the C-type [18]. This phase separation at low temperature is not obtained in $Pr_{0.15}Sr_{0.85}MnO_3$.

The interesting common magnetic characteristic of $Sm_{0.15}Ca_{0.85}MnO_3$ and $Sm_{0.10}Ca_{0.90}MnO_3$ is indeed the G-type AFM phase which structure is still not totally understood. $Sm_{0.10}Ca_{0.90}MnO_3$ crystallizes in a Pnma space group at low temperature (its structure is kept unchanged as T is swept through T_N) and exhibits a G-type AFM structure (Fig.3) below $T_N \approx 115K$ [19]. However, concomitantly a FM component develops at T_N so that its magnetic state can be viewed as the coexistence of AFM and FM at low temperature.

The FM component of $Sm_{0.10}Ca_{0.90}MnO_3$ is thus induced by the 10% Mn^{3+} which creates some FM Mn^{3+}-O-Mn^{4+} pathways in the matrix of majority Mn^{4+}. At the present time the arrangement of the FM and AFM components is not known. But , since $Sm_{0.10}Ca_{0.90}MnO_3$ is metallic below T_N (Fig.2) whereas $CaMnO_3$ is an insulator, the FM component is responsible for both metallicity and unconventional FM of the former. By considering the magnetic and electronic properties of $Sm_{0.10}Ca_{0.90}MnO_3$, one can propose an explanation of the CMR properties for "x~0.15" compositions such as in $Sm_{0.15}Ca_{0.85}MnO_3$. As shown in Fig.2, by field cooling this compound, the resistive behavior is much more metallic compared to the insulating behavior obtained in zero field cooling. This may be connected to the absence of percolation between the metallic clusters in the absence of field ; the magnetic field application may favor a size increase of these clusters (P2$_1$/m,C→Pnma,F field induced transition) so that a percolative path is achieved. NPD vs magnetic field and small angle neutron scattering experiments are now in progress to confirm this hypothesis.

The magnetic structure established for the non CMR compound $Pr_{0.1}Sr_{0.9}MnO_3$ gives some indications concerning the role of the FM-type phase in the CMR properties. The low temperature NPD data refinements show the coexistence of both the G and C-type AFM phases in a ratio close to 50 : 50. But in contrast with the "SmCa" samples, the FM component connected to the G-type phase cannot be detected. Accordingly, as shown in Fig.2, this sample exhibits an insulating behavior for the lowest temperature without any magnetoresistance.

Finally, this first set of data for Mn^{4+} rich compositions of the $Sm_{1-x}Ca_xMnO_3$ and $Pr_{1-x}Sr_xMnO_3$ series, emphasizes the role of the A-site average cationic size upon the CMR properties. As $<r_A>$ decreases, the coexistence of FM and AFM for a narrow range of compositions is favored. A magnetic field induced coupled structural and magnetic transition from the monoclinic AFMI high volume phase into the orthorhombic low volume (AFMI+FMM) phase is probably responsible for the CMR properties for $Sm_{0.85}Ca_{0.15}MnO_3$. The presence of G-type and FM clusters in the C-type matrix may act as nucleation seeds for this phase transition.

magnetoresistance in the layered manganites $Ln_{2-x}Ca_xMnO_4$ (x>1.8)

The lack of magnetoresistance for the hole-doped first member (n=1) of the (Ln AE)$_{n+1}$Mn$_n$O$_{3n+1}$ layered manganites is connected to the lack of ferromagnetism for this compound. Its bidimensional K_2NiF_4 structure reduces the 3d-bandwith with respect to the perovskites (n=∞) so that a spin-glass behavior without any magnetoresistance properties has been reported for this material [30-31].

The existence of ferromagnetism, although unconventional, in Mn^{4+} rich perovskite characterized by small $<r_A>$ suggested the possibility to increase the FM contribution in the K_2NiF_4 type manganite by reducing the thickness of the separating rock-salt type layer. Several series of $Ln_{2-x}Ca_xMnO_4$ phases have been prepared with Ln=Pr, Sm, Gd and Ho and $1.8 \leq x \leq 2.0$. The structural study confirmed the absence of perovskite defects [20].

In a similar way to the Mn^{4+} rich perovskite, the introduction of Mn^{3+} in Ca_2MnO_4 induces a weak ferromagnetic behavior, as shown in (Fig.5a). For $Pr_{2-x}Ca_xMnO_4$ and $1.80 \leq x \leq 1.96$, the magnetization values at 5K go through a maximum for x=1.94–1.90. Here again, in contrast with the $<r_A>$ effect on T_C in Mn^{3+} rich 3D manganites, the smaller cations induce higher magnetization (M) values, reaching $0.7\mu_B/$mol of Mn for Ln=Ho and x=1.92 against $0.45\mu_B/$mol of Mn for Ln=Pr and x=1.92 in Fig.5a, but T_C keeps a fixed value. This induced FM component is connected to the appearance of an inflexion point on the $\rho(T)$ curves for x=1.94 and x=1.92 (Fig.5b). However the small bump on the $\rho(T)$ curves does not coincide with the increase of M as the ferromagnetic component increases ($T_C \sim 115K$ in Fig. 5b). The magnetoresistance properties are illustrated by the $\rho(T)$ curves registered in 0 an 7T for $Pr_{0.08}Ca_{1.92}MnO_4$ (Fig.6). Roughly below T_C, the $\rho(T)_{H=0}$ and $\rho(T)_{H=7T}$ curves start to be separated showing the existence of negative magnetoresistance (MR) properties. From the field dependence of ρ (inset of Fig.6), 40% of MR

$(\%MR=100 \times (1 - \frac{\rho_{7T}}{\rho_0}))$ is obtained at 10K.

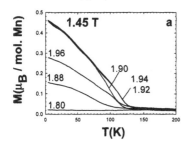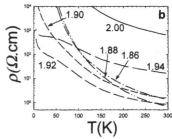

Fig. 5 : T dependence of the magnetization (a) and resistivity (b) for $Pr_{2-x}Ca_xMnO_4$ with $1.96 \geq x \geq 1.80$.

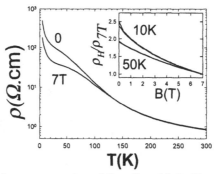

Fig. 6 : Magnetoresistance properties of $Pr_{0.08}Ca_{1.92}MnO_4$ illustrated by the $\rho_{H=0}(T)$ and $\rho_{H=7T}(T)$ curves. Inset : the $\rho(H)$ curves are also shown.

These results demonstrate that MR and some ferromagnetism can also be found in the $Ln_{2-x}Ca_xMnO_4$ layered manganites for electron doped compositions lying close to Ca_2MnO_4. Based on the Mn^{3+} rich compositions results, this behavior was not expected. The small size of the (Ln, Ca) cations in the rock salt-type layer is probably at the origin of the different nature of the interlayer magnetic coupling between two successive MnO_2 planes, in comparison with the compositions based on larger cations such as lanthanum and strontium.

B-site substitution effects in $Sm_{1-x}Ca_xMnO_3$ with x>0.5

One remarkable demonstration of this kind of chemical action upon the CMR properties was previously reported for the CO CE-type AFM $Pr_{0.5}Ca_{0.5}MnO_3$ insulating manganite [21]. Only 3% of Cr, Ni, Co substituted for Mn were sufficient to restore a FMM state and thus to induce CMR properties in this type of manganite. Recent experiments have shown that the doping is a tool to induce in a controlled way the phase separation between FMM regions and CO AFMI ones [32-33]. A global metallic state is thus realized when the content of FMM regions overcomes the percolation threshold (~16% [33]).

In the Mn^{4+} rich perovskite manganites, the small size of $<r_A>$ tends to favor the CO phenomena as shown above for $Sm_{1-x}Ca_xMnO_3$ with $0.8 \geq x \geq 0.5$ (Fig.1b). T_{CO} reaches a maximum value for x around 0.6-0.7. In order to test the efficiency of the substitutions for the electron doped manganites, substitutions by the same foreign elements (Cr, Co, Ni) on the magneto-transport properties have been studied.

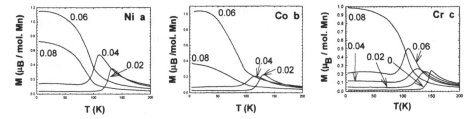

Fig. 7 : M(T) curves registered is 1.45T after a zfc process for $Sm_{0.2}Ca_{0.8}Mn_{1-x}M_xO_3$ with M=Ni (a), Co (b) and Cr (c).

This effect of B-site doping on the CO is illustrated in Fig.7 for $Sm_{0.2}Ca_{0.8}Mn_{1-x}M_xO_3$ with M=Cr, Co, Ni. It is remarkable that only 6% of foreign elements (M=Co or Ni Fig.7a and 7b) or 8% of chromium (Fig.7c) suppress the AFM transition associated to the CO and structural transition of $Sm_{0.2}Ca_{0.8}MnO_3$ and induce a ferromagnetic state. However, the magnetization magnitude at 5K reaches only ~1 μ_B per mole of Mn which is reminiscent of the unconventional FM discussed previously in $Sm_{0.10}Ca_{0.90}MnO_3$. For lower content of doping elements, such as 2% and 4% Ni and Co (Fig.8), the samples are insulating below T_N, in agreement with the small value of the magnetization at 5K (Fig.7). But, since T_N is strongly depressed for 4% in comparison with the 2% doped $Sm_{0.2}Ca_{0.8}MnO_3$ compounds (by ~30K), one can expect that small ferromagnetic regions already exist in these samples similarly to $Sm_{0.15}Ca_{0.85}MnO_3$ which exhibit the same type of transport and magnetic properties. The application of a magnetic field of

7T on the sample dramatically reduces the resistivity below T_N as shown in Fig. 8 for the 4% doped Ni and Co doped samples. Consequently, CMR properties with resistance ratios larger than 10^5 are obtained at low temperature.

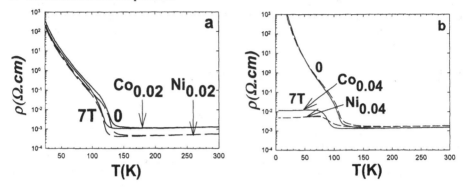

Fig. 8 : $\rho_0(T)$ and $\rho_{7T}(T)$ curves of (a) $Sm_{0.20}Ca_{0.80}Mn_{0.08}M_{0.02}O_3$ and (b) $Sm_{0.2}Ca_{0.8}Mn_{0.96}M_{0.04}O_3$ with M=Co and Ni.

Starting from CO compounds, the realization of CMR has been shown to be possible for all x values in the range $0.3<x\leq0.8$ in $Sm_{1-x}Ca_xMnO_3$ by using these foreign magnetic cations [25]. As one can see on the $Sm_{1-x}Ca_xMnO_3$ phase diagram of figure 1b, the CO temperature is maximum for x~0.60. Accordingly, the restoration of CMR properties by weakening of the CO requires more Cr for these compositions. Finally, since the ferromagnetism of the Mn^{4+} rich compositions is favored by the smaller $<r_A>$ values, the efficiency of the doping effect on the B-site increases as $<r_A>$ decreases as shown in Fig.9, where different Ln have been used to vary $<r_A>$ in $Ln_{0.3}Ca_{0.7}Mn_{0.90}Cr_{0.10}O_3$. The higher M values are reached for the smallest lanthanides Ln=Gd and Ho.

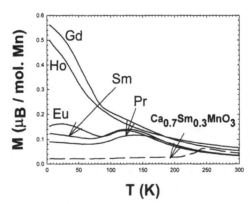

Fig. 9 : M(T) curves of the $Ln_{0.3}Ca_{0.7}Mn_{0.9}Cr_{0.1}O_3$ samples showing the increasing ferromagnetism as r_{Ln} decreases.

CONCLUSIONS

The electron doped perovskite manganites characterized by sufficiently small average A-site cationic radius exhibit metallic properties in the paramagnetic state but their transport and magnetic properties at low temperature are very sensitive to the Mn^{3+} concentration. The study of the magnetic structures shows that their CMR properties exist at the frontier between the C-type CO manganites ($x \leq 0.8$ in $Ln_{1-x}Ca_xMnO_3$) and the G-type ones ($x \sim 0.10$). The intriguing feature of the latter is the coexistence of FM and AFM which is connected to the unconventional magnetic properties of these materials such as the cluster-glass signature, revealed by AC-susceptibility, and the metallicity. The magnetic phase separation observed below T_N in $Ln_{0.15}Ca_{0.85}MnO_3$ between C and G type AFM indicates that the phase separation scenario is also valid in the case of Mn^{4+} rich manganites.

Though the lowering of dimensionality of the structure, as in $Ln_{2-x}Ca_xMnO_4$, reduces the MR magnitude, the small Ln and A cations are also favorable to the appearance of FM in the Mn^{4+} manganites contrasting with the Mn^{3+} rich ones.

Finally, B-site substitutions can also be used to induce CMR properties in electron doped manganites. The partial destabilization of the CO state by foreign B cations is thought to be responsible for the appearance of a weak FM state in these AFM compounds. By inducing the coexistence of magnetic phases, the Mn-site doping provides thus a way to control the CMR properties.

REFERENCES

1. P. Schiffer, A.P. Ramirez, W. Bao and S.W Cheong, Phys. Rev. Lett. **75**, p. 3336 (1995).

2. A.P. Ramirez, S.W. Cheong and P. Schiffer, J. Appl. Phys. **81**, p. 5337 (1997).

3. C. Martin, A. Maignan, M. Hervieu and B. Raveau, to appear in Phys. Rev. B vol.**60** (1999).

4. C. Zener, Phys. Rev. **82**, p. 403 (1951).

5. P.G. de Gennes, Phys. Rev. **118**, p. 141 (1960).

6. J.B. Goodenough, Prog. Solid. State Chem. **5**, p. 149 (1971).

7. A.J. Millis, P.B Littlewood an B.I Shraiman, Phys. Rev. Lett. **74**, p. 5144 (1995).

8. R. Mahesh, R. Mahendiran, A.K. Raychaudhuri and C.N.R. Rao, J. Solid. State Chem. **11**, p 297 (1995) ; ibid. **120**, p. 204 (1995).

9. H.Y. Hwang, S.W. Cheong, P.G. Radaelli, M. Marezio an B. Batlogg, Phys. Rev. Lett. **75**, p. 914 (1995).

10. A. Maignan, Ch. Simon, V. Caignaert and B. Raveau, Solid State Comm. **96**, p. 623 (1995).

11. L.M. Rodriguez-Martinez and J.P. Attfield, Phys. Rev. **B54**, p. 15622 (1996).

12. F. Damay, C. Martin, A. Maignan and B. Raveau, J. Appl. Phys. **82**, p. 6181 (1997).

13. A. Moreo, S. Yunoki and E. Dagotto, Science **283**, p. 2034 (1999).

14. H. Chiba, M. Kikuchi, K. Kusaba, Y. Muraoka, Y. Syono, Solid State Comm. **99**, p. 499 (1996).

15. I.O. Troyanchuk, N.V Samsonenko, H. Szymczak, A. Nabialek, J. Solid State Chem. **131**, p. 144 (1997).

16. C. Martin , A. Maignan, F. Damay, M. Hervieu, B. Raveau, J. Solid State Chem. **134**, p. 198 (1997).

17. A. Maignan, C. Martin, F. Damay, B. Raveau, Chem. Mater. **10**, p. 950, (1998).

18. C. Martin, A. Maignan, M. Hervieu, B. Raveau, Z. Jirak, A. Kurbakov, V. Trounov,
G. André and F. Bourée, J. Magn. Magn. Mater **205**, p. 184 (1999).

19. C. Martin, Z. Jirak, M. Hervieu, A. Maignan and B. Raveau (unpublished).

20. A. Maignan, C. Martin, G. Van Tendeloo, M. Hervieu and B. Raveau, J. Mater.Chem. **8**, p. 2411 (1998).

21. B. Raveau, A. Maignan and C. Martin, J. Solid State Chem. **130**, p. 162 (1997).

22. A. Barnabé, A. Maignan, M. Hervieu, F. Damay, C. Martin and B. Raveau, Appl. Phys. Lett. **71**, p. 3907 (1997).

23. F. Damay, C. Martin, A. Maignan, M. Hervieu, B. Raveau, F. Bourée and G. André, Appl. Phys. Lett. **73**, p. 3772 (1998).

24. F. Damay, C. Martin, A. Maignan and B. Raveau, J. Magn. Magn. Mater. **183**, p. 143 (1998).

25. A. Maignan, C. Martin, F. Damay, M. Hervieu and B. Raveau, J. Magn. Magn. Mater **188**, p. 185 (1998).

26. A. Maignan, C. Martin and B. Raveau, Mater. Res. Bull. **34**, p. 345, (1999).

27. M. Hervieu, A. Barnabé, C. Martin, A. Maignan, F. Damay and B. Raveau, Eur. Phys. J. B **138**, p.31 (1999).

28. A. Maignan, C. Martin, F. Damay, B. Raveau and J. Hejtmanek, Phys. Rev. B **58**, p. 2758 (1998).

29. R. Mahendiran, A. Maignan, C. Martin, M.R. Ibarra and B. Raveau (unpublished).

30. C.N.R. Rao, P. Ganguly, K.K. Singh and R.A. Mohan Ram, J. Solid State Chem. **72**, p. 14 (1988).

31. Y. Moritomo, Y. Tomioka, A. Asamitsu and Y. Tokura, Phys. Rev. B **51**, p. 3297, (1995).

32. Y. Moritomo, A. Machida, S. Mori, N. Yamamoto and A. Nakamura, Phys. Rev. B **60**, p. 9220 (1999).

33. T. Katsufugi, S.W. Cheong, S. Mori and C.H. Chen, J. Phys. Soc. Jap. **68**, p. 1090 (1999).

Two-Phase Coexistence
in the Manganites

Inhomogeneous Magnetic Structures of $La_{0.7}Ca_{0.3}MnO_3$ Investigated by ESR and Magnetization Measurements

Keon Woo Joh[*], Chang Hoon Lee[*], Cheol Eui Lee[*], Yoon Hee Jeong[**], C. S. Hong[+],
Nam Hwi Hur[+]
* Department of Physics, Korea University, Seoul 136-701, Korea
**Department of Physics, Pohang University of Science and Technology, Pohang
790-784, Korea
+Korea Research Institute of Standards and Science, Taejon 305-600, Korea

ABSTRACT

Comprehensive measurements of electron spin resonance and magnetization of
$La_{0.7}Ca_{0.3}MnO_3$ in the ferromagnetic as well as paramagnetic phases were carried out.
From the quantitative analysis of the ESR signal, attributed to itinerant spins and
localized spins in the bottleneck regime, evidences for the inhomogeneous nature of both
phases, consisting of clusters of one phase embedded in the other, were found. It is
suggested that the microscopic local magnetic structures above and below T_c are
qualitatively similar except that the phase below T_c carries long range order.

INTRODUCTION

Perovskite manganites with mixed manganese valence, $La_{1-x}D_xMnO_3$, with D for a
divalent metal, have attracted a great deal of attention due to the anomalously large
negative ``colossal" magnetoresistance (CMR) for $0.2 < x < 0.5$ [1-5]. In this
composition range, the system shows the simultaneous appearance of metallic conduction
and ferromagnetism at low temperatures while the end materials ($x = 0, 1$) are insulating
antiferromagnets. The underlying physics of this system has been traditionally explained
in terms of the double exchange (DE) mechanism due to Zener [6-8]; however, recent
works have shown that the DE alone may not be sufficient to explain the phenomenon.
Other effects such as lattice polarons due to the Jahn-Teller (JT) distortion [9], magnetic
polarons, and electron localization [10] were advanced as an additional physics. One
interesting proposal is that the phase separation of carriers, already seen in other systems
[11], is an essential ingredient in understanding the CMR materials [12].
Several computational works indicated that a homogeneous state is unstable against
phase separation (PS), as charge carriers are doped into manganites [13-15]. It was also
shown that the inclusion of Coulomb interactions leads to a charge-inhomogeneous state
with clusters of one phase (high carrier density) embedded in the other (low carrier
density). Since inhomogeneity in the carrier density also gives rise to corresponding
inhomogeneity in the magnetic structures, the PS state consists of the ferromagnetic
metallic clusters surrounded by either the antiferromagnetic regions or the paramagnetic
insulators [12]. Thus, a detailed microscopic study of magnetic properties of perovskite
manganites will shed light on the PS model.
Currently evidences for the inhomogenous nature of perovskite manganites are
offered by several studies, including NMR [16-18]. However, as will turn out, electron
spin resonance (ESR), which directly probes the electron spins, is a most suitable means
for this purpose. A few of ESR studies have already been carried out, and the spin
clusters and the inhomogeneous nature of the doped manganites have been supposed by
some ESR works [19-21]. Up to now, most of the ESR studies have been carried out
on powder or thin film samples, as a sample dimension greater than the skin depth at
the microwave frequency gives rise to asymmetric ESR lineshapes [22,23]. While the
absolute values of the g_{eff}-value and the linewidth may be less reliable, the asymmetric
lineshape can give more information on the inhomogeneity of the sample [23]. Rettory
et al. have performed an ESR experiment on a $La_{0.7}Ca_{0.3}MnO_3$ single crystal sample
whose size was larger than the skin depth at the microwave frequency, and obtained
asymmetric lineshapes [24]. However, they did not fully discuss the asymmetry of the
lineshape. Furthermore, within our knowledge, there is no report on a comprehensive

159

ESR study of the single crystal samples of the CMR materials covering a wide temperature range both above and below the magnetic transition temperature T_c.

In order to carefully examine the local magnetic properties related with the inhomogeneous nature of the CMR systems, we have made systematic ESR measurements and quantitative analysis for a single crystal sample of a representative CMR material, $La_{0.7}Ca_{0.3}MnO_3$ (LCMO), covering a wide temperature range both above and below the magnetic transition temperature (T_c). A careful dc magnetization measurement was also carried out just above T_c in order to find a relation between the macroscopic and the microscopic inhomogeneous magnetic properties. As a result of the comprehensive study of this unique system, we observed for the first time that the microscopic magnetic structures are very similar above and below T_c, except for the long range ordering. Furthermore, an antiferromagnetic correlation as well as the ferromagnetic clusters were explicitly revealed in this work.

EXPERIMENT

The $La_{0.7}Ca_{0.3}MnO_3$ (LCMO) single crystal sample used in this work was prepared by the floating zone melting method and ground into a spherical shape for the ESR measurements in order to minimize the demagnetization factor. The x-ray diffraction analysis showed a single phase perovskite structure of pseudo-cubic symmetry with a lattice constant a = 3.867 Å. The resistivity was measured with the standard four probe method. In order to examine the physical properties associated with the spin ordering transition, the dc magnetization was measured with a superconducting quantum interference device (SQUID) magnetometer as a function of temperature, and as a function of magnetic field at a temperature, 240 K, in the paramagnetic regime. The magnetic ordering temperature is conventionally defined to be the onset of the ferromagnetic magnetization, which gave the T_c of 225 K for the single crystal sample, in good agreement with the conductivity measurements as shown in Fig. 1. The sample dimension of single crystal was greater than the calculated skin depth at room temperature, and the ESR measurements were made at 9.4 GHz employing a Bruker ESP 300E spectrometer using a microwave power of 10 μ W and a modulation amplitude of 5 G.

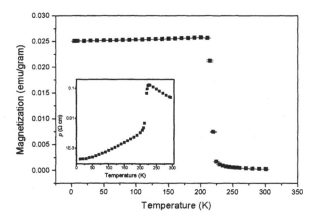

Fig. 1. Temperature dependence of the magnetization measured at the external field of 10 mT. The inset shows the temperature dependence of the resistivity.

RESULTS & DISCUSSION

A typical differential ESR spectrum for our sample is shown in Fig. 2. As can be seen, the ESR spectra are of an asymmetric lineshape both above and below the T_c.

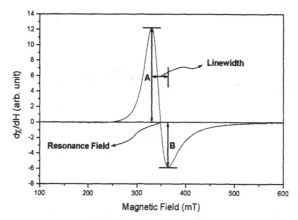

Fig. 2. A typical differential ESR spectrum above T_c (300 K) showing how the parameters were defined.

Three main parameters were derived as depicted in Fig. 2; the asymmetry parameter (AP), the g_{eff}-value, and the linewidth. The AP, defined as the ratio of the positive to the negative maximum (A/B) in the differential ESR lineshape, serves as a clue to determine the homogeneity in the sample [22,23]. The resonance field is determined from the zero crossing point of the ESR spectrum, and the g_{eff}-value is obtained from the relation

$$g_{eff} = \frac{h\nu}{\mu_B H_{res}}, \tag{1}$$

where ν is the microwave frequency, μ_B is the Bohr magneton, and H_{res} is the resonance field. However, the ESR parameters could not be determined unambiguously near the T_c, which is attributed to the combined effect of the Q-factor change in the cavity due to the large magnetoresistance during the ESR measurements [23], and the coexistence of the signals characteristic of the high temperature and low temperature regimes well above and well below T_c, respectively.

The minimum AP value for a homogeneous sample is 2.7, which corresponds to the slow diffusion limit, except for the case of the sample dimension smaller than the skin depth, which is not the case here. On the other hand, Fig. 3 shows that the AP value is about 2.0 above T_c, without a temperature dependence. For an inhomogeneous sample, localized spins confined to the surface of a sample will give rise to AP = 1, whereas the AP is expected to be 2 for those randomly distributed throughout the volume [23]. Using a combination of volume thermal expansion, magnetic susceptibility, and small-angle neutron scattering, De Teresa et al. indicated the existence of the ferromagnetic microregions in the paramagnetic regime [13]. When the ferromagnetic moments coexist with the paramagnetic moments, the ferromagnetic moments make the major contribution to the ESR intensity. Thus, the temperature-independent AP value of about 2 indicates that the ferromagnetic microregions, acting as the ESR centers, are randomly distributed throughout the volume of the LCMO. According to the PS model, the LCMO is separated into two regions; one of them the hole-rich metallic region, the other the hole-poor insulative region. The hole-rich metallic regions are expected to have ferromagnetic properties and to play the role of the magnetic impurities above T_c.

161

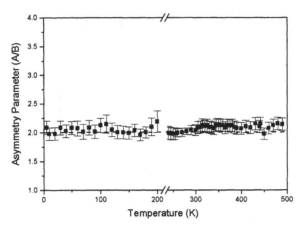

Fig. 3. Temperature dependence of the ESR lineshape asymmetry parameter. Essentially the same asymmetry parameter both above and below T_c strongly suggests that the microscopic local magnetic structures of those above and below T_c are qualitatively similar except that the latter carries long range order.

It is noticed that the AP values obtained in this work are in good agreement with the existence of the magnetic clusters, which follows from the PS model. The PS model also suggests that the phase separation persists below T_c, presumably with the hole-poor regions remaining paramagnetic or antiferromagnetic, which can be supported by the same AP values below T_c as those above T_c. Indeed, as shown in Fig. 2, it is remarkable to note the AP values below the T_c, which are also about 2 within errors. Thus, the AP values essentially the same both above and below T_c are in excellent agreement with the PS model.

Figure 4 shows the temperature dependence of the effective g-value (g_{eff}), corresponding to the magnetic field of the maximum microwave absorption. While the absolute values may not be quite reliable, the relative values are reliable because of the temperature independent AP. The g_{eff}-values above T_c are not much deviated from that of the free electron value of about 2, which is indicative of relatively weak interactions between the ferromagnetic microregions. On the other hand, the g_{eff} exhibits a marked increase with decreasing temperature below T_c, corresponding to rapidly growing interactions between the ferromagnetic microregions leading to their alignment in a direction. As the temperature is lowered below T_c, the ferromagnetic microregions can be expected to grow into the nearby hole-poor regions, random in their shape and distribution, besides their alignment with each other. In this scenario, the local magnetic fields would become more inhomogeneous with lowering the temperature, the ESR linewidth thus becoming broader. A prominent increase of the peak-to-peak linewidth with decreasing temperature below T_c, which can be seen in Fig. 5, supports this scenario. Furthermore, using the neutron scattering experiment, Lynn et al. indicated that the ferromagnetic transition in LCMO is not driven by the thermal excitation of spin waves as for conventional ferromagnetic transition, but rather is driven by the spin diffusion mechanism [16]. This unconventional ferromagnetic transition also is compatible with our scenario. Thus, we suggest that the isolated ferromagnetic microregions start to grow and start to interact with each other near T_c. Upon increasing sizes of the metallic regions and their alignment with applied external magnetic field, the conductivity will dramatically increase as well as the magnetization, resulting in the colossal magnetoresistance.

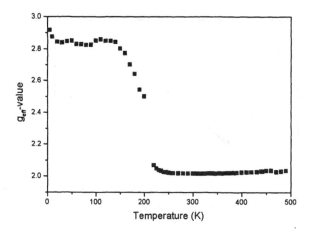

Fig. 4. Temperature dependence of the effective g-value. The error bars are within the size of the data points. Above T_c the g_{eff}-value is not much deviated from the free electron value of about 2, indicative of relatively weak interaction between the ferromagnetic microregions. On the other hand, the marked increase with decreasing temperature below T_c is attributed to the rapidly growing interaction between the ferromagnetic microregions and their alignment with each other.

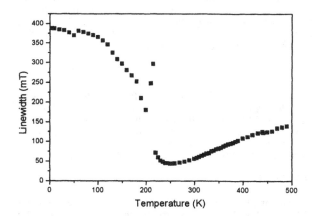

Fig. 5. Temperature dependence of the peak-to-peak ESR linewidth. The error bars are within the size of the data points. The prominent increase in the linewidth with decreasing temperature below T_c corresponds to the more inhomogeneous local magnetic field distribution, with the ferromagnetic microregions growing more inhomogeneous as the temperature is lowered.

In order to elucidate the nature of the magnetic structures, we have made a careful measurement of the field dependent magnetization at a temperature, 240 K, above T_c (225 K). While a linear dependence is seen for low magnetic fields in Fig. 6, a nonlinearity is evident for high fields. In particular, it is quite interesting to note an upward inflection at a critical field of about 1 T, which can only be explained by

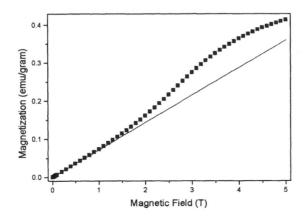

Fig. 6. Magnetic field dependence of the magnetization at 240 K, above T_c (225 K). The error bars are within the size of the data points. The inflection at around 1 T indicates an antiferromagnetic correlation. This magnetic field dependent magnetization suggests that the structure of the phase separation state consists of the ferromagnetic microregions surrounded by insulating antiferromagnetic regions.

supposing an antiferromagnetic correlation in the system. In other words, this upward inflection can naturally be attributed to spin-flips in the antiferromagnetic insulating regions, which are induced by the magnetic fields greater than a critical strength. For high enough magnetic fields, the magnetization shows a saturation behavior as would readily be expected. Given the ferromagnetic regions in the LCMO, it is quite reasonable to interpret the structure of the PS state as consisting of the ferromagnetic microregions surrounded by insulating antiferromagnetic regions.

CONCLUSION

In conclusion, we have studied the nature of spin systems in a $La_{0.7}Ca_{0.3}MnO_3$ single crystal sample undergoing the colossal magnetoresistance by means of the electron spin resonance and magnetization measurements. Our data provided convincing experimental microscopic evidences for the the phase separation model, and revealed the detailed magnetic structure of the ferromagnetic microregions surrounded by antiferromagnetic insulating regions both above and below the T_c.

ACKNOWLEDGEMENTS

This work was supported by the Korea Science and Engineering Foundation through the RCDAMP at Pusan National University and by the Korea Research Foundation (1998-015-D00113). Measurements at the Korea Basic Science Institute are acknowledged.

REFERENCES

1. S. Jin, T. H. Tiefel, M. McCormack, R. A. Fastnacht, R. Ramesh, and L. H. Chen, Science **264**, 413 (1994).

2. M. McCormack, S. Jin, T. H. Tiefel, R. M. Fleming, J. M. Phillips, and R. Ramesh, Appl. Phys. Lett. **64**, 3045 (1994).

3. P. Schieffer, A. P. Ramirez, W. Bao, and S-W. Cheong, Phys. Rev. Lett. **75**, 3336 (1995).

4. H. Y. Hwang, S-W. Cheong, P. G. Radaelli, M. Marezio, and B. Batlogg, Phys. Rev. Lett. **75**, 914 (1995).

5. M. F. Hundley, M. Hawley, R. H. Heffner, Q. X. Jia, J. J. Neumeier, J. Tesmer, J. D. Thompson, and X. D. Wu, Appl. Phys. Lett. **67**, 860 (1995).

6. C. Zener, Phys. Rev. **81**, 440 (1951).

7. P. W. Anderson and H. Hasegawa, Phys. Rev. **100**, 615 (1955).

8. P.-G. de Gennes, Phys. Rev. Lett. **118**, 141 (1960).

9. A. J. Millis, P. B. Littlewood, and B. I. Shraiman, Phys. Rev. Lett. **74**, 5144 (1995).

10. C. M. Varma, Phys. Rev. B **54**, 7328 (1996).

11. E. L. Nagaev, Physics-Uspekhi **38**, 497 (1995).

12. A. Moreo, S. Yunoki, and E. Dagotto, Science, **283**, 2034 (1999)

13. S. Yunoki, J. Hu, A. L. Malvezzi, A. Moreo, N. Furukawa, and E. Dagotto, Phys. Rev. Lett. **80**, 845 (1998).

14. S. Yunoki, A. Moreo, and E. Dagotto, Phys. Rev. Lett. **81**, 5612 (1998).

15. Shun-Qing Shen and Z. D. Wang, Phys. Rev. B **58**, R8877 (1998).

16. J. W. Lynn, R. W. Erwin, J. A. Borchers, Q. Huang, A. Santoro, J-L. Peng, and Z. Y. Li, Phys. Rev. Lett. **76**, 4046 (1996).

17. J. M. De Teresa, M. R. Ibarra, P. A. Algarabel, C. Ritter, C. Marquina, J. Blasco, J. Garcia, A. del Moral, and Z. Arnold, Nature **386**, 256 (1997).

18. G. Allodi, R. De Renzi, G. Guidi, F. Licci, and M. W. Pieper, Phys. Rev. B **56**, 6036 (1997).

19. S. E. Lofland, S. M. Bhagat, C. Kwon, S. D. Tyagi, Y. M. Mukovskii, S. G. Karabashev, and A. M. Balbashov, J. Appl. Phys. **81**, 5737 (1997).

20. M. T. Causa, M. Tovar, A. Caneiro, F. Prado, G. Ibanez, C. A. Ramos, A. Butera, B. Alascio, X. Obradors, S. Pinol, F. Rivadulla, C. Vazquez-Vazquez, M. A. Lopez-Quintela, J. Rivas, Y. Tokura, and S. B. Oseroff, Phys. Rev. B **58**, 3233 (1998).

21. O. Chauvet, G. Goglio, P. Molinie, B. Corraze, and L. Brohan, Phys. Rev. Lett. **81**, 1102 (1998).

22. F. J. Dyson, Phys. Rev. **98**, 349 (1955).

23. G. Feher and A. F. Kip, Phys. Rev. **98**, 337 (1955).

24. C. Rettory, D. Rao, J. Singley, D. Kidwell, S. B. Oseroff, M. T. Causa, J. J. Neumeier, K. J. McClellan, S-W. Cheong, and S. Schultz, Phys. Rev. B. **58**, 14490 (1998).

25. D. P. Choudhury, H. Srikanth, S. Sridhar, and P. C. Canfield, Phys. Rev. B. **58**, 14490 (1998).

SPECTROSCOPIC STUDIES OF INHOMOGENEOUS ELECTRONIC PHASES IN COLOSSAL MAGNETORESISTANCE AND CHARGE-ORDERING COMPOUNDS

S. L. COOPER, H. L. LIU, AND S. YOON,
Dept. of Physics and Frederick Seitz Materials Research Laboratory
U. of Illinois at Urbana-Champaign, Urbana, Illinois 61801 USA

S-W. CHEONG
Bell Laboratories, Lucent Technologies, Murray Hill, New Jersey 07974
and Dept. of Physics and Astronomy, Rutgers University, Piscataway, New Jersey USA

Z. FISK
Dept. of Physics and National High Magnetic Field Laboratory,
Florida State University, Tallahassee, Florida 32306 USA

ABSTRACT

Infrared reflectance and Raman spectroscopy measurements have been performed through the metal-semiconductor (or -semimetal) transitions of various "colossal magnetoresistance" and charge-ordering materials, including EuB_6 and the manganese perovskites. As described in this paper, our results demonstrate that these systems have phase regions characterized by electronic inhomogeneity of various types, including: (1) pure spin polarons, which form in a narrow temperature range above T_C in the "colossal magnetoresistance" system EuB_6; (2) magnetoelastic polarons, and an inhomogeneous ferromagnetic phase comprised of both metallic and insulating regions, in the "colossal magnetoresistance" regime of $La_{1-x}(Sr,Ca)_xMnO_3$ (x < 0.5); and (3) phase separation behavior involving coexisting ferromagnetic and antiferromagnetic domains in the intermediate-temperature (CO) phase of the charge-ordering system $Bi_{1-x}Ca_xMnO_3$ (x > 0.5).

INTRODUCTION

In the past few years there has been an enormous amount of experimental and theoretical interest in "Complex Materials," such as the ternary transition metal oxides and rare-earth hexaborides, which exhibit metal-insulator transitions involving complex magnetic phase changes and, in many cases, inhomogeneous electronic phases. The remarkable propensity for electronic inhomogeneity in various phase regimes of Complex Materials is indeed evolving into one of the most significant problems in condensed

167

matter physics, as there is growing evidence that the physics associated with this phenomenon is at the root of some of the most dramatic properties discovered in this field over the past few decades, including (a) "electronic phase separation" behavior, i.e., the coexistence of meso-scale magnetic phase regions [1,2], (b) magnetic polaron formation and "colossal magnetoresistance" (CMR) [3,4]; and (c) "charge-ordering" (CO) behavior, i.e., the spontaneous ordering of charge in periodic patterns on the lattice [5,6,7]. An understanding of these important phenomena, and the relationship among the diverse ground states exhibited by these systems, demands an understanding of the processes by which electronic phase separation, polaron formation, and charge ordering occurs in various doped insulators.

In this paper, we juxtapose three different forms of electronic inhomogeneity in various strongly-coupled systems: (a) pure spin polaron development in EuB_6, a ferromagnetic metal exhibiting CMR-type behavior with weak electron-lattice coupling; (b) small magnetoelastic polarons, and their persistence into the ferromagnetic metal phase, in the strongly electron-lattice coupled CMR system $La_{1-x}(Sr,Ca)_xMnO_3$ ($x < 0.5$); and (c) electronic phase separation into coexisting ferromagnetic and antiferromagnetic regions in the CO system $Bi_{1-x}Ca_xMnO_3$ ($x > 0.5$).

OPTICAL PROBES OF THE METAL-INSULATOR TRANSITION

Both infrared and Raman spectroscopy were used in these studies to investigate metal-insulator (MI) transitions, and inhomogeneous electronic phases, in colossal magnetoresistance and charge-ordering systems. Infrared spectroscopy is the typical spectroscopic means by which metal-insulator transitions have been studied in the past: infrared spectroscopy provides a measure of the optical spectral weight, $\sim \omega_p^2$ ($\sim n/m^*$), which is the natural order parameter for the Mott transition, as it approaches a zero value for either of the two paths by which a Mott transition is achieved: a diminution of the carrier density n, or a divergence of the carrier effective mass m^*. However, there are several strong limitations to probing metal-insulator transitions using conventional techniques such as infrared spectroscopy and transport, some of which reveal themselves particularly clearly in strongly spin-lattice-charge coupled systems. In particular, infrared spectroscopy is primarily sensitive to the charge response, and hence provides little information about changes in the spin degrees of freedom through metal-insulator transitions. This is a severe limitation in systems with complex magnetic, structural, as well as electronic phase diagrams.

On the other hand, Raman scattering has evolved as a powerful probe of metal-insulator transitions in strongly correlated systems [4,8,9]. Raman scattering distinguishes itself from more conventional probes of phase transitions such as infrared reflectivity and neutron scattering in that it (a) affords a means of simultaneously probing changes in the spin-, lattice-, and electronic-degrees-of-freedom, (b) is sensitive to low carrier density phases, which is of particular utility for studying mixed-phase regimes of various complex materials, (c) provides specific details about the nature of the ground state, (d) allows measurements of very small samples (~ 80 μm) and does not require data processing (e.g., Kramers-Kronig transformations) that can introduce errors into the fundamental physical quantities of interest (e.g., $\sigma(\omega)$), and (e) lends itself to sophisticated techniques for exploring the low frequency charge dynamics across metal-insulator transitions, including pressure, magnetic-field, and time-resolved studies.

SPIN POLARON FORMATION IN EuB$_6$: CMR-TYPE BEHAVIOR IN THE WEAK ELECTRON-LATTICE COUPLING REGIME

It is instructive to juxtapose CMR-type behavior in "conventional" ferromagnetic metals with that observed in the manganese perovskites. Like the manganese perovskites, many "conventional" ferromagnetic metals exhibit a large negative magnetoresistance near the paramagnetic semiconductor (or semimetal) to ferromagnetic metal transition, but unlike the manganites, these systems generally exhibit much weaker electron-lattice coupling, as well as Fisher-Langer type scaling between the carrier density and the size of the magnetoresistance [10]. As a representative of this class of ferromagnetic metals, in this study we consider EuB$_6$, which previous studies have shown is a paramagnetic semimetal at high temperatures, but which develops short-range magnetic correlations and "colossal magnetoresistance" behavior at low temperatures before transiting into a ferromagnetic metal phase near $T_C \sim 16$ K [11]. Because electron-phonon coupling effects are much weaker in EuB$_6$ than in the manganese perovskites, studies of the former system afford an interesting and important means of investigating colossal magnetoresistance behavior absent strong lattice effects.

As shown in Fig. 1, in the high temperature paramagnetic phase, the Raman scattering spectrum of EuB$_6$ is characterized by a collision-dominated scattering response [4],

$$S(\omega) \sim A\omega\Gamma/(\omega^2 + \Gamma^2), \tag{1}$$

where A is the scattering amplitude and Γ is the carrier scattering rate. This scattering response is associated with diffusive hopping of the carriers. The carrier scattering rate Γ in EuB$_6$ decreases precipitously with temperature in the high temperature PM phase, but levels-off and then increases again upon entering a short-range magnetic order (SRMO) regime just above the Curie temperature T_C. As shown in Fig. 1, the PM semimetal → FM metal transition in EuB$_6$ is revealed in the electronic Raman scattering spectrum by an abrupt change from a collision-dominated scattering response (Eq. 1) to a flat continuum response typical of that observed in strongly-correlated metals such as the cuprates and titanates; this change reflects a crossover from a diffusive carrier scattering regime to a strongly-correlated metal regime with frequency-dependent carrier scattering [4].

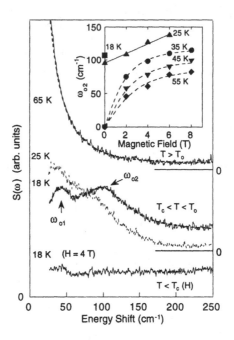

Figure 1. The Raman scattering spectra of EuB$_6$ in various temperature regimes: (top) a diffusive response (Eq. 1) in the paramagnetic phase $T > T_o$; (middle) an inelastic scattering response associated with spin polaron formation in the intermediate temperature regime $T_C < T < T_o$; and a flat continuum response in the low temperature ferromagnetic regime $T < T_C$. Inset: Field-dependence of the spin-flip Raman scattering energy associated with the spin polaron, illustrating the evolution of the spin polaron size with decreasing temperature.

The most interesting phase regime of EuB$_6$ is evident in a narrow temperature range just above $T_C(H)$, in which an inelastic scattering response develops in the Raman spectrum due to the formation of magnetic polarons (Fig. 1). We attribute this inelastic response to spin-flip scattering from carriers bound to ferromagnetic clusters of Eu^{2+} spins [4]. This is best evidenced by the field-dependence of the inelastic peak ω_{o2}, which is nicely described by the functional form (inset, Fig. 1)

$$\hbar\omega_o = xJ_{df} \langle S_z \rangle \qquad (2)$$

where x is the concentration of polarons, J_{df} is the d-f exchange constant, and $<S_z>$ is the thermal average of the Eu^{2+} spins. This result illustrates that the transition from the PM semimetal to the FM metal phase is preceded by the binding of some of the carriers into spin polaron states, i.e., ferromagnetic clusters. Notably, by measuring the field- and temperature-dependent energy of the polaron peak at ω_{0s}, one can map out the magnetization of these polarons as a function of temperature and magnetic field. As reflected in the increasing saturation magnetization of the polarons as a function of decreasing temperature (see inset of Fig. 1), the size (or number) of these clusters increases with decreasing temperature until a percolation threshold is reached at T_C, precipitating the transition to the ferromagnetic metal phase. These results provide direct evidence that the PM semimetal $\rightarrow$ FM metal transition in EuB_6 is initiated by the development of an inhomogeneous electronic phase comprised of magnetic polarons, and that the highly field- and temperature-dependent size (and overlap) of these polarons is responsible for both the dramatic negative magnetoresistance near T_C and the transition into the ferromagnetic metal phase.

Thus, magnetic cluster formation near T_C of itself is not unique to the CMR phase manganese perovskites, but is present even in low carrier density metallic ferromagnets. Indeed, on a fundamental level the physics associated with magnetic cluster formation in these systems is similar, reflecting the competition between the carrier kinetic energy and the "localizing" energies associated with both the static and "dynamic" (e.g., lattice polaron) disorder potentials, and the ferromagnetic exchange interaction, which favors localizing carriers around ferromagnetically-aligned clusters of moments. However, while rather large "localizing" energies (chiefly associated with Jahn-Teller and breathing mode distortions) are essential to localizing the charge in the relatively high carrier density manganites, in low carrier density metallic ferromagnets the confining energies can be substantially weaker, and indeed appear to be principally associated with static disorder and the relatively small exchange couplings in these systems ($J \sim 0.1$ eV in EuB_6). As we will argue in the next section, an interesting byproduct of the large confining potentials in the manganites appears to be a robustness of the magnetic clusters as a function of temperature and magnetic field; this is manifested most dramatically as a persistence of these clusters even into the ferromagnetic metal phase.

MAGNETOELASTIC POLARONS AND THE INHOMOGENEOUS FERROMAGNETIC METAL PHASE IN THE CMR REGIME OF $La_{1-x}(Sr,Ca)_xMnO_3$ ($x < 0.5$)

As in the case of EuB_6 (see above), the PM semimetal $\rightarrow$ FM metal transition in $La_{1-x}(Sr,Ca)_xMnO_3$ ($x < 0.5$) is characterized in the Raman scattering response by a change from a diffusive carrier scattering response

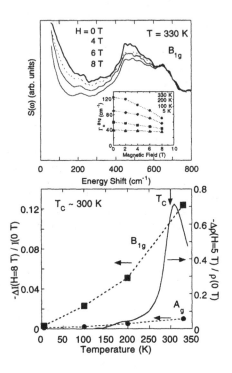

Figure 2. (Top) Diffusive (Eq. 1) B_{1g}-symmetry Raman scattering response of a $T_C \sim 300$ K manganese perovskite single crystal as a function of magnetic field. Inset: Field- and temperature-dependence of the carrier scattering rate as extracted from fits to the B_{1g} Raman response using Eq. 1. (Bottom) Temperature-dependence of the fractional change in the B_{1g} Raman intensity between $H = 8$ T and $H = 0$ T (squares), compared with the temperature-dependent magnetoresistivity measured on the same sample (solid line).

with a frequency-independent scattering rate to a flat electronic continuum response typical of a strongly correlated metal [12].

There is also a strong field-dependence associated with the diffusive scattering in both EuB_6 and $La_{1-x}(Sr,Ca)_xMnO_3$, which is attributable to a reduction in carrier-spin scattering with increasing applied field; indeed, in what amounts to "optically-detected magnetotransport," a careful analysis of these data allow us to estimate the field- and temperature-dependence of the carrier scattering rate by fitting the measured spectra using a collision-dominated scattering response function (Eq. 1). The results are illustrated in the inset of the top of Fig. 2. Most significantly, our Raman measurements illustrate that the magnetic-field dependence of the scattering response persists well into the ferromagnetic phase, and well below the temperature at which the magnetoresistivity goes to zero (see bottom, Fig. 2). This is strong evidence for an inhomogeneous ferromagnetic metal phase in $La_{1-x}(Sr,Ca)_xMnO_3$, wherein insulating (small polaronic) and metallic regions coexist [12] - by contrast, in EuB_6 there is absolutely no field dependence associated with the electronic Raman scattering response below T_C, suggesting a homogeneous ferromagnetic metal phase in this system. Note that magneto-transport measurements of $La_{1-x}(Sr,Ca)_xMnO_3$ reveal an abrupt transition near T_C which may be ascribed to a percolative transition (solid line in lower Figure 2) - this difference between the optical and transport measurements reflects the fact that below the percolation threshold, transport measurements are not sensitive to secondary insulating phases, as they are shorted-out by the majority metallic component. One interesting possibility is that the strong electron-lattice coupling in the $La_{1-x}(Sr,Ca)_xMnO_3$ system allows the localization of some charge even in the presence of predominantly metallic phase regions, although a more detailed understanding of such an unusual state of matter demands further theoretical investigation.

PHASE SEPARATION BEHAVIOR IN THE CHARGE-ORDERING REGIME OF $Bi_{1-x}Ca_xMnO_3$ (x > 0.5)

A great deal of recent interest has been focused on the antiferromagnetic (AFM) insulating ground state observed in the manganese perovskites at higher doping $x \geq 0.5$, which involves long-range "stripe" ordering of polarons below the Neel temperature T_N. In particular, there has been much effort devoted to clarifying the process by which small polarons in the high temperature phase, which hop with predominantly ferromagnetic correlations via the double-exchange mechanism, organize at low temperatures into charge stripes with AF long-range order. We have investigated this issue through infrared optical studies of single-crystal $Bi_{1-x}Ca_xMnO_3$ (x > 0.5), which has a phase diagram essentially identical to that of $La_{1-x}Ca_xMnO_3$ [13].

Our optical studies show that the temperature-dependent optical conductivity $\sigma_1(\omega)$ of $Bi_{1-x}Ca_xMnO_3$ (x > 0.5) has three regimes of behavior [14] (see Fig. 3). In the high temperature phase $T > T_{co}$ (top spectrum, Fig. 3), which is dominated by FM correlations [13], the optical conductivity below 1.5 eV is characterized by a small polaron response (photoionization energy ~ 0.25 eV) similar to that observed in the PM phase of $R_{1-x}Ca_xMnO_3$ for x < 0.5. However, below the Neel temperature $T < T_N$ (bottom spectrum, Fig. 3), the optical conductivity is characterized by a distinct charge gap $2\Delta \sim 0.4$ eV and a

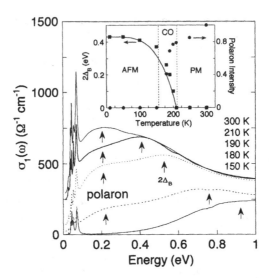

Figure 3. The optical conductivity $\sigma_1(\omega)$ below 1 eV as a function of temperature for $Bi_{1-x}Ca_xMnO_3$ ($x \sim 0.82$). Inset: Compares the temperature-dependence of the optical gap energy $2\Delta_B$ (filled squares) to the small polaron oscillator strength normalized to its value at T = 300 K (open circles), illustrating the coexistence of high- and low-temperature electronic phases in the CO regime. The solid line is a fit to a BCS function.

splitting of the optical Mn-O stretch mode, indicative of charge-density-wave-"like" gap formation and long-range charge ordering.

Perhaps most interestingly, the intermediate temperature regime $T_{co} < T < T_N$ is characterized by two distinct electronic contributions to the optical conductivity below 1.5 eV (middle curves, Fig. 3): (a) a remnant of the small polaron absorption, which decreases in spectral weight with decreasing temperature due to the progressive freezing-out of small polarons via charge ordering; and (b) a higher-energy conductivity contribution that gradually evolves into the charge gap below T_N with a BCS-like temperature-dependence (inset, Fig. 3). The coexistence of these two conductivity contributions is direct evidence for electronic phase separation into hole-rich FM domains and hole-poor AFM domains in the intermediate temperature regime. Indeed, these results suggest a picture of the charge-ordering transition in which mobile small polarons in the high temperature phase gradually order below T_{co}, forming charge stripe "fragments" that grow in length with decreasing temperature until long-range charge-ordering occurs below T_N [14].

SUMMARY

The optical spectroscopic results presented here demonstrate that electronically inhomogeneous phases appear in different guises in various strongly spin/charge/lattice coupled systems such as the rare-earth hexaborides and the manganese perovskites, reflecting the remarkable manner in which these systems transition between disparate magnetic and electronic phases - in all these systems the evolution of the magnetic ground state appears to originate with "nucleation" process, wherein "clusters" or "domains" characteristic of the ground state first evolve in mesoscale regions, then eventually grow in size with decreasing temperature until a percolation threshold is reached.

In CMR materials, we observed that magnetic clusters tend to persist only over a narrow temperature range around T_C in systems with weak electron-lattice coupling (e.g., hexaborides), whereas in systems with strong electron-lattice coupling such as the manganites, the enhanced stability of the clusters seems to result in a persistence of clusters well into the low temperature phase, i.e., "phase coexistence." The similarity between cluster formation in the manganites and that observed in other metallic ferromagnets reflects a similar competition between the carrier kinetic energy and the confining energies associated with static disorder, lattice distortions, and the ferromagnetic exchange interaction. However, while rather large "localizing" energies (chiefly associated with Jahn-Teller and breathing mode distortions) are essential to localizing the charge in the relatively high carrier density manganites, in low carrier density metallic ferromagnets the confining energies are likely to be primarily associated with static disorder combined with ferromagnetic exchange.

In the CO manganites, we found evidence that the charge stripes in the low temperature antiferromagnetic phase form by a gradual process in which mobile small polarons in the high temperature phase gradually order below T_{co}, forming charge stripe "fragments" that grow in length with decreasing temperature until long-range charge-ordering occurs below T_N. This intermediate phase of CO systems is reminiscent of a "nematic"-like phase in which there is orientational order of the charge "fragments" but no long-range stripe order [15].

ACKNOWLEDGEMENTS

This work was supported by NSF grant DMR97-00716, Dept. of Energy grant DEFG02-96ER45439, and by the NHMFL through grant NSF DMR 90-16241.

REFERENCES

1. V. J. Emery, S. A. Kivelson, and O. Zachar, Phys. Rev. B **56**, 6120 (1997).
2. A. Moreo, S. Yunoki, and E. Dagotto, Science **283**, 2034 (1999).
3. M. Jaime, H. T. Hardner, M. B. Salamon, M. Rubinstein, P. Dorsey, and D. Emin, Phys. Rev. Lett. **78**, 951 (1997).
4. P. Nyhus, S. Yoon, M. Kauffman, S. L. Cooper, Z. Fisk, and J. Sarrao, Phys. Rev. B **56**, 2717 (1997).
5. V. J. Emery, S. A. Kivelson, and J. M. Tranquada, Proc. Natl. Acad. Sci. **96**, 8814 (1999).
6. C. H. Chen, S.-W. Cheong, and A. S. Cooper, Phys. Rev. Lett. **71**, 2461 (1993).
7. C. H. Chen and S.-W. Cheong, Phys. Rev. Lett. **76**, 4042 (1996).
8. P. Nyhus, S. L. Cooper, and Z. Fisk, Phys. Rev. B **51**, 15626 (1995).
9. P. Nyhus, S. L. Cooper, Z. Fisk, and J. Sarrao, Phys. Rev. B **55**, 12488 (1997).
10. P. Majumdar and P. B. Littlewood, Nature **395**, 479 (1998).
11. C. N. Guy, S. von Molnar, J. Etourneau, and Z. Fisk, Solid State Commun. **33**, 1055 (1972).
12. H. L. Liu, S. Yoon, S. L. Cooper, S-W. Cheong, P. D. Han, and D. A. Payne, Phys. Rev. B **58**, R10115 (1998).
13. W. Bao, J. D. Axe, C. H. Chen, and S-W. Cheong, Phys. Rev. Lett. **78**, 543 (1997).
14. H. L. Liu, S. L. Cooper, and S-W. Cheong, Phys. Rev. Lett. **81** 4684 (1998).
15. S. A. Kivelson, E. Fradkin, and V. J. Emery, Nature **393**, 550 (1998).

CHARGE INHOMOGENEITIES IN THE COLOSSAL MAGNETORESISTANT MANGANITES FROM THE LOCAL ATOMIC STRUCTURE

S. J. L. BILLINGE*, V. PETKOV*, TH. PROFFEN*, G. H. KWEI**, J. L. SARRAO**, S. D. SHASTRI*** and S. KYCIA****
*Center for Fundamental Materials Research and Department of Physics and Astronomy, Michigan State University, East Lansing, MI 48824.
**Los Alamos National Laboratory, Los Alamos, NM 87545
***Advanced Photon Source, Argonne National Laboratory, Argonne, IL
****Cornell High Energy Synchrotron Source, Ithaca, NY

ABSTRACT

We have measured atomic pair distribution functions (PDF) of $La_{1-x}Ca_xMnO_3$ using high energy x-ray diffraction. This approach yields accurate PDFs with very high real-space resolution. It also avoids potential pitfalls from the more usual neutron measurements that magnetic scattering is present in the measurement, that the neutron scattering length of manganese is negative leading to partial cancellation of PDF peaks, and that inelasticity effects might distort the resulting PDF. We have used this to address the following questions which do not have a satisfactory answer: (1) What are the amplitudes and natures of the local Jahn-Teller and polaronic distortions in the CMR region. (2) Is the ground-state of the ferromagnetic metallic phase delocalized or polaronic. (3) As one moves away from the ground-state, by raising temperature or decreasing doping, towards the metal insulator transition, how does the state of the material evolve?

INTRODUCTION

A very large magnetoresistance is observed in the, so-called, colossal magnetoresistant manganites [1]. This coupling of magnetic field to transport is qualitatively understood using the concept of double exchange [2]. That this description underlies the CMR phenomenon appears not to be in question [3]. However, the DE model has difficulty to account quantitatively for the observed T_c's and the magnitude of the resistivity change at T_c, and is totally incapable of explaining the complex phenomenology away from the ideal CMR compositions [4,5]. A number of authors [4,6–10] have suggested that the extra contribution to the model should be an electron-lattice coupling term. Extensive experimental evidence exists for lattice effects in these materials. For example, anomalous increases in crystallographic Debye-Waller factors [11] and, equivalently, PDF peak-widths [12] and XAFS Debye-Waller factors [13,14] at the MI transition; an oxygen isotope dependence to T_c [15,16]; thermal conductivity [17,18] and transport measurements above T_c interpreted in terms of polaron dynamics [19,20]. These results can be consistently interpreted if charges localize as small polarons above T_c.

Early work using a number of different techniques [11–13,15,21] established the importance of the lattice to the colossal magnetoresistance phenomenon. More recently attention has been focussed on understanding the precise *nature* of the polaronic state and the related Jahn-Teller distortions. We have concerned ourselves with a number of questions: (1) What does the polaronic distortion look like? (2) What is the ground-state of the ferromagnetic metallic (FM) phase? (3) How do the polarons and Jahn-Teller (JT) distortions evolve with compositional and temperature changes?

Mat. Res. Soc. Symp. Proc. Vol. 602 © 2000 Materials Research Society

The principle result which we have obtained as a result of these studies is that the ferromagnetic metallic phase of $La_{1-x}Ca_xMnO_3$ is *microscopically* inhomogeneous [22,23]. This is in accord with a number of other experimental and theoretical observations [7]. It is particularly interesting in light of the observation of *macroscopic* phase coexistence in materials with a smaller A-site ion size, $\langle r_A \rangle$ [24,25]. The temperature dependence of the local structure in $La_{1-x}Ca_xMnO_3$ is characteristic of regular phase separation phenomena with two phases coexisting and the proportion of each phase evolving with changing temperature or doping. However, in this system *no macroscopic phase coexistence is observed*: it is a purely local structural phenomenon. The relative proportion of the sample in the localized and delocalized states can be quantified using the PDF as we describe below. The results are compared to earlier neutron PDF results.

EXPERIMENTAL

Samples of $La_{1-x}Ca_xMnO_3$ with $x = 0.0, 0.12, 0.21, 0.25$ and 0.33, were synthesized by standard solid-state reaction. Details are reported elsewhere [22]. X-ray powder diffraction data were collected out at the A2 24 pole wiggler beam line at Cornell High Energy Synchrotron Source (CHESS) and at beamline BM1 at the Advanced Photon Source (APS). The measurements were made in flat-plate symmetric-transmission geometry. Finely ground powders of uniform thickness ~ 5 mm were supported between thin kapton foils on the cold-finger of a helium displex refrigerator. The sample thickness was determined so that the absorption of the sample was $\mu t \sim 1$. Xrays of energy 60 keV and 65 keV were used at CHESS and APS, respectively. The raw data were corrected for experimental effects such as absorption and multiple scattering. Unwanted scattering such as backgrounds and Compton scattering were removed, and the data were normalized to obtain the total scattering factor, $S(Q)$, where Q is the magnitude of the momentum transfer $(Q = 4\pi \sim \theta/\lambda)$. $S(Q)$ is defined as

$$S(Q) = 1 + \frac{I_{el}(Q) - \sum_i c_i f_i^2(Q)}{[\sum_i c_i f_i]^2} \tag{1}$$

where I_{el} is the measured elastic part of the spectrum, c_i and $f_i(Q)$ are the atomic concentration and scattering factor of the atomic species of type i ($i =$ La, Ca, Mn and O), respectively [26]. The data processing was carried out using the program RAD [27]. The PDF is obtained from $S(Q)$ by a direct Fourier transform according to

$$G(r) = \frac{2}{\pi} \int_0^\infty Q[S(Q) - 1]\sin(Qr)\,dQ, \tag{2}$$

The $F(Q) = Q[S(Q) - 1]$ and $G(r)$ functions from the $x = 0.33$ sample collected at 20 K are shown in Fig. 1. The low-r regions of the PDF from this, and various other, data sets are superimposed on the phase diagram in Fig. 2

Data are modeled by calculating the PDF from a model structure. Model parameters are then varied in such a way as to optimize the agreement between the calculated and data-PDFs [29]. The program PDFFIT was used [30]. The solid line in Fig. 1(b) shows the PDF obtained from the average crystal structure refined to the $x = 0.33$, 20 K data-set.

RESULTS

Nature of the Jahn – Teller distortion :

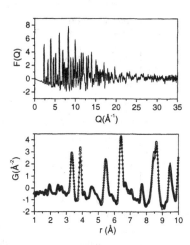

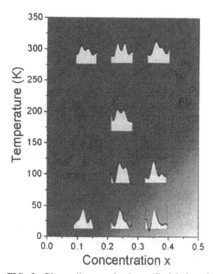

FIG. 1. (a) Total structure function, $S(Q)$, and (b) Reduced radial distribution functions, $G(r)$, for $La_{1-x}Ca_xMnO_3$, $x = 0.33$ at 20 K. In (b) the data are shown as open circles. The solid line is a fit of the average structure to the data. [28]

FIG. 2. Phase diagram for $La_{1-x}Ca_xMnO_3$ with high-resolution x-ray PDFs superimposed. The region of the PDF showing the Mn-O nearest neighbor bonds is shown. The speckled pattern indicates the sample is microscopically inhomogeneous in this region [22].

Undoped $LaMnO_3$ is insulating. The Mn^{3+} ions are orbitally ordered and the full Jahn-Teller distortion can be measured crystallographically [31–33]. We have verified that the local Jahn-Teller distortion is *quantitatively* the same as the average one [28]. This establishes nature of the local Jahn-Teller distortion around Mn^{3+} in the absence of strain. It also establishes the accuracy of our fitting programs. We have repeated this fit on the present room temperature x-ray data for $x = 0$. The results are summarized in Fig. 3 and Table I. The reproducibility of the results from the x-ray and neutron measurements is excellent as can be seen in Tab. I. The Jahn-Teller distorted octahedra have four short bonds, two each of length 1.92 Å and 1.97 Å, respectively, and two long-bonds of length 2.15 Å.

Evolution of polarons with T and x :

The presence of JT distorted octahedra implies the presence of Mn^{3+} ions whereas doped charges which delocalize result in all the MnO_6 octahedra being undistorted [34]. Thus, using the PDF we can search for evidence of locally Jahn-Teller distorted octahedra which indicate that, in that region of the sample, doped charges are absent or are localized as distinct Mn^{4+} species.

Within the sensitivity of our measurement there is no detectable structural disorder and no local Jahn-Teller distortions are present deep in the FM phase [22,23]. However, as the MI transition is approached, small polarons do appear. We can identify two kinds of phase: a charge localized (Jahn-Teller) phase and a delocalized (Zener) phase which is structurally ordered. We have made a qualitative survey of the phase diagram of $La_{1-x}Ca_xMnO_3$ to search for the presence of JT long-bonds indicating localized material. In Fig. 2 we have superimposed on the electronic phase diagram the region of the PDFs showing the Mn-O short and long bonds [22]. Deep in the FM

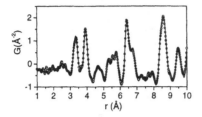

	Rietveld	IPNS (neutron)	APS (X-ray)
a	5.542(1)	5.5422(7)	5.5241(9)
b	5.732(1)	5.7437(8)	5.6769(11)
c	7.6832(2)	7.690(1)	7.6865(11)
x(La)	-0.0068(3)	-0.0073(2)	-0.0069(3)
y(La)	0.0501(3)	0.0488(2)	0.04345(17)
$\langle u^2 \rangle$(La)	0.0022(4)	0.00199(4)	0.00564(5)
$\langle u^2 \rangle$(Mn)	0.0011(6)	0.00067(7)	0.00713(10)
x(O1)	0.0746(4)	0.0729(3)	0.077(3)
y(O1)	0.4873(4)	0.4857(3)	0.494(3)
$\langle u^2 \rangle$(O1)	0.0031(5)	0.00233(7)	0.0333(18)
x(O2)	0.7243(3)	0.7247(3)	0.7249(22)
y(O2)	0.3040(3)	0.3068(3)	0.2952(17)
z(O2)	0.0390(2)	0.0388(3)	0.0463(17)
$\langle u^2 \rangle$(O2)	0.0030(4)	0.00378(5)	0.0201(6)
R_{wp}	12.1	16.2	10.9

FIG. 3. PDF from $LaMnO_3$ at 300 K from x-ray data taken at APS. Solid line is a fit of the crystallographic model

TABLE I. Structural data of $LaMnO_3$ (*Pbnm*) from Rietveld and PDFFIT refinements of neutron powder diffraction data at 20 K [28] and a PDF-FIT refinement of the room temperature x-ray data from APS. La and O1 are on $(x,y,\frac{1}{4})$, Mn is on $(0,\frac{1}{2},0)$ and O2 is on (x,y,z). The units for lattice parameter are Å and for $\langle u^2 \rangle$, Å^2.

phase ($x = 0.33$, $T = 10$ K) there is no long bond: the refined local structure agrees with the average structure, there is no detectable structural disorder and no JT distortion [22]. There may be a suggestion of peak asymmetry on the Mn-O PDF peak at $r = 2.0$ Å (Fig. 1(b) and Fig. 2), and we cannot rule out the presence of large polarons [35]. However, there is no peak intensity around 2.15 Å, the position of the JT-long-bond in JT distorted material. As the MI transition is approached as a function of T and x, JT long-bonds appear, even in the FM phase. The long-bonds appear, not as a continuous elongation of the Mn-O long-bonds, but are apparent always at 2.15 Å and merely grow in intensity as the MI transition is approached. This is in qualitative agreement with our own neutron data and those of others [21], though Louca *et al.* present a different interpretation. It is also in agreement with XAFS data [13,14]. It is qualitatively, though not quantitatively, similar to the XAFS data of Lanzara *et al.* [35]. This is strong evidence for a microscopically inhomogeneous state in the ferromagnetic metallic region of the phase diagram, with a *microscopic* coexistence of localized JT and delocalized (Zener) material.

The PDF can *quantify* the relative proportion of the sample in each of these phases, even when there is no macroscopic phase separation. This is possible by monitoring the presence or absence of a Mn-O long-bond, or equivalently, by monitoring the width of the peak in the PDF originating from the O-O correlations in the MnO_6 octahedra. The PDF peak width is inversely proportional to its height and it is easier to monitor changes in peak-height. We can determine an excess peak height, h_{xs}, by subtracting an extrapolation to low temperature of the PDF peak-height from above T_c. This is shown in Fig. 4 for two data-sets. This preliminary result indicates that in $La_{1-x}Ca_xMnO_3$ the proportion of the sample in the delocalized phase may scale universally with reduced temperature close to T_c (Fig. 4, inset). From h_{xs} we obtain the proportion of the sample in the delocalized phase, $c(H, T)$, referred to as the "mixing factor" in Ref. [36]. Jaime *et al.* [36] make a prediction for the field and temperature dependence of $c(H, T)$, reproduced in Fig. 5, from their phenomenological model. This can be compared with the excess-PDF peak-height variation shown in Fig. 4(inset).

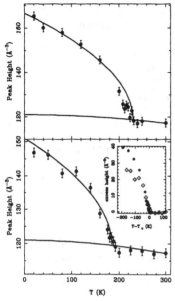

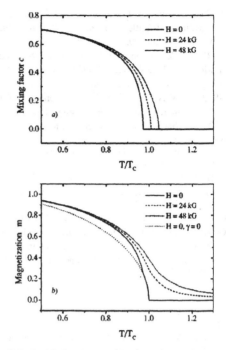

FIG. 4. O-O PDF Peak-height vs T from La$_{1-x}$Ca$_x$MnO$_3$. Top panel $x = 0.25$, bottom panel $x = 0.21$. Debye behavior has been fit to the high-temperature region. Excess-peak height is plotted vs. reduced temperature in the inset for $x = 0.25$ (circles) and $x = 0.21$ (crosses).

FIG. 5. Mixing factor $c(H, T)$ predicted for the proportion of the sample in the delocalized phase below T_c in the theory of Jaime et al. [36]

CONCLUSIONS

Using the PDF we can study the size and shape and presence and absence of the local Jahn-Teller distortions. We have carried out such an analysis on high real-space resolution x-ray diffraction data. We find that, deep in the ferromagnetic metallic phase there is no Jahn-Teller distortion and the sample is fully delocalized. As the MI transition is approached both with T and x, JT long-bonds appear indicating a microscopic coexistence of localized and delocalized material. The PDF can quantify the proportion of the sample in each phase. Preliminary results suggest that the quantity of delocalized sample seems to scale universally with reduced temperature.

ACKNOWLEDGEMENTS

This work was supported by NSF through grant DMR-9700966. CHESS is supported by NSF through grant DMR-9713424 and APS by DOE through grant W-31-109-Eng-38.

REFERENCES

1. A. P. Ramirez, J. Phys: Condens. Matter **9**, 8171 (1997).
2. C. Zener, Phys. Rev. **82**, 403 (1951).

3. A. Chattopadhyay, A. J. Millis, and S. D. Sarma, Unpublished. (preprint: cond-mat/9908305).

4. A. J. Millis, P. B. Littlewood, and B. I. Shraiman, Phys. Rev. Lett. **74**, 5144 (1995).

5. A. J. Millis, Nature **392**, 147 (1998).

6. H. Röder, J. Zang, and A. R. Bishop, Phys. Rev. Lett. **76**, 1356 (1996).

7. A. Moreo, A. Yunoki, and E. Dagotto, Science **283**, 2034 (1999).

8. A. J. Millis, B. I. Shraiman, and R. Müller, Phys. Rev. Lett. **77**, 175 (1996).

9. J-S. Zhou, W. Archibald, and J. B. Goodenough, Nature **381**, 770 (1996).

10. A. S. Alexandrov and A. M. Bratkovsky, Phys. Rev. Lett. **82**, 141 (1999).

11. P. Dai, J. Zhang, H. A. Mook, S.-H. Liou, P. A. Dowben, and E. W. Plummer, Phys. Rev. B **54**, R3694 (1996).

12. S. J. L. Billinge, R. G. DiFrancesco, G. H. Kwei, J. J. Neumeier, and J. D. Thompson, Phys. Rev. Lett. **77**, 715 (1996).

13. C. H. Booth, F. Bridges, G. J. Snyder, and T. H. Geballe, Phys. Rev. B **54**, R15606 (1996).

14. C. H. Booth, F. Bridges, G. H. Kwei, J. M. Lawrence, A. L. Cornelius, and J. J. Neumeier, Phys. Rev. B **57**, 10440 (1998).

15. G. Zhao, K. Conder, H. Keller, and K. A. Müller, Nature **381**, 676 (1996).

16. G. M. Zhao, M. B. Hunt, and H. Keller, Phys. Rev. Lett. **78**, 955 (1997).

17. D. W. Visser, A. P. Ramirez, and M. A. Subramanian, Phys. Rev. Lett. **78**, 3947 (1997).

18. J. L. Cohn, J. J. Neumeier, C. P. Popoviciu, K. J. McCellan, and Th. Leventouri, Phys. Rev. B **56**, R8495 (1997).

19. M. Jaime, M. B. Salamon, M. Rubinstein, R. E. Treece, J. S. Horwitz, and D. B. Chrisey, Phys. Rev. B **54**, 11914 (1996).

20. M. Jaime, H. T. Hardner, M. B. Salamon, M. Rubinstein, P. Dorsey, and D. Emin, Phys. Rev. Lett. **78**, 951 (1997).

21. D. Louca, T. Egami, E. L. Brosha, H. Röder, and A. R. Bishop, Phys. Rev. B **56**, R8475 (1997).

22. S. J. L. Billinge, Th. Proffen, V. Petkov, J. Sarrao, and S. Kycia, submitted to Phys. Rev. B. (Preprint: cond-mat/9907329.

23. S. J. L. Billinge, in *Physics of Manganites*, page 201, Klewer Academic/Plenum, 1999.

24. M. Uehara, S. Mori, C. H. Chen, and S.-W. Cheong, Nature **399**, 560 (1999).

25. D. E. Cox, P. G. Radaelli, M. Marezio, and S.-W. Cheong, Phys. Rev. B **57**, 3305 (1998).

26. Y. Waseda, *The structure of non-crystalline materials*, McGraw-Hill, New York, 1980.

27. V. Petkov, J. Appl. Crystallogr. **23**, 387 (1989).

28. Th. Proffen, R. G. DiFrancesco, S. J. L. Billinge, E. L. Brosha, and G. H. Kwei, Phys. Rev. B **60**, 9973 (1999).

29. S. J. L. Billinge, in *Local Structure from Diffraction*, edited by S. J. L. Billinge and M. F. Thorpe, page 137, New York, 1998, Plenum.

30. Th. Proffen and S. J. L. Billinge, J. Appl. Crystallogr. **32**, 572 (1999).

31. J. F. Mitchell, D. N. Argyriou, C. D. Potter, D. G. Hinks, J. D. Jorgensen, and S. D. Bader, Phys. Rev. B **54**, 6172 (1996).

32. Q. Huang, A. Santoro, J. W. Lynn, R. W. Erwin, J. A. Borchers, J. L. Peng, and R. L. Greene, Phys. Rev. B **55**, 14987 (1997).

33. J. Rodríguez-Carvajal, M. Hennion, F. Moussa, A. H. Moudden, L. Pinsard, and A. Revcolevschi, Phys. Rev. B **57**, R3189 (1998).

34. W. E. Pickett and D. J. Singh, Phys. Rev. B **53**, 1146 (1996).

35. A. Lanzara, N. L. Saini, M. Brunelli, F. Natali, A. Bianconi, P. G. Radaelli, and S. Cheong, Phys. Rev. Lett. **81**, 878 (1998).

36. M. Jaime, P. Lin, S. H. Chun, M. B. Salamon, P. Dorsey, and M. Rubinstein, Phys. Rev. B **60**, 1028 (1999).

Strain Effects in
Manganite Thin Films

ANOMALOUS MAGNETORESISTANCE EFFECT IN STRAINED MANGANITE ULTRATHIN FILMS

QI LI, H. S. WANG, Y. F. HU, AND E. WERTZ

Department of Physics, Pennsylvania State University, University Park, PA 16802

ABSTRACT

Magnetotransport properties of strained epitaxial $Pr_{2/3}Sr_{1/3}MnO_z$ (PSMO) ultrathin films have been studied. The strains are controlled by growing PSMO films on different lattice-mismatched substrates which impose biaxial compressive-, tensile-, and very little strain in the films, respectively. Distinctive magnetoresistance effects in low magnetic field (LFMR) have been observed in films with different types of strain. The films with compressive strain show very large negative LFMR (> 1000 % at 2500 Oe) and significant magnetoresistance hysteresis when a magnetic field is applied perpendicular to the film plane. In a parallel field, the LFMR is much smaller with almost no hysteresis. In contrast, the tensile-strained ultrathin films show positive LFMR at low temperatures in a perpendicular field and a negative LFMR in a parallel field. In comparison, the almost strain-free ultrathin films show very small LFMR (< 2 %) as in single crystal samples at similar temperatures and magnetic fields. In the compressive-strained samples, the large LFMR decreases rapidly with increasing film thickness. These results are interpreted based on spin dependent scattering at the domain walls and domain-rotation.

INTRODUCTION

Recently, the doped manganese oxides have been the focus of intensive research due to the colossal magnetoresistance (CMR) effect observed in the system [1, 2, 3, 4]. The manganites are perovskite oxides with a composition $Re_{1-x}A_xMnO_3$, where Re is a rare earth and A is a divalent alkali element. The material is ferromagnetic in the doping range of about $x \sim$ 0.2-0.5, and the CMR effect occurs near the ferromagnetic transition temperature T_c. The ferromagnetic properties of manganites are traditionally interpreted by the double exchange picture between Mn^{3+} and Mn^{4+} [5, 6, 7] and are very sensitive to the lattice distortions, such as the external hydrostatic pressure in the bulk samples [8, 9, 10], the "internal pressure" due to rare earth substitution [11, 12]. It is proposed theoretically that the CMR effect is related to the localization effect due to a strong dynamical electron-lattice coupling involving

Jahn-Teller vibration mode [13]. The Jahn-Teller mode is associated with a uniaxial volume-preserving lattice distortion. Therefore, the uniaxial lattice strain in thin films has been shown to affect the magnetotransport properties [14, 15, 16, 17].

Strain also affects the magnetic anisotropy of the thin film samples. In manganites, experiments show that the intrinsic magnetocrystalline anisotropy is very weak [18]. There is essentially no MR anisotropy in single crystals. In thin films, on the other hand, the strain-induced anisotropy has been found to play a dominant role in the magnetic anisotropy energy [19, 20]. Magnetic anisotropy can have dramatic influence on the MR properties, in particular the low field MR (LFMR) effect[21, 22]. Since most proposed applications of the CMR materials involve thin films, it is very important to understand how the strain affects the magnetotransport properties in thin films.

In this paper, we present results on the LFMR and MR anisotropy for ultrathin (50 - 150 Å,) $Pr_{2/3}Sr_{1/3}MnO_z$ (PSMO) films epitaxially grown on different substrates which induce different types of biaxial strain in the thin films due to lattice mismatch with the films. Distinctive low-field MR effects have been observed in films with different types of strain. The films with compressive strain show very large negative LFMR and MR hysteresis when a magnetic field is applied perpendicular to the film plane, while those with tensile strain show positive LFMR for the same field orientation. All samples show negative MR in a parallel magnetic field. These results are closely related to spin-dependent scattering at domain boundaries and domain movement.

EXPERIMENT

The PSMO thin films with thickness ranging from 5 to 400 nm were grown epitaxially *in situ* by pulsed-laser deposition. The samples were made with a substrate temperature of 750-800 C, an oxygen pressure of 0.75 mbar, and the laser energy density of ~ 2 J/cm^2. After deposition, the films were cooled down to room temperature in ~ 1 atmosphere oxygen. Using these conditions, the *in situ* grown PSMO films with the thickness larger than 100 nm show similar T_c to the bulk materials, indicating no oxygen deficiency as commonly observed in the *in situ* PSMO films. Details of the film preparation and MR measurement have been described previously [21, 22, 17]. We will focus our attention in this paper on ultrathin films (50 to 150 Å) in which the strain distribution is relatively uniform. LaAlO$_3$ (001) (LAO), SrTiO$_3$ (001) (STO), and NdGaO$_3$ (110) (NGO) have been used as the substrates which impose biaxial compressive, tensile, and nearly no strains in the films, respectively. LAO is peudocubic with a lattice constant of $a \sim 3.79$ Å, STO is cubic with $a \sim 3.905$ Å, and NGO is orthorhombic with the lattice parameters of 3.862 Å and 3.854 Å. Bulk PSMO is orthorhombic with equivalent lattice parameters of $a \sim 3.879$ Å, $b \sim 3.866$ Å, and $c \sim 3.856$ Å. Therefore, among the three types of films studied, the PSMO/NGO films have the smallest film-substrate lattice mismatch ($\sim - 0.3$ %), and hence the smallest lattice distortion. Both PSMO/LAO ($\sim - 2$ %) and PSMO/STO (~ 1.0 %) films have large distortions with biaxially

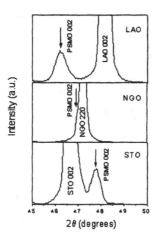

Figure 1: XRD patterns of three 30 nm thick PSMO films grown on LAO (a), NGO (b), and STO (c) substrates, respectively. The arrows indicate the PSMO 002 peaks.

compressed and expanded in-plane lattice parameters respectively. X-ray diffraction (XRD) experiments reveal that the films are c-axis oriented. The c-axis lattice parameter of the PSMO/NGO film is essentially thickness independent and very close to the bulk value, while those of the ultrathin PSMO/LAO and PSMO/STO films are enlarged to 3.95 Å and reduced to 3.82 Å respectively due to the in-plane strain as shown in Fig.1. Thickness-dependence studies show that, except for the PSMO/NGO samples, the c-axis lattice parameter changes gradually to the bulk value when the film thickness is increased above ~ 20 nm, indicating a gradual relief of strain. For 10 nm thick samples, in- plane X-ray diffraction data show that the films on STO are fully strained, while those on LAO are partially strained with the in-plane lattice parameter of 3.81 Å. The diffraction peaks are broader for the strained films than for the strain-free films. The magnetoresistance was measured using a Quantum Design PPMS system with a maximum magnetic field of 9 Tesla.

RESULT

The resistance of all the PSMO films show a crossover at T_p from a high temperature semiconducting state to a low temperature metallic state, typical for manganites. At a fixed thickness, among the three types of samples, the T_p of the PSMO/STO film is always the lowest and that of the PSMO/NGO film the highest. Fig. 2 shows the T_p as a function of the film thickness. Compared to thick PSMO films, T_p is usually reduced in the ultrathin films even for strain-free samples. This can be due to the finite size effect for ferromagentic

187

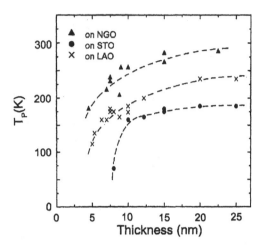

Figure 2: T_p as a function of film thickness for samples on different substrates.

ordering and surface and interface related disorder [17]. The qualitative features of CMR at high magnetic fields for the strained films are similar to that of the thick films, but the CMR ratio is orders of magnitude larger in strained films than in strain-free films [17]. However, the MR at low magnetic field for strained films is dramatically different from that of single crystal or thick films.

The LFMR in manganite single crystals or thick films is usually very small. Strikingly, in the compressive-strain ultrathin films, unusually large low-field MR has been observed. Fig. 3 shows the normalized resistance as a function of magnetic field applied perpendicular to the film plane for a 7.5 nm thick sample when the sample was first cooled in zero magnetic field. The resistance is reduced by more than an order of magnitude in a magnetic field H = 2.5 KOe. The resistance did not return to the virgin R(0) value when the magnetic field was removed. In the subsequent magnetic field scan, large MR hysteresis curves were obtained and reproducible upon field cycling as shown in Fig. 4. The MR hysteresis is observed in a wide temperature range below T_p and the LFMR shows a maximum value at a temperature which is different and well below the maximum CMR temperature. Shown in the inset of Fig. 3 is the magnetization measurement in parallel and perpendicular fields. The result indicates that the easy magnetization axis is perpendicular to the film surface due to the compressive strain [22, 23]

The large low-field MR is observed only in the films under compressive strain on LAO substrate. For both the strain-free (on NGO substrate) and the tensile-strain (on STO substrate) samples, the MR ratios are all very small. This indicates that the observed effect is mainly due to the compressive strain rather than the reduced dimensionality of the very thin films.

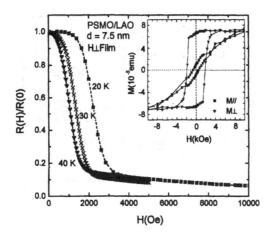

Figure 3: Normalized resistance as a function of magnetic field for a PSMO/LAO sample when the sample was cooled in zero magnetic field. Inset shows the magnetization measurements in different magnetic field directions

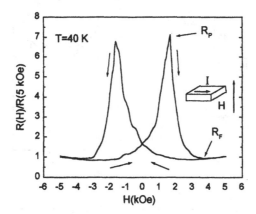

Figure 4: Normalized resistance as a function of magnetic field scan for a PSMO/LAO sample.

The observed anomalous LFMR effects are anisotropic. Fig. 5 shows the MR measurements for a 7.5 nm thick film with the field scan parallel and perpendicular to the film plane at 50 K. In the perpendicular-field geometry, a very large MR with pronounced hysteresis is observed. In contrast, in the parallel-field geometry, the MR hysteresis is much weaker and the MR ratio is smaller.

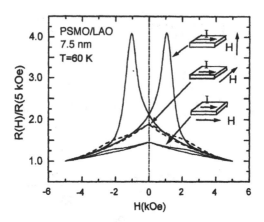

Figure 5: MR hysteresis loops of a 7.5 nm thick PSMO/LAO film measured in parallel and perpendicular magnetic fields.

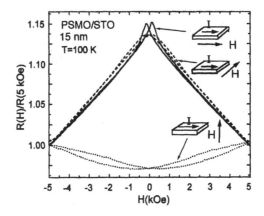

Figure 6: MR hysteresis loops of a 15 nm thick PSMO/STO film measured in parallel and perpendicular magnetic fields.

For tensile-strain samples, the LFMR properties are very different from the compressive-strain samples. Shown in Fig. 6 are the MR results of a 15 nm thick PSMO/STO film. Strikingly, the MR has different signs for different field directions. It is positive in perpendicular magnetic field and negative in parallel magnetic field. Also, the MR ratios are different for the two field directions and both are orders of magnitude smaller than those of the PSMO/LAO samples. The positive MR effect in perpendicular fields is only present in small magnetic fields. When the applied field is larger than 1 - 2 teslas, the MR becomes

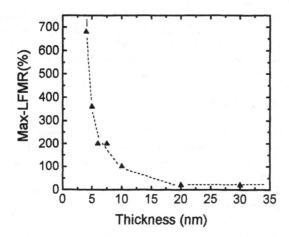

Figure 7: T_p as a function of film thickness for samples on different substrates.

negative. The positive MR also has a strong temperature dependence. It occurs at low temperatures and when the temperature is close to T_p, the positive MR crosses over to a negative MR.

With the exception of the PSMO/NGO samples, the measured MR properties depend strongly on the film thickness. For PSMO/LAO films, the MR ratio decreases rapidly when the film thickness increases. The dependence of the LFMR as a function of film thickness is shown in Fig.7. For a 20 nm thick sample, the maximum low-field MR is less than 10 %. In thicker PSMO/STO films, the positive MR effect also disappears. These effects are probably due to the gradual release of strain in thicker films as well as changes in domain size and the domain wall width as the thickness of the sample increases.

DISCUSSIONS

We suggest that spin-dependent scattering at domain walls and domain wall rotation are responsible for the observed anomalous LFMR effect in our samples. In the PSMO/LAO films, the strain- induced anisotropy field, which favors an out-of-plane magnetization, is larger than the demagnetization field, resulting in an out-of-plane easy axis [23]. This is consistent with the Magnetic anisotropy in $La_{0.7}Ca_{0.3}$-MnO_3 and $La_{0.7}Sr_{0.3}$-MnO_3 films [24, 20, 19]. In our samples, when the domains are all aligned in a perpendicular field, the resistance is low. As the field is swept from one direction to another, domain reversal occurs at its switching field. Due to the difference in the switching fields of different domains, a mixture of oppositely aligned domains is present in a moderate field range near the coercive field. Due to spin dependent scattering at the domain walls, the resistance is high. This gives rise to a large MR. The observed peak resistance field from the R vs. H curve is very close to the coercive field obtained from the magnetization measurement. For the parallel field, because the magnetization hard axis is in the film plane, a much larger field is required to align all the domains in the plane, therefore the MR ratio at low field is smaller than that in a perpendicular field. The lack of hysteresis in the parallel field can be explained by the flipping of the magnetization of the domains from the in-plane to the out-of-plane direction (easy-axis) when the applied field approaches zero. In this case, there is no memory of the history of the in-plane magnetization. Generally, scattering at domain walls in ferromagnetic materials does not contribute to a large MR because the domain walls are usually broad and the magnetization direction changes continuously across the wall regions. Since a large MR due to spin-dependent scattering relies on a high density of unaligned ferromagnetic domains, it suggests that in very thin PSML/LAO films, the perpendicular domain sizes are very small with extremely thin walls.

In the PSMO/STO samples, the tensile strain induces a magnetic ani- sotropy which favors an in-plane magnetization. When a perpendicular field is applied, the magnetization will rotate out of the plane and be perpendicular to the film plane, when H is larger than a threshold field of $H_c = H_K + H_D$, where H_K is the strain-induced anisotropy field, and $H_D = 4\pi M$ is the demagnetization field. As discussed by Eckstein et al. [25], the resistance may increas with field due to the increasing angle between the measuring current and the magnetization. On the other hand, we have found that in strained films, there is also a large MR anisotropy when the magnetic field (magnetization) is applied along different crystal directions. The two types of MR anisotropy together resulted in the positive MR in PSMO/STO films in perpendicular fields. The positive MR crosses over to a negative MR due to the intrinsic negative CMR when H is larger than H_c and the film will be in a state of a single domain. In the temperature ranges of 60 - 100 K, the measured H_c for a 10 nm thick PSMO/STO film is about 1 tesla, comparable to the sum of the demagnetization field and the anisotropy field measured. In a parallel field, the LFMR should be due to the domain rotation and movement within the film plane. In this case, unlike the perpendicular domains in the PSMO/LAO samples, the domain wall is conventional and the domain rotations do not cause a large MR, as in the usual anisotropic MR of ferromagnetic materials.

It should be noted that in strained thin films, disorders due to strain, surface, and interface are all present. They will also contribute to the MR behaviors, especially the MR ratio.

In summary, we have studied the strain effects on the magnetoresistance of the ultrathin PSMO films. Dramatic differences in the low-field MR properties have been observed in films with different types of strains. The anomalous anisotropic MR effects are mainly attributed to the spin-dependent scattering at the domain wall. We have observed very large low-field MR effect and MR hysteresis in compressively-strained ultrathin PSMO films. The result demonstrates a large LFMR in single-layer epitaxial manganite films. By improving the T_p value of the samples, the effect should also be seen at higher temperatures which is technologically desirable.

ACKNOWLEDGMENTS

The authors wish to thank X. X. Xi, R. Willis, X. W. Wu, M. Rzchowski, K. Liu, C. L. Chien, D. Schlom, T. Malouk, and M. Rubinstein for collaborations in the experiment and stimulating discussions about the results. This work is supported in part by NSF under Grant DMR-9707681, DMR-9623315, and Petroleum Research Fund (Award 31863-G5).

References

[1] R. von Helmholt, J. Wecker, B. Holzapfel, L. Schultz, and K. Samwer, Phys. Rev. Lett. 71, 2331 (1993).

[2] K. ichi Chahara, T. Ohno, M. Kasai, and Y. Kozono, Appl. Phys. Lett. 63, 1990 (1993).

[3] S. Jin, T. Tiefel, M. McCormack, R. Fastnacht, R. Ramesh, and L. Chen, Science 264, 413 (1994).

[4] G. C. Xiong, Q. Li, H. L. Ju, S. N. Mao, L. Senapati, X. X. Xi, R. L. Greene, and T. Venkatesan, Appl. Phys. Lett. 66, 1427 (1995).

[5] C. Zener, Phys. Rev. 82, 403 (1951).

[6] P. W. Anderson and H. Hasegawa, Phys. Rev. 100, 675 (1955).

[7] P.-G. de Gennes, Phys. Rev. 118, 141 (1960).

[8] Y. Moritomo, A. Asamitau, and Y. Tokura, Phys. Rev. B 51, 16491 (1995).

[9] J. J. Neumeier, M. F. Hundley, J. D. Thompson, and R. H. Heffner, Phys. Rev. B 52, 7006 (1995).

[10] K. Khazeni, Y. X. Jia, L. Lu, V. H. Crespi, M. L. Cohen, and A. Zettl, Phys.Rev. Lett. 76, 295 (1996).

[11] H. Y. Hwang, S.-W. Cheong, P. G. Radaelli, M. Marezio, and B. Batlogg, Phys. Rev. Lett. 75, 914 (1996).

[12] J. Fontcuberta, B. Martinez, A. Seffar, S. Pinol, J. L. Garcia-Munoz, and X. Obradors, Phys. Rev. Lett. 76, 1122 (1996).

[13] A. J. Millis, B. I. Shraiman, and R. Mueller, Phys. Rev. Lett. 77, 175 (1996).

[14] A. J. Millis, Nature 392, 147 (1998).

[15] A. J. Millis, T. Darling, and A. Migliori, J. Appl. Phys. 83, 1588 (1998).

[16] T. K. Nath, R. A. Rao, D. Lavric, C. B. Eom, L. Wu, and F. Tsui, Appl. Phys. Lett. 74, 1615 (1999).

[17] H. S. Wang, E. Wertz, Y. F. Hu, and Q. Li, J. Appl. Phys. 2212 (1999).

[18] T. M. Perekalina, I. E. Lipinski, V. A. Timofeeva, and S. A. Cherkezya, Sov. Phys. Solid State 182, 1828 (1990).

[19] J. O'Donnell, M. S. Rzchowski, J. N. Echstein, and I. Bozovic, Appl. Phys. Lett. 72, 1 (1998).

[20] Y. Suzuki, H. Y. Hwang, S.-W. Cheong, and R. B. V. Dover, Appl. Phys. Lett. 71, 140 (1997).

[21] H. S. Wang and Q. Li, Appl. Phys. Lett. 73, 2360 (1998).

[22] H. S. Wang, Q. Li, K. Liu, and C. L. Chien, Appl. Phys. Lett. 2360 (1998).

[23] X. W. Wu, M. Rzchovski, H. S. Wang, and Q. Li, to be published in Phys. Rev. B (2000).

[24] C. Kwon, Q. Li, I. Takeuchi, C. D. P. Warburton, S. N. Mao, X. X. Xi, and T. Venkatesan, Physica C 266, 75 (1996).

[25] J. N. Eckstein, I. Bozovic, J. O. Donnell, M. Onellion, and M. S. Rzchowski, Appl. Phys. Lett. 69, 1312 (1996).

STRAIN ANALYSIS IN THIN La$_{1-x}$Ca$_x$MnO$_3$ FILMS BY GRAZING INCIDENCE X-RAY SCATTERING

M. PETIT*, L. J. MARTINEZ-MIRANDA*, M. RAJESWARI**, A. BISWAS**, D. J. KANG** AND T. VENKATESAN**, ***

*Dept. of Materials and Nuclear Eng. And NSF-MRSEC, University of Maryland, College Park, MD, 20742, **martinez@eng.umd.edu**
**NSF-MRSEC and Center for Superconductivity, University of Maryland, College Park, MD 20742
***Dept. of Physics and Dept. of Electrical Eng., University of Maryland, College Park, MD 20742

ABSTRACT

We have performed depth profile analyses of the lattice parameters in epitaxial thin films of La$_{1-x}$Ca$_x$MnO$_3$ (LCMO), where x = 0.33 or 0.3, to understand the evolution of strain relaxation processes in these materials. The analyses were done using Grazing Incidence X-ray Scattering (GIXS) on films of different thicnesses on two different substrates, (100) oriented LaAlO$_3$ (LAO), with a lattice mismatch of ~ 2% and (110) oriented NGO, with a lattice mismatch of less than 0.1%. Films grown on LAO can exhibit up to three in-plane strained lattice constants, corresponding to a slight orthorhombic distortion of the crystal, as well as near-surface and columnar lattice relaxation. As a function of film thickness, a crossover from a strained film to a mixture of strained and relaxed regions in the film occurs in the range of 700 Å. The structural evolution at this thickness coincides with a change in the resistivity curve near the metal-insulator transition. The in-plane compressive strain has a range of 0.2 – 1.5%, depending on the film thickness for filsm in the range of 400 - 1500 A.

INTRODUCTION

Heteroepitaxial films are of great industrial interest because of their specific functional properties (magnetic, electronic) which can be exploited in device applications. However, the growth of a thin film on a substrate almost always induces a state of structural strain due to, among others, lattice and thermal expansion mismatch. For La$_{0.7}$Ca$_{0.3}$MnO$_3$ (LCMO) films grown on different substrates, this strain results in a variation of electrical and magnetic properties related to the sensitivity of the electronic state of Mn^{3+} ion to the biaxial Jahn-Teller strain. [1]. Changes in the electrical properties are more notable for films thinner than 800 Å. Since the physical properties of materials are strongly dependent on the structural stacking, it is imperative to understand the role of in-plane stress induced by the lattice mismatch between the film and the substrate and the resulting evolution of structure within the films.

We have studied the structural evolution in LCMO films grown on both NdGaO$_3$ (NGO) and LaAlO$_3$ (LAO) substrates by pulsed laser deposition (PLD), using grazing incidence x-ray diffraction (GIXS) [2,3]. This method allows us to measure the in-plane structure of the films as a function of depth non-destructively. In thin films it is also possible to measure the structure of the vicinity of the buried susbstrate-film interface [4]. By performing a mapping on the in-plane scattering profile, we have successfully measured strains, strain distribution and strain evolution in these films, and we have been able to compare the in-plane strain of films for samples of different thickness.

LCMO has a cubic structure with unstrained bulk lattice parameters $a = 3.86$-3.87 Å. The LAO substrate has a pseudo-cubic structure with a lattice parameter $a = 3.78$-3.79 Å and NGO is orthorhombic with lattice parameters 5.43 Å, 5.5 Å and 7.71 Å. `Since the surface is (110) oriented for deposition, the in-plane lattice parameters of the NGO substrate are 3.855 Å and 3.864 Å. As is clear from the lattice parameters, the LCMO films on LAO are expected to be uinder a large (~2%) compressive strain while for the films on NGO the compressive strain magnitude is significantly smaller. We investigated four films $La_{0.7}Ca_{0.3}MnO_3$ deposited on LAO, 400 Å, 700 Å, 850 Å and 1500 Å thick, in order to study the evolution of stress as a function of thickness.

EXPERIMENTAL

The samples were prepared using pulsed laser deposition, as described in Ref. 1. The GIXS technique, illustrated in figure 1, combines the principle of total external reflection, sample absorption and diffraction to probe different depths within a film [2,3]. The diffraction condition can be satisfied for crystallographic planes nearly perpendicular to the sample surface in this geometry. With appropriate x-ray energies, regions close to the buried film-substrate interface can also be probed. By rocking the sample about the azimuthal angle φ, the in-plane evolution of the lattice parameters can be determined, as well as the scattering distribution about the main peaks and their in-plane orientation.

The experiments were performed at the National Synchrotron Light Source at Brookhaven National Laboratories, at beamline X18A using 1.24 Å (energy = 10 keV) wavelength X-rays with a wavevector resolution of $\Delta q = 0.003q_0$. The incident beam has a horizontal convergence of 6 mrad and a vertical convergence of 1.2 mrad. The X-ray beam had a spot size of 1×1 mm^2. Samples were mounted on a 4-circle Huber goniometer (2θ, θ, χ, φ). The film surface was adjusted perpendicular to the rotation axis φ with a Huber head goniometer and was moved in the middle of the beam with a translation axis. A rocking curve in the total reflection region (scan θ at $2\theta = 0.2°$ for example) was performed before the in-plane measurements and served to align the surface of the film with respect to the X-ray beam, defining therefore the zero value of the rotation axis θ. The zero value of the χ axis was already defined, so that the surface plane of the film is now perfectly vertical at $\theta = 0°$ and $\chi = 0°$, and as a consequence is parallel to the plane defined by the incoming and the outgoing beams (k_i, k_d), as shown in Figure 1.

Depth within the films was controlled by varying the X-ray incidence angle α which is defined by a combination of the two axes θ and χ on the 4-circle and given by the relation,

$$\sin \alpha = \sin \theta \cdot \sin \chi. \qquad (1)$$

First the depth within the films was calibrated by a measurement of the critical angle α_c of the films obtained from the reflectivity curve. This measurement determined both the density of the films as well as the values of the incidence angle used in the measurements. This could be realized by varying either the θ angle or the χ angle, according to equation 1. Our in-plane measurements were done by varying χ at fixed θ. The previous range was chosen for χ to vary α considering the critical angle, and the θ angle was fixed and selected as follows: The film and substrate peaks were optimized at χ = 90 (out-of plane conditions), for a $2\theta_0$ value and a corresponding θ_0 value which is not necessary the half of $2\theta_0$, but depends if the out-of-plane wavevector is exactly parallel or not to the film surface. The θ_0 value relative to the (001) film peak is chosen to fix the θ axis for the in-plane measurements. The penetration angle α has then an accuracy of:

$$\Delta\alpha = [\Delta\chi \cos\chi + 0.003\sin\chi]\sin\theta_0 /\cos\alpha , \qquad (2)$$

with $\Delta\chi$ the uncertainty of the χ rotation axis.

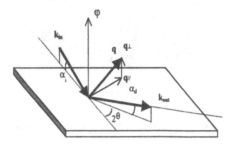

Figure 1. Geometry of the GIXS experiment.

RESULTS AND DISCUSSION

At each penetration depth, mappings of the reciprocal space were realized. Each map is a two-dimensional 2θ-φ scan, where the 2θ interval is centered around the 2θ value corresponding to the unstrained known parameter, and with a φ interval centered around one of the two in-plane substrate axis directions. One direction is located around φ = 0° and the other one around φ = 90°, since the substrate has a cubic symmetry. Figure 2 shows the map for the 1500 Å on LAO sample at α = 0.4° (or χ = 1.25°).

To analyze these diffraction maps, a diffraction profile is extracted for each peak at its maximum φ value. These direct (without any correction of intensity) diffraction data are fit with a Gaussian function or a sum of Gaussians to determine the location of each peak, 2θ. The result is used to determine the in-plane parameters. Figure 3 shows the evolution of the in-plane parameters for each sample. The error bars represented are calculated considering Equation 1. These values are higher than the ones given by the fitting function. As the penetration depth increases, regions closer to the interface with the substrate are probed. Films below 1500 Å exhibit compressive strain through the thickness of the film. There is a relaxation near the surface of the films. The strain is

197

highest for the 400 and 700 Å thick films where it has a value of –1.2%. A closer look at the lattice parameter of these films shows that the 700 Å film consists of three main lattice parameters – two along $\phi = 0°$, and one along $\phi = 90°$. The two parameters along $\phi = 0°$ are shown in Figure 4a. A comparison of this figure with Figure 3 shows that the relaxed axis is comparable to the in-plane axes of the 850 Å film.

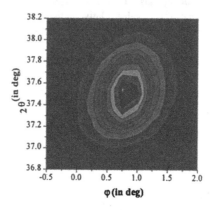

Figure 2. In-plane 2θ-φ scan for a 1500 Å film, at an incidence angle of 0.4°.

The above result suggests the presence within the film of regions where the lattice parameter relaxes toward the bulk value of LCMO. Evidence for this appears in Figure 4b, which shows a plot of the integrated intensity as a function of depth for the relaxed (a1) and strained (a2) axes observed in the 700 Å film. Figure 4b is a plot of the integrated intensity for the peaks corresponding to the relaxed and strained lattices. Near the surface, the intensity of the strained lattice parameter is close to two order of magnitudes larger than the relaxed lattice parameter. The apparent increase of the relaxed peak intensity beyond a depth of 300 Å is due to the fact that the incidence angle causes the in-plane scan to be slightly off-center for the epitaxial strained component. The evolution of the intensity for the relaxed lattice component of the film is consistent with that predicted from optical theory [3]. This suggests that at this thickness, the film begins to relax by forming columns in the film, consistent with TEM observations [5]. The development of relaxed columns in the films coincides with the change in the resistivity curve mentioned above [1].

CONCLUSION

We have presented the results of a depth analysis of the in-plane structure of LCMO films of varying thickness grown on LAO. Films below 1500 Å thick are strained in the plane of the film. This strain is relaxed near the surface of the films, and is constant throughout most of the bulk of the film. We have observed a cross-over from a uniformly strained epitaxial film to a film consisting of strained regions and relaxed columns in the thickness range 400 – 700 Å. This crossover coincides with a change in the shape of the resistivity curves near the metal-insulator transition in these films.

Future measurements may include temperature dependent measurements in order to fully understand the structure-resistivity relation in the films.

ACKNOWLEDGEMENTS

This work is supported by a NSF-MRSEC Grant NO. DMR-9632521. Work at the National Synchrotron Light Source is partially supported by the US Department of Energy.

REFERENCES

1. A. Millis et al., to be published.

2. M.F. Toney, T.C. Huang. S.Brennan, Z. Rek, J. Mater. Res. 3, 351 (1988).

3. Marra, W. C., Eisenberger, P. and Cho, A. Y., J. Appl. Phys., 50, 6927 (1979).

4. For example, see L. J. Martínez-Miranda, Y. Hu and T. K. Misra, Moleq. Cryst. Liq. Cryst., 329, 121 (1999); L. J. Martínez-Miranda, Y. Li, G. M. Chow and L. K. Kurihara, NanoStructured Materials, 12, 653 (1999); L. J. Martínez-Miranda, submitted to PR B, 1999.

5. Y. H. Li, private communication, 1999.

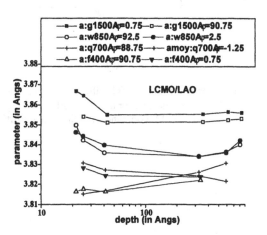

Figure 3. In-plane lattice evolution as a function of depth and thickness for films of LCMO grown on LAO.

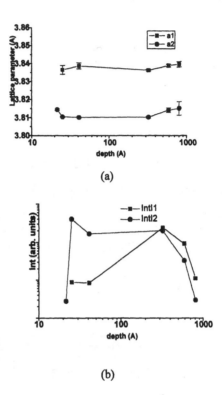

(a)

(b)

Figure 4. a. Lattice parameters around φ = 0 for the 700 Å film. The larger lattice parameter is comparable to that measured for the 850 Å film. b. Integrated intensity for the strained (IntI1) and relaxed (IntI2) components of the in-plane diffraction peak for the 700 Å film.

Substrate-induced anomalous electrical transport and magnetic transitions in epitaxial La$_{0.66}$Sr$_{0.33}$MnO$_3$ films grown on (001) BaTiO$_3$ substrates

T.K. Nath, M.K. Lee, and C.B. Eom
Department of Mechanical Eng. and Materials Science, Duke University, Durham, NC 27708.
M. Smoak, P.A. Ryan, and F. Tsui
Department of Physics and Astronomy, University of North Carolina, Chapel Hill, NC 27599

ABSTRACT

Effects of drastic change of 3-dimensional films' lattice strain by substrate structural transformations on the transport and magnetic properties of tensile strained epitaxial La$_{0.66}$Sr$_{0.33}$MnO$_3$ thin films were studied using (001) BaTiO$_3$ as the templates. These films exhibit dramatic jumps in electrical resistance and magnetization at temperatures very close to different crystallographic structural change of BaTiO$_3$ substrate. Application of high magnetic field (50 kOe) suppresses both the electrical transport and magnetization jumps. The observed low-field sharp transitions correspond sudden changes in magnetic anisotropy, caused by the structural transformation of the substrate. The magnetic easy axis lies in the film plane along [100] direction having a strong correlation with the substrate-induced 3-dimensional strain states as determined by the normal and grazing incidence x-ray diffraction. The sudden substrate structural transformations imposes an additional drastic sharp change of the film strain which is found to reorient the magnetic easy axis within the film plane.

INTRODUCTION

Ever since the discovery of colossal magnetoresistance (CMR) effect in the epitaxial thin films of doped manganite perovskite materials Re$_{1-x}$Ae$_x$MnO$_3$ [trivalent rare-earth ions (Re) and divalent alkaline-earth ions (Ae)], extensive research has been focussed in these materials for possible device applications [1-3]. The occurrence of CMR behavior has been attributed to the presence of lattice strain and disorder in epitaxial films. It has been shown that substrate-induced lattice strain plays a dominant role in CMR behavior, T$_P$, T$_c$, magnetic domain structure and associated magnetic anisotropy [4-11] of these epitaxial manganite films. In recent time the subject of modulation of substrate-induced lattice strain has become a critical issue in thin film CMR materials to understand their 3D lattice strain states, crystallographic domain structure and magnetic anisotropy both theoretically [12-13] and experimentally [4-11] for the implementation of magnetic and electronic device applications.

In this paper, we report the effects of substrate structural transformations on the transport and magnetic properties of epitaxial La$_{0.66}$Sr$_{0.33}$MnO$_3$ (LSMO) thin films using (001) BaTiO$_3$ (BTO) as the templates. BTO is a stable ferroelectric material having Curie temperature of 393 K. The symmetry of the non-polar phase is cubic (m $3m$) having a *perovskite-type* structure [14]. The symmetry of the polar phase below 393 K is tetragonal (4 mm). Below 278 K it makes a transition from tetragonal to orthorhombic (mm) structure changing sharply the c-axis lattice parameter by –0.6% and a-axis lattice parameter by –0.1%. At 183 K, a third phase transition occurs and the symmetry changes, upon cooling, from orthorhombic to rhombohedral (3 m). The c-axis lattice parameter changes sharply by –0.4% and a-axis lattice parameter by +0.5% across this transition. Generally, in any epitaxial thin film elastic film strain due to epitaxy provided by the substrate-film lattice mismatch gradually changes with temperature depending on the temperature coefficient of lattice constants for both film and substrate. The sudden substrate structural transformations at the transition temperatures impose an additional drastic sharp change of the 3-dimensional lattice strain in the film. The effect of this additional sudden change of film strain on electrical and magnetic behavior of epitaxial CMR film has been studied in this paper.

Mat. Res. Soc. Symp. Proc. Vol. 602 © 2000 Materials Research Society

EXPERIMENTAL DETAILS

We have grown 500 Å $La_{0.66}Sr_{0.33}MnO_3$ epitaxial films on (001) BTO substrate by 90^0 off-axis rf-magnetron sputtering using a 2 in. diameter stoichiometric target. A rf (13.56 MHz) power of 100 W was applied. The process was carried out in a flowing gas mixture of argon with oxygen in the 12 : 8 volumetric ratio with 200 mTorr total partial pressure. The growth temperature for the substrate was optimized at 750^0 C. After deposition the films were cooled under ambient oxygen of 300 Torr.

The in-plane and out-of-plane lattice parameters and the associated 3D strain states in the films were determined by normal and grazing incidence diffraction (GID) using a four-circle x-ray diffractometer with Cu $K\alpha$ radiation. The bulk LSMO target has a pseudocubic perovskite structure with lattice parameters a^p = 3.884 Å. The tilting of MnO_6 octahedra results in a rhombohedral structure with a relation a^{Rhom} = $\sqrt{2}a^p$ and α = 60^0 .The pseudocubic lattice parameters are rotated 45^0 from the rhombohedral lattice parameters (a^{Rhom}). In this paper we will describe the Miller indices of the LSMO planes based on pseudocubic perovskite unit cell structure.

The temperature-dependent electrical and magneto transport measurements were carried out using a 4-point DC technique employing Oxford Instruments' MagLab System 2000™. Magnetic measurements were carried out using a Quantum Design superconducting quantum interference device (SQUID) magnetometer. Special sample holders were used to position the samples along chosen crystallographic directions.

RESULTS

The LSMO films are grown on (001) BTO substrate. At film deposition temperature the substrate is in cubic phase. At room temperature in tetragonal phase, the a-axis and c-axis lattice parameters of BTO substrate are 3.997 Å and 4.037 Å (c/a =1.01), respectively. After a thermal cycle through the film growth temperature the single domain BTO substrate becomes a mixture of two different kind of domains of (001) (~60%) and (100) (~40%) normal orientations. The lattice mismatch (+2.9 %) between film and substrate induces in-plane biaxial tensile stress in the LSMO films. Figure 1 shows the normal $\theta - 2\theta$ x-ray diffraction scan of 500 Å LSMO film on BTO

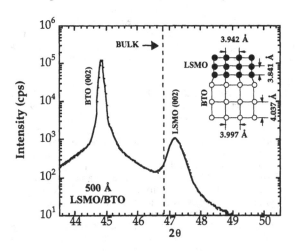

Fig. 1. X-ray normal θ - 2θ scans for 500 Å LSMO epitaxial films grown on (001) $BaTiO_3$ substrates. The dotted line indicate bulk 2θ value of LSMO. Inset shows the schematic diagrams of cross-sectional view of the strained lattice.

substrate. The out-of-plane pseudocubic lattice spacing (d_{001}) of the film is obtained to be (3.841 ± 0.001) Å. The in-plane lattice spacing of the film from grazing incidence x-ray diffraction

(GID) is found to be (3.942±0.006) Å. This indicates that the 500 Å LSMO film grown on BTO has an in-plane biaxial tensile strain with $\varepsilon_{xx} = \varepsilon_{yy} = +1.5$ % and a corresponding out-of-plane uniaxial compression with $\varepsilon_{zz} = -1.1$ %. The film is not fully coherent (partially strain relaxed) with the substrate because of the large 2.9% biaxial tensile stress imposed by BTO. The thin film perovskite unit cell volume gets expanded by 1.8%.

Figure 2(a) shows the temperature dependent electrical transport of 500 Å LSMO/BTO film in the temperature range of 4.2 – 380 K in both warming up and cooling down cycle. The metal-insulator transition peak (T_p) of this epitaxial thin film has been observed at 340 K which is much suppressed from the bulk value (360 K).

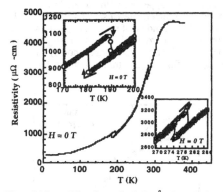

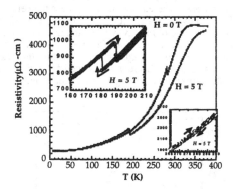

Fig. 2(a). ρ(T) plot of 500 Å LSMO/BTO epitaxial film in both warming up and cooling down cycle at H=0T. Inset : Expanded view of sharp change in ρ and thermal hysteresis effect.

Fig. 2(b). Comparative plot of ρ(T) at H = 0T and 5T of LSMO/BTO film showing high field effect. Inset : Field effect of thermal hysteresis loop (expanded view).

Anomalous dramatic jumps in electrical resistance of the film on BTO has been observed at 277 K and 180.5 K which are very close to the transitions where the BTO substrate makes different crystallographic structural transformations. At 277 K and 180.5 K, the sharp changes of resistivity have been observed to be about +10% and +12%, respectively (Fig.2(a)). Around these temperatures the changes of c-axis lattice parameter of BTO substrate are about -0.6% and -0.4%, respectively. The changes of a-axis lattice parameters are about −0.1% and +0.5% respectively [14]. At both the transitions the BTO cell volume gets contracted by about 0.15% . The observed sharp changes of resistivity of the LSMO film around transitions are at least one to two order of magnitude higher than structural changes of BTO template. The observation of sharp change in resistivity near the substrate structural transformation could be associated with number of effects. Several recent theoretical work [15,16] argued that in CMR manganites conduction mechanism involves the interplay of hopping of conduction band Mn e_g electrons from site to site (within the framework of double-exchange model) and the localizing effect of a strong electron-lattice coupling (Jahn-Teller strain). The substrate induced lattice strain in epitaxial manganite film can be decomposed into bulk strain (ε_B) and volume preserving Jahn Teller strain (ε_{JT}) [7]. The sudden change of substrate-imposed film lattice strain will result in changes in ε_B and ε_{JT} leading to abrupt changes in electrical transport behaviour; it could be of magnetic origin due to sudden changes in magnetic anisotropy caused by the sharp structural transformation.

Thermal hysteresis effect has been observed around both the sharp jumps of resistivity. In the warming up cycle the transition temperature for the resistivity jump around 277 K is enhanced by ($\Delta T = +5$ K) while it is enhanced by $+9.5$ K (ΔT) around 180.5 K (inset of Fig.2(a)). The magnitude of the sharp change of resistivity remains same around both the transitions as observed in cooling cycle. The observation of these thermal hysteresis effects of resistivity is possibly due to the thermal hysteresis effect of the substrate itself as the structural transition temperatures get shifted at least by 10 K (ΔT) in the cycle [14].

In order to explore the effect of magnetic field on these resistivity jumps the electrical transport measurement was carried out in the temperature range of 4.2 - 380 K in presence of 5 T dc field. Figure 2(b) shows the comparative plot of temperature dependent electrical resistivity at H= 0 and 5 T applied dc magnetic field. T_p gets enhanced to very high value with the application of 5 T field as normally observed in LSMO films grown epitaxially on any single crystal substrates. In presence of 5 T field the resistivity jump at higher temperature (277 K) completely disappeared whereas the jump at lower temperature (180.5 K) appeared with reduced magnitude. Thermal hysteresis effect also had been observed around this temperature (inset of Fig.2(b)). The disappearance of resistivity jump in presence of magnetic field could be of magnetic origin, such as effect of enhancement of effective band width of the charge carriers due to spin alignment in presence of high magnetic field (within the framework of double-exchange model) is more compared to the disorder introduced by sudden change of film lattice strain changing the Jahn-Teller distortion and charge - lattice coupling.

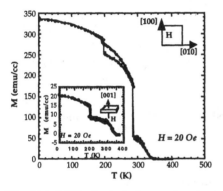 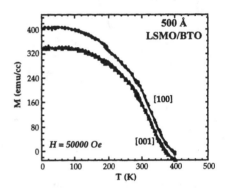

Fig. 3(a). M(T) plot in both warming up and cooling down cycle at 20 Oe applied field along in-plane [100] direction. Inset : Out-of-plane magnetization along [001] at 20 Oe field.

Fig. 3(b). M(T) plot along in-plane [100] and out-of-plane [001] at 50000 Oe applied field in both warming up and cooling down cycle.

In order to understand the magnetic behavior of the film properties induced by substrate structural transformations, temperature dependent dc magnetization measurements had been carried out in the temperature range of 5 K- 400 K in presence of several static magnetic fields : 5, 10, 20, 50, 100 and 50000 Oe. Figure 3(a) shows the M(T) plot in both warming up and cooling down cycle at 20 Oe applied magnetic field along the in-plane [100] direction. The ferromagnetic transition temperature (T_c) of this strained LSMO/BTO film has been observed at 340 K (< bulk T_c of 360 K). The saturation magnetization (M_s) in this tensile strained film (2.9 μ_B/Mn) is significantly reduced (M_s^{bulk} =3.6 μ_B/Mn). The sharp magnetization jumps are observed at 292 K and 200 K in both the cycle. Around 200 K and 292 K the magnitude of sharp changes of magnetization are about 15 % and 70%, respectively. Small thermal hysteresis of magnetization have been observed around both the transitions. The inset in Fig. 4(a) shows temperature dependent out-of-plane magnetization with relatively reduced magnitude at 20 Oe applied field along [001]. Upon increasing the measuring field the magnitude of the sharp changes of magnetization gets diminished.

Figure 3(b) shows the M(T) plot of the film in presence of 50000 Oe applied magnetic field along both in-plane [100] and out-of-plane [001] of the film. Both the sudden jumps almost disappeared with this applied high field in each magnetization curve. When one takes a closer look, the jump at 200 K with very small magnitude still can be distinguished unlike the jump at 292 K which is absolutely disappeared with this field. The observed facts indicate that there could be sudden changes in magnetic anisotropy resulting a sharp drop in magnetization caused by structural transformation of the substrate. The magnetization drops due to change in magnetic anisotropy becomes tiny in presence of very high magnetic field as the magnetic spins are almost pinned (saturation) along the direction of high field.

In order to probe the effects of magnetic anisotropy in the LSMO/BTO film, temperature-dependent magnetic hysteresis loops were measured with the applied field along both in-plane and perpendicular to the film plane at several temperatures just above and below the substrate structural transformations. The behaviors at 290 K and 282 K which are the temperatures above and below one of the observed magnetization jumps (Fig.3(a)) are shown in Fig. 4. In-plane magnetization loop at 282 K with field applied along [100] direction are nearly square and it saturates much faster (H_s ~ 250 Oe). The observed saturation magnetization (Ms) is about 2 μ_B /Mn and the coercive field is about 50 Oe. The magnetization along [100] at 290 K exhibits relatively high saturation field (~800 Oe) and nearly zero remanent magnetization M_r . The magnetization perpendicular to the film at both the temperatures, after it has been corrected for the demagnetization effect, exhibits very high saturation field H_s (~ 10 kOe) and nearly zero remanent magnetization M_r , indicating that the magnetic hard axis along [001]. The higher remanence for the behavior at 282 K along [100] (nearly equal to M_s) compared to that of the behavior at 290 K indicates that the magnetic easy axis at 282 K is along [100] directions. At 290 K, [100] direction becomes a relatively hard axis in the film plane after the film passes through substrtate stuuctural transformation from orthorhombic to tetragonal phase. However, the film plane still remains the easy plane reorienting the easy-axis in some other direction within the plane.

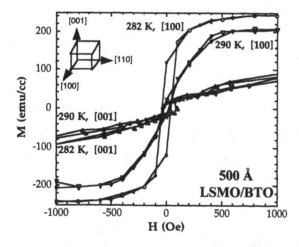

Fig. 4. Magnetic hysteresis loop at 282 K and 290 K (below and above orthorhombic to tetragonal structural transfor -mation of the substrate) along in-plane [100] and out-of-plane [001] of 500 Å LSMO epitaxial film grown on [001] BaTiO₃ substrate.

Detailed orientation dependent magnetotransport and magnetic anisotropy results of this LSMO/BTO film will be discussed elsewhere [15].

In summary, we have studied the effect of substrate structural transformation on electrical and magnetic properties of epitaxial LSMO films grown on BTO. The sudden substrate structural transformations at the transition temperatures impose an additional drastic sharp change of the 3-dimensional lattice strain in the film. In both electrical transport and magnetic measurements dramatic sharp jumps have been observed near those substrate structural transformations. With th application of high magnetic field (50 kOe) both these electrical and magnetic jumps get suppressed. We show that sudden change of magnetic anisotropy in large magnetostrictive LSMO film imposed by sharp change of substrate structural transformations is the best explanation of the observed facts. The magnetic easy axis is found to lie in the film plane along [100] direction having a strong correlation with the substrate-induced 3-dimensional strain states. The sudden substrate structural transformations imposes an additional drastic sharp change of the film strain which is found to reorient the magnetic easy axis within the film plane. Temperature dependent x-ray diffraction data of the films and the substrates will provide a detailed quantitative information of the 3-dimensional strains states at those temperatures where the anomalous jumps have been observed.

ACKNOWLEDGMENTS

This work was supported by the NSF Grant No. DMR-9802444, the NSF Young Investigator Award (CBE) and the David and Lucile Packard Fellowship (CBE)

REFERENCES

1. S. Jin, T.H. Tiefel, M. McCormack, R.A. Fastnacht, R. Ramesh, and L.H. Chen, *Science* **264**, 413 (1994); S. Jin, T.H. Tiefel, M. McCormack, H.M. O'Bryan, L.H. Chen, R. Ramesh, and D. Schurig, Appl. Phys. Lett. **67**, 557 (1995).

2. J. Z. Sun, W. J. Gallagher, P. R. Duncombe, L. Krusin-Elbaum, R. A. Altman, A. Gupta, Y. Lu, G. Q. Gong, and G. Xiao, Appl. Phys. Lett. **69**, 3266 (1996).

3. S. Jin, M. McCormack, T. H. Tiefel, and R. Ramesh, J. Appl. Phys. **76**, 6929 (1994).

4. Y. Suzuki, H.Y. Hwang, S.-W. Cheong, and R.B. van Dover, Appl. Phys. Lett. **71**, 140 (1997).

5. J. O'Donnell, M.S. Rzchowski, J.N. Eckstein and I. Bozovic, Appl. Phys. Lett. **72**, 1775 (1998).

6. C. Kwon, M.C. Robson, K.-C. Kim, J.Y. Gu, S.E. Lofland, S.M. Bhagat, Z. Trajanovic, M. Rajeswari, K. Ghosh, R. Shreekala, R.L. Greene, R. Ramesh, J. Magn. Magn. Mater. **172**, 229 (1997).

7. A. Millis, T. Darling, and A. Migliori, J. Appl. Phys. **83**, 1588 (1998).

8. A. Millis, A. Goyal, M. Rajeswari, K. Ghosh, R. Shreekala, R. L. Greene, R. Ramesh, and T. Venkatesan (unpublished).

9. T.K. Nath, R.A. Rao, D. Lavric, C.B. Eom, L. Wu, and F. Tsui, Appl. Phys. Lett. **74**, 1615 (1999).

10. R.A. Rao, D. Lavric, T.K. Nath, C.B. Eom, L. Wu, and F. Tsui, Appl. Phys. Lett. **73**, 3294 (1998).

11. Yan Wu, Y. Suzuki, U. Rudiger, J. Yu, A.D. Kent, T.K. Nath, and C.B. Eom, Appl. Phys. Lett. **75**, 2295 (1999).

12. A. Millis, Nature **392**, 147 (1998).

13. A. B. Shick, Phys. Rev. B **60**, 6254 (1999).

14. F. Jona and G. Shirane in *Ferroelectric crystals* , (Dover publications, Inc., New Yorrk, 1993) p.126.

15. T.K. Nath, M.K. Lee, C.B. Eom, M. Smoak, and F. Tsui, (unpublished).

16. A.J. Millis, P.B. Littlewood, and B.I. Shraiman, Phys. Rev. Lett. **74**, 5144 (1995).

17. A.J. Millis, R. Muller, and B.I. Shraiman, Phys. Rev. **B54**, 5405 (1996).

LATTICE DISTORTIONS AND DOMAIN STRUCTURE
IN EPITAXIAL MANGANITE THIN FILMS

Y. SUZUKI, YAN WU
Dept. of Materials Science and Engineering, Cornell University, Ithaca, NY 14853

U. RÜDIGER, J. YU, A.D. KENT
Dept. of Physics, New York University, New York, NY 10003

T.K. NATH, C.B. EOM
Dept. of Mechanical Engineering and Materials Science, Duke University, Durham, NC 20008

ABSTRACT

Lattice distortions, be they in the form of chemical and hydrostatic pressure in bulk or lattice mismatch between film and substrate, have significant effects on the transport as well as the magnetic properties of colossal magnetoresistance (CMR) materials. We summarize here our results on tensilely and compressively strained $La_{0.7}Sr_{0.3}MnO_3$ (LSMO) thin films that indicate the important role of lattice distortions due to the lattice mismatch between the film and substrate. The strain due to lattice distortions can be used to tune the magnetic domain structure, magnetization, magnetic anisotropy and magnetotransport of LSMO thin films.

INTRODUCTION

The magnetism and magnetotransport of colossal magnetoresistance materials has been shown to be highly sensitive to lattice distortions both in bulk and in thin films. Magnetic domain structure may also lead to distinctive magnetotransport effects in thin films. Several groups have shown that properties such as Curie temperature, resistivity and magnetoresistance effect are extremely sensitive to chemical as well as hydrostatic pressure. Others have identified domain wall scattering as an additional source of magnetoresistance. [1-11]. Studies of bulk polycrystalline pellets, thin films of varying polycrystallinity and isolated grain boundaries have shown that the magnetoresistance is profoundly affected by transport across grain boundaries [2,12,13].

In epitaxial manganite films, the substrate imposes a strain on the film. Such strain is manifest in lattice distortions measured by diffraction techniques. These distortions, in turn, affect the magnetic anisotropy, magnetic domain formation and anisotropy of magnetoresistance. We have conducted a detailed investigation of the evolution of structure, magnetic anisotropy, magnetic domain configurations and magnetoresistance (MR) of smooth epitaxial $La_{0.7}Sr_{0.3}MnO_3$ (LSMO) thin films. Films grown on (100) and (110) $SrTiO_3$ substrates are tensilely strained while (100) oriented films grown on $LaAlO_3$ substrates are compressively strained and exhibit a strong perpendicular anisotropy. In films grown on $SrTiO_3$, the magnetic anisotropy and the anisotropic magnetoresistance are dominated by tensile strain effects. In thin films on $LaAlO_3$, the magnetic anisotropy is more than sufficient to overcome the film demagnetization factors and results in perpendicularly magnetized domains with fine scale ~ 200 nm domain subdivision, which we image directly at room temperature using magnetic force microscopy. The main MR effects can be understood in terms of bulk colossal MR and anisotropic MR. We also find evidence for a small but measurable domain wall (DW) contribution to the MR, which is significantly larger than that expected based purely on a double exchange picture.

EXPERIMENT SETUP

The epitaxial CMR thin films and heterostructures were deposited in a pulsed laser deposition (PLD) system with a KrF excimer laser (248nm) operating at 10 Hz with an energy density of ~3 J/cm^2. The films were deposited on (100) and (110) SrTiO$_3$ and (001)/(110) LaGaO$_3$ substrates as well as (100) LaAlO$_3$ substrates. In this paper, we will describe the orientation of the manganites in terms of the pseudocubic lattices parameters a'$_{bulk}$= b'$_{bulk}$= c'$_{bulk}$= 3.875Å. These lattice parameters are rotated 45 degrees from the rhombohedral, nearly cubic lattice parameters a$_{bulk}$= b$_{bulk}$= c$_{bulk}$≈ 5.48Å. The former two substrates place the overlying LSMO films under tensile strain while the latter places the overlying LSMO film under compressive strain (see table 1 below). The (110) orientation of the LSMO films is particularly useful as it gives rise to uniaxial anisotropy in the plane of the film. The LSMO films were deposited at 700°C in 320 mTorr atmosphere of 100% O$_2$. We grew CMR layers with thicknesses varying from 250~3000Å and found no degradation of crystal quality in this range. Structural characterization of these thin films was performed by normal and grazing incidence diffraction, Rutherford backscattering spectroscopy (RBS). Magnetization data were taken on a Lake Shore vibrating sample magnetometer (VSM) at fields of up to 1.4kOe at room temperature and in a SQUID magnetometer at fields up to 50kOe from 4.2K to 350K.

Substrate	Lattice parameter
SrTiO$_3$	a=b=c= 3.911Å
LaGaO$_3$	a=5.496Å, b=5.524Å, c=7.787Å
LaAlO$_3$	a=b=c= 3.79Å
La$_{0.7}$Sr$_{0.3}$MnO$_3$	a$_{bulk}$= b$_{bulk}$= c$_{bulk}$≈ 5.48Å (rhombohedral) a'$_{bulk}$= b'$_{bulk}$= c'$_{bulk}$= 3.875Å (pseudocubic)

Table 1. Lattice parameters of substrates and CMR thin film material.

RESULTS

STRUCTURE

We have performed x-ray diffraction and Rutherford backscattering spectroscopy in order to determine the epitaxy, stoichiometry and crystalline quality of the films. From normal incidence x-ray diffraction measurements, we observe that (100) LSMO grows on (100) STO and LAO while (110) LSMO grows on (110) STO. In-plane phi scans of the (110) and the (100) LSMO peaks respectively for the two orientation films indicate that there is in-plane alignment of the crystal axes. These oriented single crystalline films have excellent crystallinity with full width half maximum values of Δω=0.1°. RBS channeling of the LSMO films reveals excellent crystallinity with typical figures of merit of 11%. We have probed surface morphology using

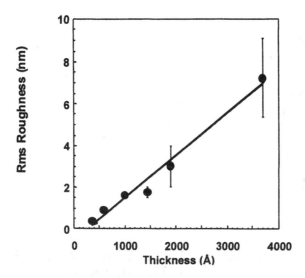

Figure 1. Rms roughness versus LSMO film thickness indicates a roughening trend for our PLD deposited films.

atomic force microscopy (AFM). We observe that thinner films are, on average, smoother with an rms roughness as low as 0.5nm for very thin films. A plot of the rms roughness versus thickness of LSMO films indicates that there is a roughening mechanism at work (figure 1). Possible explanations include large scale inhomogeneities which propagate upward with growth or an uphill Ehlrich- Schwoebel type growth mechanism. A typical AFM image is shown in figure 2.

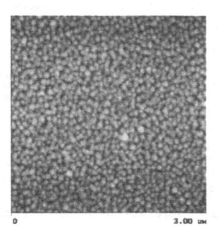

Figure 2. Atomic force microscopy image of 1100Å thick LSMO thin film with rms roughness of 0.97nm.

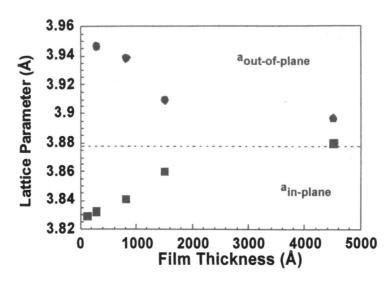

Figure 3. Relaxation of lattice parameters in and out of the plane of the film for compressively strained LSMO thin films on (100) LAO.

From table 1, we can see that the single crystal substrates described above place biaxial strain on the films. The biaxial compressive and tensile strains are directly measured via normal and grazing incidence x-ray diffraction. Grazing incidence diffraction (GID) probe the crystal structure to a depth of about 100Å below the surface. In LSMO films on LAO substrates, the LSMO are grown under compressive strain. As the films are grown thicker, the in-plane lattice parameters relax toward the bulk value of 3.875Å (figure 3). The strain in these films can be written down in terms of a bulk strain ($\varepsilon_B = \varepsilon_{xx} + \varepsilon_{yy} + \varepsilon_{zz}$), associated with a hydrostatic volume distortion, and a volume preserving Jahn-Teller strain ($\varepsilon_{JT} = \frac{1}{\sqrt{6}}\left(2\varepsilon_{zz} - \varepsilon_{xx} - \varepsilon_{yy}\right)$), associated with the lattice mismatch between film and substrate[14]. Since we have direct measurements of the lattice parameters both in and out of the plane, we can easily separate these two contributions. In agreement with the results of Rao et al. [15] and Millis et al. [14], we observe an increasing Jahn-Teller strain with decreasing thickness while the bulk strain contribution becomes more negative with decreasing thickness (table 2). The volume preserving Jahn-Teller strain is a dominant factor in the epitaxial thin LSMO films.

Thickness (Å)	$a_{\text{out-of-plane}}$ (Å)	$a_{\text{in-of-plane}}$ (Å)	$\varepsilon_{\text{bulk}}$ (%)	$\varepsilon_{\text{Jahn-Teller}}$ (%)	T_c
280	3.947	3.832 ± 0.005	-0.36	2.42	334
600	3.939	3.841 ± 0.003	-0.104	2.07	343
1150	3.910	3.860 ± 0.005	0.0129	1.05	360

Table 2. Calculation of bulk and Jahn-Teller strain for a series of LSMO/LAO thin films.

MAGNETIC PROPERTIES

The effect of these strains dominates the magnetic anisotropy in these epitaxial LSMO films. Magnetization measurements of LSMO/LAO films, taken with a SQUID magnetometer as well as a vibrating sample magnetometer, establish the Curie temperature T_c around 360 K for films of nominal thickness greater than 1000Å . Measurements of the magnetization as a function of temperature indicate that T_c varies from 334K in the thinnest films to 360K in the thickest films (see table 2). In these CMR materials, as the Mn-O-Mn bond is decreased, the carrier hopping probability decreases and in turn the magnetic ordering temperature T_c decreases [see for example 2]. These results are consistent with the fact that our thinnest compressively strained LSMO films exhibit the lowest T_c.

(110) LSMO/STO films exhibit an in-plane uniaxial anisotropy while (100) LSMO/STO films exhibit in-plane fourfold anisotropy. In these tensilely strained films, the in-plane direction is an easy direction while the normal direction is a hard one (figure 4(a)). To understand further the stress effects, we also grew LSMO films on (001)/(110)LGO substrates. Since the lattice mismatch of the LSMO with the LGO (1%) is smaller than that with STO (1.4%), we expect the stress effects to be smaller in the films on LGO. Indeed we observe anisotropy fields a factor of four smaller in films on LGO compared to those on STO. (100) LSMO/LAO thin films are compressively strained and give rise to a magnetically hard direction in the plane of the film (figure 4(b)). These results can be understood self-consistently using magnetoelastic theory[5].

We have also characterized the surface domain structure of these LSMO thin films. Since magnetic force microscopy (MFM) detects force gradients between the ferromagnetic sample and the magnetic tip, imaging in-plane domains is more difficult than imaging perpendicular domains. We have characterized the surface magnetic domain structure of LSMO/LAO thin films of varying thicknesses. In our case, the exchange thickness ($A/2\pi M^2$) dictates that the surface domain structure is a good measure of most of the film thickness.

Magnetic force microscopy (MFM) measurements reveal small scale "stripe" domains consistent with a perpendicular anisotropy. Fig. 5 shows images taken in zero field at room

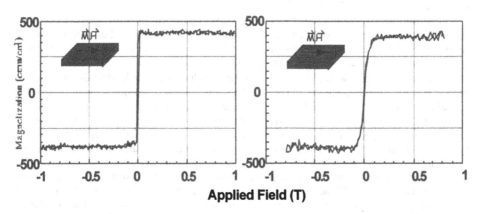

Figure 4. Magnetization loops in the plane of the film for (a) tensilely strained LSMO/STO films and (b) compressively strained LSMO/LAO thin films.

temperature after the films have been demagnetized with an in-plane magnetic field. Stripe domains form due to a competition between the exchange, magnetostatic and magnetic anisotropy energies. The magnetic anisotropy must be sufficient to overcome the magnetostatic energy of the stripe domain state, which is significantly less than the maximum magnetostatic energy of $2\pi M^2$ (where M = magnetization) for a perpendicular magnetized single domain film. In the weak stripe domain limit, $Q = K/2\pi M^2 < 1$ (where K is the magnetic anisotropy), relevant to our films (Q~0.2), this condition amounts to the film thickness being greater than approximately the domain wall width ($= \pi(A/K)^{1/2}$ ~300Å where A is the exchange stiffness) [16]. We observe a trend toward increasing domain size with film thickness (Fig. 5), which reflects the basic scaling expected for the magnetic interactions with film thickness as well as a reduction in the magnetic anisotropy with increasing film thickness, due to reduced average strain in thicker films [17].

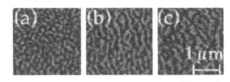

Figure 5. Magnetic force microscopy in zero applied field of films of thickness: (a) 800Å, (b) 1450Å and (c) 3750Å after a magnetic field had been applied in-plane (vertically in image). The domains are aligned in stripes along the direction of the demagnetization field. Domain size is seen to increase with increasing film thickness.

MAGNETOTRANSPORT

We have studied the magnetotransport of tensilely and compressively strained films and have found the magnetoresistance to be affected by the magnetic anisotropy which, is in turn, dominated by strain effects. In compressively strained films with perpendicular anisotropy, we expect a contribution to the magnetotransport from the intrinsic CMR material as well as microstructural defects and domain walls.

Anisotropy of magnetotransport is particularly prominent in (110) LSMO/STO films. We have patterned our (110) LSMO films with bars along the [001] and [1$\bar{1}$0] directions on the same substrate. The bars were 500μm wide, 2000~4000Å thick and 2mm long. In the absence of applied field, the resistance along the [001] direction is, to within the error of the experiment, $1/\sqrt{2}$ of the resistance along the [1$\bar{1}$0] direction. This results suggests that while the macroscopic current may be flowing along the [1$\bar{1}$0] direction, the microscopic current paths follow the crystal axis directions. We carried out four probe measurements along the two bars with fields applied in three different orientations: (i) perpendicular to the current flow and to the plane of the film (A), (ii) perpendicular to the current flow and in the plane of the film (B) and (iii) parallel to the current flow (C). In all cases the sample was magnetically saturated at fields of 3 Tesla before the MR measurements were performed.

Figure 6 shows resistivity versus magnetic field for the three different field configurations. In configuration A, the MR curve exhibits demagnetizing effects around 5.0 kOe as well as irreversibility at ±1.3 kOe. The irreversibility corresponds to domain wall motion that is observed at ±50 Oe in the magnetization loops. In other words, the conduction electrons "feel" a field of 50 Oe when the applied field is 1.3 kOe because of the field contribution from the moments. When the current is flowing along the easy [001] direction and the field is applied in configuration B, we observe a positive magnetoresistance at low fields which crosses over to negative magnetoresistance at higher fields. The cross over field is too high to be the shape effect of the transport bar. It is in fact the uniaxial magnetic anisotropy field of 1kOe (@300K) measured in the magnetization data. The evidence suggests that below the anisotropy field, the MR with current parallel to the moments is less than the MR with current perpendicular to the moments. This is in agreement with the conclusions drawn by O'Donnell et al. on (001) $La_{0.7}Ca_{0.3}MnO_3/STO$ films. They attribute the sublinear deviation from linearity of MR, with current parallel to the field, to anisotropic MR effects of transverse domains nucleating during the magnetization process. They observe negative MR relative to zero field at all times. In our films, we see a distinctly positive magnetoresistance at temperatures from 300K down to 10K. We believe that they are both manifestations of the same AMR in these doped manganites but in films with differing magnetic anisotropies- biaxial anisotropy in O'Donnell's samples versus uniaxial anisotropy in our samples.

In configuration C, the MR exhibits a linear dependence on field and an irreversibility corresponding to the coercive field of the LSMO. The irreversibility in the MR curves at ±50Oe can be explained in terms of a rapid reversal from moments antiparallel to parallel to the applied field (figure 6). At 300K the anisotropy field is about 1 kOe. However since the field is applied in an easy direction, we would not expect any features at 1000 Oe nor would we expect the appearance of transverse magnetic domains.

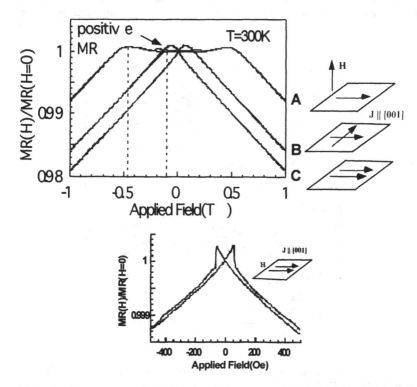

Figure 6. Magnetoresistance versus applied field in three different current-field configurations. The lower figure is a detailed view of MR measured in configuration C.

We have also performed magnetotransport measurements on compressively strained LSMO films. We measured the MR of films of varying thicknesses and found that the main MR effects can be understood in terms of bulk colossal MR and anisotropic MR. Figure 7 shows MR curves for a 800Å thick LSMO/LAO film from room temperature down to 4K with the magnetic field applied in the plane of the film and perpendicular to the current. The MR curves exhibit a low field hysteretic behavior and a sublinear deviation from the linear high field behavior. In the limit of extremely high magnetic field, we would expect the MR to approach a constant value. However our MR measurements up to 6 Teslas, as well as those of other groups, [2] indicate that we continue to suppress spin fluctuations even at these high magnetic field values. Below magnetic fields corresponding to saturation in the MH loop (figure 7), the MR deviates from linearity. At these fields, moments rotate out from the applied field direction and out of the film plane direction as observed in the MFM data. Due to demagnetization effects the internal field is reduced as the moments rotate, which by CMR alone would lead to a superlinear deviation of the MR. However, the resistivity anisotropy in our films is such that the resistivity is smallest when the magnetization is perpendicular to the film plane and for this reason the MR deviation is sublinear [18].

As the temperature is reduced below room temperature, we observe an increase in the coercive field that directly correlates with the hysteretic behavior in the MR measurements, thus confirming that the MR hysteresis is associated with domain wall motion. With decreasing temperature, the remnant magnetization increases significantly. This remnant magnetization indicates that the moments have a nonnegligible projection of magnetization in the plane of the

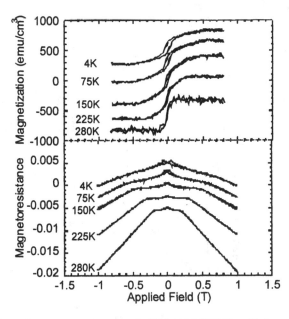

Figure 7. Magnetization and magnetotransport curves of a 800Å thick LSMO film with the magnetic field applied in the plane of the film and perpendicular to the current.

214

film. Therefore moments in the demagnetized state do not exhibit true perpendicular anisotropy but are most likely canted out of the film normal direction.

Now let us turn to MR measurements where the applied magnetic field is perpendicular to the current and the plane of the film. We observe a negative MR with a high field linear dependence on the applied field. We attribute this negative MR to the suppression of spin fluctuations at high fields (CMR). At lower fields, there is a sublinear deviation of the MR from the linear high field behavior, with the resistivity approximately field independent. The field at which the deviation occurs corresponds to the magnetic saturation field, (figure 8). As the field is lowered below the saturation field, reversed magnetic domains form and, while the applied field is varying, the internal magnetic field B is approximately constant and the CMR is suppressed.

As a function of temperature, we observe very little change in the MR and MH loops (figure 8). With decreasing temperature, the perpendicular anisotropy associated with the biaxial compressive strain increases faster than the magnetization, thus giving rise to remanence. At lower temperatures, we do observe an increase in the coercive field which is consistent with an increasing in-plane component of magnetization.

The difference between the in-plane transverse and perpendicular MR at H=0 reflects the effect of domain configurations on film resistivity. However we need to take into account the fact that in the "maze" configuration (figure 5) a portion of the current is shunted by the domains and need not cross domain walls, thus producing a smaller resistivity. With this consideration in mind, we have estimated the DW MR from the MR data to be 3×10^{-3} at room temperature [19].

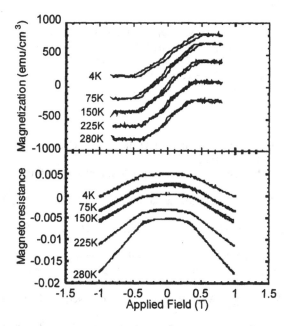

Figure 8. Magnetization and magnetotransport curves of a 800Å thick LSMO film with the magnetic field applied normal to the plane of the film.

A simple double exchange picture predicts a DW MR of 1.5×10^{-4}. As the temperature is lowered, the DW contribution to the MR remains the same order of magnitude. An increasing in-plane magnetization component renders the DW resistivity estimate a limiting case, thus emphasizing even more the discrepancy between experiment and a simple exchange model. A more detailed model of the domain wall MR, in double exchange ferromagnets, predicts a 1 to 2% effect which is consistent with our measured MR [20].

CONCLUSIONS

We have studied the role of lattice distortions in the form of lattice mismatch between the film and substrate on the magnetics and magnetoresistance of LSMO thin films. We observe the effect of tensile strain in LSMO/STO thin films in the magnetic anisotropy and low field magnetoresistance. The effect of the compressive strain in LSMO/LAO films is evident in the shape of the magnetization loops and the magnetotransport measurements at various temperatures. Although the domain wall contribution to the magnetoresistance is significantly larger than that predicted from a simple double exchange picture, the contribution is a small fraction of the measured magnetoresistance.

ACKNOWLEDGMENTS

We thank Jonathan Sun, Harold Hwang, Maura Weathers and Peter Revesz for useful discussions. This work is supported in part by a David and Lucile Packard Foundation Fellowship (Y.S.) and an NSF CAREER Award (Y.S.). Structural characterization was performed in the Cornell Center for Materials Research (NSF-MRSEC).

REFERENCES

1. S. Jin, T.H. Tiefel, M. McCormack, H.M. O'Bryan, L.H. Chen, R. Ramesh and D. Schurig, Appl. Phys. Lett. **67** 557 (1995).
2. H.Y. Hwang, T.T.M. Palstra, S.-W. Cheong and B. Batlogg, Phys. Rev. B **52** 15046 (1995).
3. M.R. Ibarra, P.A. Algarabel, C. Marquina, J. Basco and J. Garcia, Phys. Rev. Lett. **75** 3541 (1995).
4. J.J. Neumeier, M.F. Hundley, J.D. Thompson and R.H. Heffner, Phys. Rev. B **52** R7006 (1995).
5. J. O'Donnell, M.S. Rzchowski, J.N. Eckstein, I. Bosovic, Appl. Phys. Lett. **69** 1312 (1996). J. O'Donnell, M.S. Rzchowski, J.N. Eckstein and I. Bosovic, Appl. Phys. Lett. **72** 1775 (1998).
6. Y. Suzuki, H.Y. Hwang, S-W. Cheong and R.B. van Dover, Appl. Phys. Lett. **71** 140 (1997).
7. C. Kwon, K.-C. Kim, M.C. Robson, S.E. Lofland, S.M. Bhagat, T. Venkatesan, R. Ramesh and R.D. Gomez, J.Mag. Magn. Mater. **172** 229 (1997).
8. N-C. Yeh, R.P. Vasquez, D.A. Beam, C-C. Fu, J. Huynh and G. Beach, J. Phys.: Condens. Matter **9** 3713 (1997).
9. H.S. Wang and Qi Li, Appl. Phys. Lett. **73** 2360 (1998).

10. T.K. Nath, R.A.Rao, D. Lavric, C.B. Eom, L. Wu, and F. Tsui. Appl. Phys. Lett. **74** 1615 (1999).

11. K.A. Thomas, P.S.I.P.N De Silva, L.F. Cohen, A. Hossain, M. Rajeswari, T. Venkatesan, R. Hiskes, J.L. MacManus-Driscoll, J. Appl. Phys. **84** 3939 (1998).

12. A. Gupta, G.Q. Gong, G. Xiao, P.R. Duncombe, P. Lecoeur, P. Trouilloud, Y.Y. Wang, V.P. Dravid, J.Z. Sun, Phys. Rev. B **54** R15629 (1996).

13. N.D. Mathur, G. Burnell, S. P. Isaac, T. J. Jackson, B.-S. Teo, J. L. MacManus-Driscoll, L. F. Cohen, J. E. Evetts and M. G. Blamire, Nature **387** 226 (1997).

14. A.J. Millis, A. Goyal, M. Rajeswari, K. Ghosh, R. Shreekala, R.L. Greene, R. Ramesh and T. Venkatesan, (to be published).

15. R.A. Rao, D. Lavric, T.K. Nath, C.B. Eom, L. Wu, F. Tsui, Appl. Phys. Lett. **73** 3294 (1998).

16. see, for example, A. Hubert and R. Schaefer, "Magnetic Domains," Springer, New York 1998.

17. C. Kittel, Phys. Rev. **70** 965 (1946).

18. In magnetic crystals, the spin-orbit interaction, which is at the microscopic origin of this resistivity anisotropy or anisotropic MR (AMR) effect, leads to a resistivity which depends not only on the orientation of magnetization and current and but also on the relation of these vectors to the crystal structure. For example, resistivity anisotropy of this type has recently been reported for transition metal epitaxial thin films.

19. Yan Wu, Y. Suzuki, J.Yu, U. Rudiger, A.D. Kent, T.K. Nath, C.B. Eom, Appl. Phys. Lett. **75** 2295 (1999).

20. L. Brey, Cond. Matt. Preprint (Los Alamos) #9905209 (1999).

STRAIN-INDUCED MAGNETIC PROPERTIES OF
$Pr_{0.67}Sr_{0.33}MnO_3$ THIN FILMS

X. W. Wu and M. S. Rzchowski
Physics Department, University of Wisconsin-Madison, Madison, WI 53706
H. S. Wang and Qi Li
Physics Department, Pennsylvania State Universit, University Park, Pennsylvania 16802

ABSTRACT

We report the temperature dependence of the magnetic anisotropy in both compressive and tensile strained films of $Pr_{0.67}Sr_{0.33}MnO_3$ (PSMO). Compressive strain induced by growth on $LaAlO_3$ (LAO) substrates results in a spontaneous out-of-plane magnetization, while tensile strain (grown on $SrTiO_3$) results in in-plane magnetization. The coefficient of linear proportionality between the magnetic anisotropy energy and the tetragonal strain for both compressive and tensile strained PSMO films is larger than that found previously in strained $La_{0.67}Ca_{0.33}MnO_3$ films. From the data, we estimate a 20 unit cell magnetic domain wall width for PSMO / LAO. Scattering from such a narrow domain wall could produce a potentially significant contribution to the resistivity.

Introduction

The discovery of 'colossal' magnetoresistance [1,2] in perovskite manganites, and the subsequent exploration of their doping-dependent phase diagram, has renewed interest in these materials [3,4]. The successful application of thin film growth techniques developed for high temperature superconductors has permitted careful study of epitaxial manganite thin films. Since substrates such as $SrTiO_3$ (STO) and $LaAlO_3$ (LAO) have predominantly cubic unit cells, the substrate surface presents a square template for growth of the manganite film. This template does not have the same lattice constant as bulk manganite, so that epitaxial film growth is often characterized by some degree of tensile or compressive stress on the film. The elastic response of the film leads to tetragonal unit cell, with square symmetry in the film plane. This should be contrasted with the approximately cubic cell of the ferromagnetically doped bulk material [5,6].

One directly measurable consequence of the tetragonal deformation in the thin film is the emergence of a uniaxial magnetic anisotropy [7-14]. O'Donnell *et al.* [7] have previously studied stain-induced anisotropy in pseudomorphic $La_{0.67}Ca_{0.33}MnO_3$ (LCMO) films on [100] STO substrates ($a=3.90$Å). Bulk LCMO with 33% Ca concentration is orthorhombic due to small tilts and deformations of the MnO_6 octahedral, and has a pseudo-cubic lattice constant $a=3.868$ Å. A tensile strain of 0.8% is introduced by the lattice mismatch. The magnetic anisotropy in LCMO / STO system has been shown to be uniaxial to lowest order, with the easy plane being the film plane. The anisotropy energy is dominated by strain-induced anisotropy from the lattice mismatch between film and substrate: the intrinsic magnetocrystalline anisotropy of the bulk cubic system is negligible. This strong strain-related anisotropy suggested that an appropriate substrate could result in a spontaneous out-of-plane magnetization induced by compressive strain [7]. Such spontaneous out-of-plane magnetization has recently been reported [8,9].

Here we determine the coefficient of linear proportionally between the magnetic anisotropy energy and tetragonal strain in PSMO ultra-thin films. This coefficient represents the intrinsic magneto-elastic property of the materials, and is related to microscopic materials parameters such as the spin-orbit interaction. We find this coefficient is larger than that for strained LCMO film. We attribute this difference to the different A-site cation. Two consequences of this large anisotropy coefficient are a) very little in-plane compressive strain is required to produce a spontaneous out-of-plane magnetization, and b) the domain wall width predicted from the measured anisotropy is 8 nm, potentially small enough to influence the measured resistivity.

Experimental Results

In this paper, we report the study of both tensile (grown on STO [100]) and compressive strain (grown on LAO [100]) effects in PSMO films. Both LAO and STO substrates are approximately cubic with lattice constant a=3.79 Å and 3.90Å respectively. Bulk PSMO is orthorhombic with $a/\sqrt{2} = 3.879Å$, $b/\sqrt{2} = 3.866Å$, and $c/2 = 3.856Å$ [9]. We reference our strain analysis to a pseudo-cubic cell of 3.866Å. A fully strained PSMO / LAO film would then have a lattice compression ~2.0% in the film plane, with an elongated out-of-plane c-axis determined by the film elastic constants, while the PSMO/STO films would have a 0.8% tensile strain in-plane, with a compressed c-axis.

The PSMO films were grown epitaxially by pulsed laser deposition (PLD) [10]. The energy density and the repetition rate of the laser were 2 J / cm^2 and 5 Hz.. The films were grown at 750-800^{0}C in 0.75 mbar of O$_2$. After deposition, the films were cooled to room temperature in 1 atmosphere of O$_2$. Films of 10 nm thickness were used to minimize defect-induced strain relaxation. The c-axis lattice constants for 10nm film were determined by x-ray diffraction to be 3.92 Å ($\varepsilon_{zz} = -1.4\%$) for LAO substrates, and 3.81Å ($\varepsilon_{zz} = 1.4\%$) for STO substrates. For thicker films, the c-axis relaxes back toward the bulk value. The in-plane lattice constant of PSMO / STO was determined from the $\overline{1}$03 reflection to be 3.90Å ($\varepsilon_{xx} = -0.8\%$), which indicates that the thin film grown on STO is fully strained. For fully strained films, $\varepsilon_{zz}/\varepsilon_{xx} = -2\upsilon/(1-\upsilon)$, where υ is the poisson's ratio. For a volume-preserving distortion, $\varepsilon_{zz}/\varepsilon_{xx} = -2$, or $\upsilon = 1/2$. In both fully strained La$_{0.67}$Ca$_{0.33}$MnO$_3$ and La$_{0.8}$Ca$_{0.2}$MnO$_3$ films, the $\varepsilon_{zz}/\varepsilon_{xx}$ ratio is approximately -1.25 ($\upsilon = 0.38$)[7,11]. For strained La$_{0.67}$Sr$_{0.33}$MnO$_3$ (LSMO) film, $\varepsilon_{zz}/\varepsilon_{xx}$ has been reported to be -1.42 ($\upsilon = 0.42$)[12]. For the fully strained PSMO / STO film in our study, the $\varepsilon_{zz}/\varepsilon_{xx}$ ratio is -1.75 ($\upsilon = 0.47$), slightly larger than either LCMO or LSMO films. If the PSMO / LAO films were fully strained, the in-plane strain would be $\varepsilon_{xx}^{full} \approx 2.0\%$, and the out-of plane ε_{zz}^{full} could be estimated to be 3.5% from the $\varepsilon_{zz}/\varepsilon_{xx}$ ratio. The much smaller measured c-axis strain suggests that the film grown on LAO has relaxed even at a 10nm thickness, which is consistent with the observation of LCMO / LAO [11]. The in-plane lattice constant of PSMO / LAO film was difficult to determine from the $\overline{1}$03 peak due to the peak broadening. We have estimated the in-plane lattice constant to be 3.83Å ($\varepsilon_{xx} = 0.9\%$) based on our measured c-axis lattice constant and a thickness-dependence study of lattice constants of La$_{0.8}$Ca$_{0.2}$MnO$_3$ films grown on LAO substrate [11].

Magnetization measurements of 10nm thick PSMO films were made in a SQUID magnetometer. The diamagnetic signal from the substrate was minimized by mechanically thinning the substrate to approximately 0.3mm. For accurate background correction, the susceptibility of blank substrates was measured at different temperatures. Although the LAO substrate showed a slight temperature dependence at low temperature, for T>20K the specific susceptibility had a temperature-independent value of -1.95×10^{-7} emu/gm-Oe. The susceptibility of the STO substrate showed a temperature-independent value of -1×10^{-7} emu/gm-Oe at all temperatures.

Figure 1 shows the M vs H loops of strained PSMO films at 5 K with the applied field in [100] and [001] directions (these have not been corrected for demagnetization). For tensile strained PSMO/STO films (Fig. 1a), zero remanent magnetization in out-of-plane field direction indicates a magnetic hard axis along the c-axis, which is consistent with previous studies of tensile strained LCMO. A very different behavior is observed for the compressively strained PSMO films. Zero remanent magnetization in the (in-plane) [100] direction, as shown in Fig. 1b, indicates an out-of-plane, easy axis magnetic anisotropy. The [001] easy axis magnetization shows hysteresis and an almost ideal 100% remanence at 5 K, consistent with a spontaneous out-of-plane magnetization. This general behavior is observed at all temperatures less than the Curie temperature. This is consistent with predictions based on the strain dependence of magnetic anisotropy in the tensile strained systems, where a tensile strain resulted in an out-of-plane *hard* magnetic axis.

The saturation magnetization in our strained samples is significantly reduced from the bulk value ($M_S^{bulk} = 730 emu/cc$). A similar effect has been observed in the study of strained LCMO and SrRuO$_3$ films [7,8,13]. LCMO / STO films grown by PLD show about 45% reduction in magnetization due to 1.42% tetragonal strain $\varepsilon_{zz} - \varepsilon_{xx}$, and LCMO / LAO films show a 59% reduction due to 1.91% strain [8]. For both PSMO / LAO and PSMO / STO films in our measurements (also grown by PLD), we have a reduction of about 60% due to the tetragonal strain

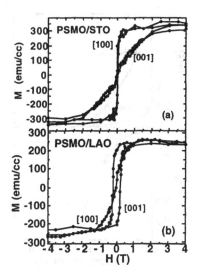

Fig.1. The magnetization loops of (a) PSMO / STO and (b) PSMO / LAO for magnetic field parallel to the film plane and perpendicular to the film plane.

of 2.2%. These results suggest an approximately linear relationship between magnetization reduction and strain. For off-axis sputtered $SrRuO_3$ / STO films [13], the relationship between magnetization reduction and the strain (about 1.2% tetragonal strain leads to ~30% magnetization reduction) is quite similar to the case of PLD-grown manganites. However, for $La_{0.67}Ca_{0.33}MnO_3$ films grown by MBE [7], a somewhat less reduction was observed (only 30% reduction for 1.8% strain). This could be due to better film quality in MBE growth.

Figure 2 shows the temperature dependence of the coercivity field (H_C) and saturation magnetization (M_s) respectively. The coercivity in both cases decreases with increasing temperature (shown in Fig.2(a)). However, the compressively strained PSMO shows a much larger H_C at 5K $(H_C=2.6kG)$ comparing with the tensile strained film $(H_C=0.38kG)$. The larger H_C could be due to a larger defect density because of film relaxation. $M_s(T)$ curves in Fig. 2(b) indicate a Curie temperature of $T_C{\sim}170$ K for both PSMO / LAO and PSMO / STO. $M_s(T)$ was determined from data just above the coercive field for the easy axis hysteresis loops.

Analysis

We quantitatively determine the anisotropy energy through an analysis of the hard axis magnetization loops, finding that it is larger for PSMO than that determined in the LCMO system [7]. In hard axis geometry, the applied field gradually rotates the spontaneous magnetization from the easy axis into the hard axis, until it saturates in the applied field direction. We can determine the anisotropy energy by fitting the magnetization curves of our samples along the magnetic hard axis from 5 K to150 K as follows [15].

In tetragonal symmetry, the total magnetic energy of a thin film in an external field $\vec{H}$ can be written as

$$E = K_1 m_z^2 + K_2 m_z^4 + K_3 m_x^2 m_y^2 - \vec{M} \bullet \vec{H} + 2\pi M^2 m_z^2 \qquad (1)$$

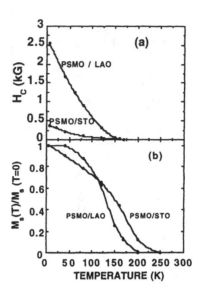

Fig. 2. The temperature dependence of (a) the coercive field and (b) the saturation magnetization $M_S(T)$.

where the $m_{x,y,z}$ are the direction cosines of the magnetization vector, and K_1 and K_2 are the second and fourth order uniaxial anisotropy constants. K_3 represents in-plane biaxial anisotropy. The last term represents the thin-film demagnetization energy. For the in-plane magnetization loops (hard axis) for PSMO / LAO, $\vec{M}$ is rotated by the applied field from the z-axis to the x-axis without substantial change in magnitude. The term $K_3 m_x^2 m_y^2$ is always zero in this case, so that the in-plane anisotropy constant K_3 is not determined. The zero-field energy difference ΔE between the z axis and [100] direction can be found from Eq. (1) to be $E_{001} - E_{100} \equiv \Delta E = K_1 + 2\pi M^2 + K_2 = \Delta E_{anis} + 2\pi M^2$, where $\Delta E_{anis} = K_1 + K_2$ is the anisotropy energy. Minimizing Eq. (1) for an in-plane field gives

$$H = \left(\frac{-2[K_1 + 2\pi M^2] - 4K_2}{M} \right) m_x + \frac{4K_2}{M} m_x^3 \qquad (2)$$

Applying a similar analysis to PSMO / STO films, we find

$$H = \left(\frac{2[K_1 + 2\pi M^2]}{M} \right) m_x + \frac{4K_2}{M} m_x^3 \qquad (3)$$

K_1 and K_2 can be obtained by fitting the hysteresis loop using these two expressions.

The temperature dependence of the anisotropy energy ΔE_{anis} using this fit procedure is shown in Fig. 3. We also calculated the anisotropy energy by direct integration of the in-plane M vs H curves [16]; the two methods agree within the error evident in Fig. 3. The diverging susceptibility as T approaches T_C (~170K), restricts us to T<170 K to ensure that the magnetization magnitude is independent of field in the fit to Eq. (2) and (3). At T=5K, the anisotropy energy ΔE_{anis} in either case ($\Delta E = 2.5 \times 10^6 erg / cc$) is larger than that of LCMO films.

We evaluate the role of strain in PSMO films by determining the value of the coupling constants that relate the magnetic anisotropy to tetragonal strain. The total internal energy of a magnetic crystal (ignoring demagnetization contributions) includes the magnetocrystalline anisotropy energy E_{mca} and elastic energy $E_{elastic}$. As argued previously [7], the mismatch stresses

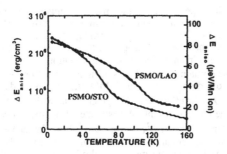

Fig. 3. The temperature dependence of the anisotropy energy $\Delta E_{aniso} = K_1 + K_2$ for PSMO / STO and PSMO / LAO films.

encountered in growth of manganites on LAO and STO are sufficiently large that the equilibrium atomic positions (strains) are determined predominantly from the elastic response of the film (i.e. by minimizing $E_{elastic}$), with only a small contribution from magnetoelastic effects (i.e. E_{mca}). This means that lattice constants determined from x-ray diffraction can be used to evaluate the strain dependence of the energy associated with rotating the magnetization (magnetic anisotropy). In our case of biaxial ($\varepsilon_{xx} = \varepsilon_{yy}$) in-plane strain, shear strains are induced only by magnetoelastic effects, and hence the terms related to shear strain have been ignored in the following discussion.

From general symmetry arguments, the magnetocrystalline anisotropy energy associated with deformations from cubic symmetry includes the energy associated with volume change of the unit cell E_{vol}, and the energy associated with the tetragonal distortion E_{tetra}, where $E_{vol} = F(\hat{m})(\varepsilon_{xx} + \varepsilon_{yy} + \varepsilon_{zz})$, and $E_{tetra} = G(\hat{m})(\varepsilon_{zz} - \varepsilon_{xx})$. Here $\hat{m}$ is a unit vector in the direction of the magnetization. Since $F(\hat{m})$ is the same for the cubic crystal axes (in particular $F(\hat{x}) = F(\hat{z})$), the E_{vol} term does not contribute to $\Delta E = E_{100} - E_{001}$. The anisotropy energy ΔE is then determined only by the tetragonal distortion, which gives

$$\Delta E = [G(\hat{m} = \hat{x}) - G(\hat{m} = \hat{z})](\varepsilon_{zz} - \varepsilon_{xx}) \equiv b'(\varepsilon_{zz} - \varepsilon_{xx})$$

Here $b' = b_1 + \frac{3}{7}b_3 + \cdots$. The b_i are magnetoelastic coupling constants (directly related to the magnetostriction constants) [17]. The coefficient b' depends on the magnetization [17], and to lowest order, it is proportional to M^2.

For strained PSMO films, the magneto-elastic coupling constants b' at 5K (determined from ΔE to be 1.13×10^8erg/cc) in both compressive (grown on LAO) and tensile (grown on STO) cases are the same due to the similar anisotropy energy and similar strain. In tensile strained LCMO films, the anisotropy energy at 5K was 1.5×10^6 erg/cm^3, while the strain $\varepsilon_{zz} - \varepsilon_{xx} = 1.8\%$[7]. Therefore, the value of b' can be obtained as 8.3×10^7erg/cm^3, which is smaller than that for PSMO films. As mentioned before, the coefficient b' increases as the magnetization increases. At T=5K, the magnetization is about 500emu/cc for LCMO / STO films [7], but only about 300emu/cc for strained PSMO films. If we compare the coefficient b' at the same magnetization in both materials, the difference between PSMO and LCMO would be even larger.

Discussion

We have measured the magnetic anisotropy constant for PSMO films (compressively strained and tensile strained) by epitaxial growth on a LAO or STO substrate. A comparison of the magnetoelastic coefficient relating magnetic anisotropy to strain with that obtained in the LCMO / STO system indicates a material-related magnitude difference due to the different A-site cations.

Our measured strain dependence of the magnetic anisotropy shows that only a small amount of compressive film strain will produce a spontaneous out-of-plane magnetization. In thin films with a magnetization of 300 emu/cc, the demagnetization energy is in the order of 5×10^5 erg/cc, 4 times smaller than the anisotropy energy, indicating that a compressive strain as small as $\varepsilon_{zz} - \varepsilon_{xx} = 0.5\%$ results in an out-of-plane magnetization.

We also estimate the domain wall thickness using the anisotropy energy. Due to the strong anisotropy, we expect narrow domain walls in PSMO films. The width of a 180° Bloch wall is $\delta = \pi (JS^2 / Ka)^{1/2}$, where J is the effective exchange constant, S the total spin at each site, K the anisotropy energy, and a the lattice constant. We estimate J from the spin-wave dispersion relation measured by neutron scattering $\hbar\omega \sim Dq^2$, where $D = 2JSa^2$ for nearest-neighbor Heisenberg interactions. Using the low-temperature [18] $D \sim 165$ meV-Å^2 for bulk PSMO, $S \sim 2$, and $a \sim$ 0.4 nm, we estimate $J \sim 2.5$ meV. Since our measured film T_C is less than half that of the bulk value, we may anticipate an exchange constant on the order of 1.2 meV. Using our measured low-temperature $K \sim 1.56$ meV/nm^3, we find a domain-wall thickness of ~ 8 nm, or about 20 unit cells. For a 180° Bloch wall, this results in an $\sim 9°$ angle between neighboring domain wall spins. This would reduce the "bandwidth" in the domain wall region (via the double-exchange mechanism), and potentially introduce a significant domain-wall scattering contribution to the measured resistance [10].

We would like to thank R. Matyi and M. Winokur for use of x-ray diffractometers and for x-ray measurement advice. We also thank J. O'Donnell and R. Joynt for useful conversations. This work at University of Wisconsin was supported by the National Science Foundation MRSEC program award No. 9632357, and at Penn State by NSF DMR-987626, DMR-9972973 and Petroleum Research Fund.

References

1. R. von Helmolt, J. Wecher, B. Holzapfel, L. Schultz, and K. Samwer, Phys. Rev. Lett. **71**, 2331 (1993).
2. S. Jin, T. H. Tiefel, M. Mccormack, R. A. Fastnacht, R. Ramech and L. H. Chen, Science **264**, 413 (1994).
3. K. Steenbech, T. Eich, K. Kirsch, K. O'Donnell, and E. Steinberb, Appl. Phys. Lett. **71**, 968 (1997).
4. J. Z. Sun, W. J. Gallagher, P. R. Duncombe, L. Krusin-Elbaum, R. A. Altman, A. Gupta, Yu Lu, G. Q. Gong, and Gang, Xiao, Appl. Phys. Lett. **69**, 3266 (1996).
5. T. M. Perekalina, I. E. Lipinski, V. A. Timofeeva, and S. A. Cherkezyan, Sov. Phys. Solid State **32**, 1827 (1991).
6. C. W. Searle and S. T. Wang, Can. J. Phys. **47**, 2703 (1969).
7. J. O'Donnell, M. S. Rzchowski, J. N. Eckstein and I. Bozovic, Appl. Phys. Lett. **72**, 1175 (1998).
8. T. K. Nath, R. A. Rao, D. Lavric, C. B. Eom, L. Wu and F. Tsui, Appl. Phys. Lett. **74**, 1615 (1999).
9. H. S. Wang, Q. Li, K. Liu and C. L. Chien, Appl. Phys. Lett. **74**, 2212 (1999).
10. H. S. Wang and Q. Li, Appl. Phys. Lett. **73**, 2360 (1998).
11. R. A. Rao, D. Lavric, T. K. Nath, C. B. Eom, L. Wu and F. Tsui, J. Appl. Phys. **85**, 4794 (1999).
12. J. Z. Sun, D. W. Abraham, R. A. Rao and C. B. Eom, Appl. Phys. Lett. **74**, 3017 (1999).
13. Q. Gan, R. A. Rao, C. B. Eom, J. L. Garrent and M. Lee, Appl. Phys. Lett. **72**, 978 (1998).
14. Y.Suzuki, H. Y. Hwang, S-W Cheong, R. B. van Dover, A. Asamitsu and Y. Tokura, J. of Appl. Phys. **83**, 7064 (1998).
15. L. D. Landau, E. M. Lifshitz, and L. P. Pitaevskii, *Electrodynamic of Continuous Media*, 2nd ed. (Butterworth Heinemann, Oxford, 1995).
16. B. D. Cullity, *Introduction to Magnetic Materials* (Addison-Wesley, Reading, MA, 1972).
17. A. E. Clark, in Ferromagnetic Materials, edited by E. P. Wohlfarth (North-Holland, Amsterdam 1980), Vol. 1.
18. J.A. Fernandez-Baca, P. Dai, H.Y. Hwang, C. Kloc, and S.-W. Cheong, Phys. Rev. Lett. **80**, 4012 (1998).

MAGNETIC MEASUREMENTS ON STRESSED AND STRESS RELIEVED La$_{0.66}$Ca$_{0.33}$MnO$_3$ THIN FILMS

H.-U. Habermeier, R. B. Praus, G. M. Gross, and F. S. Razavi
Max-Planck-Institut - FKF, Heisenbergstr.1 D-70569, Stuttgart, Germany.

ABSTRACT

In general, epitaxially grown thin films are biaxially stressed if the lattice mismatch of substrate and film is below 2%. We have used epitaxial strain as an extrinsic source of tensile stress exerted to La$_{0.66}$Ca$_{0.33}$MnO$_3$ thin films deposited on SrTiO$_3$ single crystal substrates. The average stress in the film is a function of film thickness and post deposition annealing. Thickness variation and annealing procedures have been used to explore the stress dependence of the characterisics of magnetization curves of the stressed films. It could be shown that the inflection point of the magnetization curve which determines the transition from Blochwall movements to rotations of the magnetization vector and the Rayleigh constant of the virgin magnetization curve are correlated with the stress of the films in analogy to results obtained for plastically deformed single crystals of conventional ferromagnetic metals.

INTRODUCTION

The discovery of high temperature superconductivity in hole doped copper oxide compounds has generated renewed interest in doped Mott insulators with strong electron correlation. Consequently, the electronic properties of doped rare earth manganites in perovskite structure of the type RE$_{1-x}$A$_x$MnO$_3$ [RE = La, Pr, Nd, A = Ca, Sr] have been revisited [1,2,3]. The magnetic field driven shift of a first-order metal-insulator transition in a temperature window around the ferromagnetic ordering of the Mn spins causes a colossal negative magnetoresistance (CMR),which can exceed 10^5 [4]. From a fundamental point of view the underlying physical processes leading to the metal-insulator transition and the magnetic ordering are still not fully understood. The exploration of methods in order to tailor properties such as tuning the temperature window for the magnetic ordering and the enhancement of the magnetic field sensitivity for low magnetic fields remains a challenge. Currently, doped LaMnO$_3$ is being re-examined as a possible next generation magnetoresistance sensor material [5].

The CMR materials have basically a perovskite-type crystal structure which can be rhombohedrically or orthorhombically distorted. They show a rich variety of magnetic ordering which is influenced by a cooperative tilting of the MnO$_6$ octahedra due to the ion-size mismatch of the A-site cations and the Jahn-Teller distortion as well. The LaCaMnO$_3$ system, e.g. , shows ferromagnetic ordering in the Mn-O layers of the a-b planes and a doping dependent antiferromagnetic ($x < 0.2$ and $x > 0.5$) or ferromagnetic ($0.2 < x < 0.5$) ordering along the c-axis. The first-order metal-insulator phase transition associated with the magnetic ordering can be shifted in a wide temperature range by doping. The valence state conversion of the Mn ions from Mn^{3+} to Mn^{4+} by a partial substitution of the trivalent RE^{3+} by the divalent A^{2+} leads to a mixed valence system as already proposed by Jonker and van Santen [6]. The striking correlation between ferromagnetism and metallic conductivity is due to the coupled dynamics of charge and spin essentially mediated by the double exchange interaction arising from strong on-site coupling [Hund´s rule] between itinerant e$_g$ and localized t$_{2g}$ electrons and thus expected to be largely influenced by the Mn-O-Mn bond length and bond angle [7]. Superimposed to this mechanism, more recently, models are discussed involving (i) polaron effects arising from a strong electron-phonon coupling based on the Jahn-Teller splitting of the degenerate energy levels in the Mn^{3+} ion ground state [8], (ii) antiferromagnetic superexchange interaction between local t_{2g} spins [9] and (iii) correlations of orbitals and charges [10]. It has been found empirically, that the change of resistivity upon the magnetic transition and consequently the magnetoresistance are most enhanced in a compositional region close to the charge/orbital ordering accompanied by collective Jahn Teller lattice distortions [11].

The properties of the CMR materials can be ascribed to a complex interplay of spin-, charge-, and orbital ordering superimposed to a strong electron lattice coupling via the Jahn-Teller distortion of the oxygen octahedra surrounding the Mn^{3+} ions. This suggests a strong correlation of the physical properties of the CMR materials with extrinsic parameters affecting the Mn-O-Mn bond length and bond angle. In epitaxially grown single crystal type thin films this effect has been explored systematically by modifying the biaxial strain of the film and measuring its influence on the transport properties and the magnetic ordering temperature [12-14]. Furthermore, superimposing hydrostatic pressure to biaxially strained thin films demonstrates a reversal of the effect of biaxial strain [15]. So far, however, little is known of the effect of extrinsic strain on the magnetic properties such as the characteristics of the magnetization curve. In this paper we explore the effect of strain on the magnetic properties for LaCaMnO thin films the first time. Extrinsic strain is generated by the lattice mismatch between substrate and film. In the case of $La_{0.66}Ca_{0.33}MnO_3$ films (a= 0.386 nm) deposited on SrTiO3 substrates (a = 0.3905) a biaxial tensile stress is exerted on the films whose magnitude changes systematically with film thickness [16]. Annealing of the films reduces the homogeneous epitaxial stress by the formation of a variety of lattice defects such as misfit dislocations and stacking faults with intrinsic local stress fields in their vicinity. A systematic study of the thin film properties before and after an annealing process gives more insight in the correlation of microstructure and properties of $La_{0.66}Ca_{0.33}MnO_3$ thin films.

EXPERIMENTAL

The $La_{0.66}Ca_{0.33}MnO_3$ thin films were prepared by the standard on-axis pulsed laser deposition method on 4.5 x 4.5 mm^2 SrTiO$_3$ (001) oriented single crystal substrates at a deposition temperature of 800^{0}C and oxygen pressure of 40 Pa. Immediately after the deposition the films were kept at 780^{0}C for 1 h in 10^5 Pa of flowing oxygen. The set-up allowed the deposition of four specimens in one run, thus ensuring identical preparation conditions for each set of samples. One sample was used for measurements of the magnetization curve with a Quantum Design SQUID magnetometer,subjected to an 1 hour ex-situ post annealing and measured again. Annealed and non-annealed samples were patterned by conventional lithographic techniques for four point probe resistivity measurements.

EXPERIMENTAL RESULTS AND DISCUSSION

In Fig. 1 the data for the remanent magnetization, M_R, and the magnetization at 0.1T,$M_{0.1T}$, extracted from the hysteresis loops measured at 5 K are represented. The magnetic field was applied in the film plane along the [100] direction. The measured magnetic moment is expressed in Bohr magnetons, μ_B, normalized to the volume of the unit cell. There is a systematic increase of the magnetization with film thickness in the range 40nm < t < 200nm followed by quasi-constant value of 3 μ_B to 3.1 μ_B .These values are very close to the bulk value of the saturation magnetization in ceramics and single crystals of 3.3 μ_B indicating that a magnetic field of 0.1T is sufficient to achieve single domain samples. Thermal annealing of the films results in a slight increase of $M_{0.1T}$ below 10%. In as grown films we found a systematic enlargement of the rema-

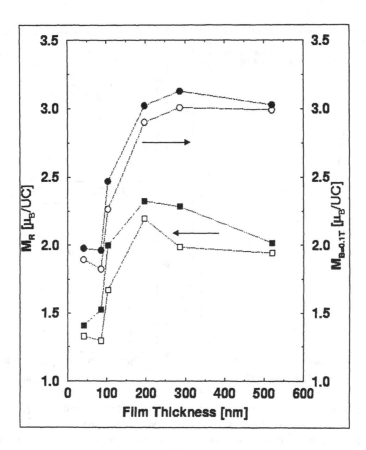

Fig. 1 Remanent magnetization (squares) and magnetization at 0.1 T (circles) La$_{2/3}$Ca$_{1/3}$MnO$_3$ thin films of different thickness before (open symbols) and after annealing (full symbols).

nent magnetization with film thickness which increase up to 20% after the annealing process. Furthermore, looking at the magnetic field B$_{rot}$ where the magnetization process turns from dominant Blochwall movement to magnetization vector rotation rather high values of 42.5 mT have been found for the thinnest films. We define B$_{rot}$ as the inflection point of the magnetization curve and determine B$_{rot}$ as the maximum in a ∂M/∂H vs. H curve. In Fig. 2 the data for B$_{rot}$ are given as a function of film thickness for specimens before and after annealing. We observe a drastic reduction of B$_{rot}$ with increasing film thickness and a decrease of B$_{rot}$ after annealing. It is evident that the shape of the curve Fig. 2 is the inverse of the plot M$_{0.1T}$, suggesting a correlation of these quantities. Qualitatively, the trends for the thickness dependence of B$_{rot}$ and B$_{0.1T}$ is similar to data reported for single crystals of conventional ferromagnetic materials with increasing plastic deformation [17] suggesting that internal stress is the dominant mechanism for generating these dependencies. A careful study of the low-field region of the virgin magnetization curve further supports this assumption. In conventional ferromagnetic systems the low-field region of the virgin magnetization curve can be described by the

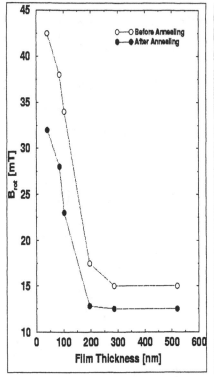

Fig.2. Transition filed B_{rot} from Blochwall movement to rotation of the magnetization vector as a function of film thickness and annealing.

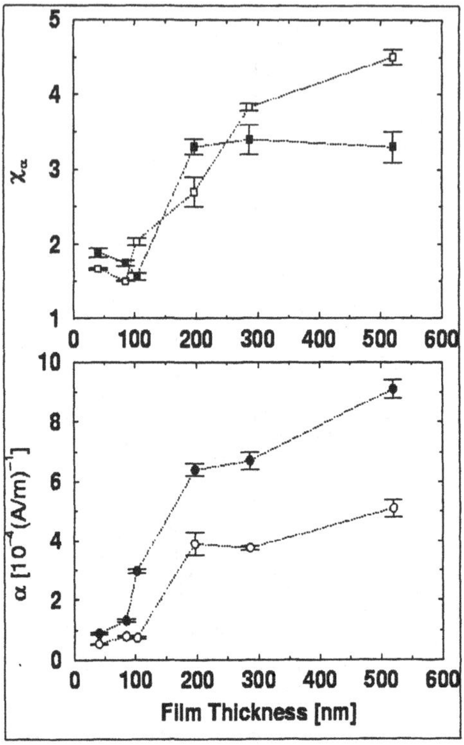

Fig.3. Initial susceptibility , χ_a, and Rayleigh constant α taken from the virgin magnetization curve as a function of film thickness and annealing.

Rayleigh law

$$M = \chi_a H + \alpha H^2 \qquad (1)$$

with χ_a being the initial susceptibility and α the Rayleigh constant. The term $\chi_a H$ describes the reversible part of the magnetization process and αH^2 the irreversible component. Both contributions show a clear increase with film thickness as shown in Fig. 3. Whereas the annealing procedure causes no clear trends in χ_a for films with $t < 200$nm, there is a pronounced increase of α after annealing.

Structural analysis of the films show a strong thickness dependence of the out-of-plane lattice parameter as a function of film thickness as demonstrated in Fig. 4. Thin films are under tensile stress in the film plane due to the larger lattice parameter of the substrate compared to the bulk material, consequently, - assuming preservation of the unit cell volume - the out-of-plane lattice parameter is reduced for films under tensile stress. With increasing film thickness this epitaxial stress is relieved by the introduction of lattice defects as demonstrated in TEM analysis [18]. Annealing of as grown films causes a partial reduction of the interfacial stress by generating

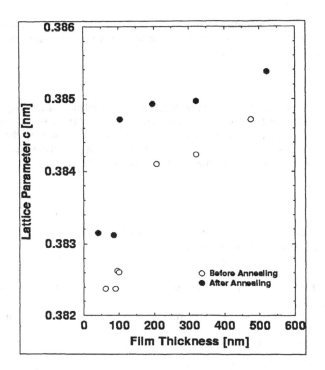

Fig 4 Out of plane lattice parameter of $La_{2/3}Ca_{1/3}MnO_3$ films before and after annealing.

dislocations, stacking faults and twin boundaries.

The experimental data presented in this paper can qualitatively be interpreted as the influence of epitaxial stress generated at the interface of substrate and film on the magnetic properties of the films. A theoretical description of the findings can be done within the frame of the micromagnetic equations describing the magnetization process of ferromagnetic materials by solving the variational problem for the total Gibbs free energy Φ_T for cubic crystals :

$$\Phi_T = \Phi_{EX} + \Phi_{CRYST} + \Phi_H + \Phi_{STRAY} + \Phi_{MAGNEL} \qquad (2)$$

with

$$\Phi_{MAGNEL} = \int [3/2 \, \lambda_{100} \, \Sigma \, \sigma_{ii} \gamma_i^2 + 3/2 \, \lambda_{1\varphi} \Sigma \, \sigma_{ij} \gamma_i \gamma_j \,] \, dV \quad (3)$$

where the different contributions to Φ_T correspond to the exchange energy, crystal energy, magnetostatic energy, stray field energy and the magnetoelastic coupling energy, respectively. The γ_i are the directional cosines, λ_{ij} corresponds to the magnetostrictive constants and σ to the stress. Equ. (2) and Equ. (3) set the frame for the physical mechanisms describing the interaction of stress fields (external as well as internal) with the strain fields surrounding Bloch walls. The results given in Fig. 4 show that the average homogeneous stress in the film decreases with increasing film thickness. The thickness dependence of $B_{0.1T}$, B_{rot} and the Rayleigh constant α

reflects average stress of the films as a function of thickness. Furthermore, comparing the data of as grown and annealed films, the strong correlation of the magnetic properties with external stress is confirmed.

SUMMARY

In our experiments we have demonstrated a close correlation of the magnetic properties of $La_{0.66}Ca_{0.33}MnO_3$ thin films with substrate-induced biaxial epitaxial stress. Qualitatively, the experimental data scale with the average stress in the film as shown by the thickness and annealing dependence of the magnetic properties. For a quantitative analysis the evaluation of the micromagnetic equations is required, this will be a task for the future.

REFERENCES

1. G. H. Jonker and J. H. Van Santen, Physica **XVI**, p. 337 (1950)
2. J. Vogler, Physica **XX**, p. 49 (1954)
3. R. M. Kusters, J. Singleton, D. A. Keen, R. McGreevy, and W. Hayes, Physica **B 155**, p. 362 (1989)
4. S. Jin, H. M. O' Brian, T. H. Tiefel, M. McCormack, and W. W. Rhodes, Appl.   Phys. Lett. **66**, p. 382 (1995)
5. Y. Tomoika, A. Asamitsu, Y. Moritomo, H. Kuwahara, and Y. Tokura, Phys. Rev. **B 53** p. R 1,689 (1996)
6. H. N. Bertram, IEEE Trans. Magn. **31**, p. 2,573 (1995)
7. C. Zener, Phys. Rev. **82**, p. 403 (1951)
8. A. J. Millis, P. B. Littlewood, and B. I. Shraiman, Phys. Rev. Lett. **74**, p.5,144 (1995)
9. G. H. Jonker, Physica **XX**   (1954)
10. J. Fontcuberta, A. Sefar-Martinez, S. Pinol, J. L. Garcia-Munoz, and X. Obradors, Phys. Rev. Lett. **76**, p. 1,122 (1996)
11. S. Jin, T. H. Tiefel, M. McCormack, R. A. Fastnacht, R. Ramesh, and L. H. Chen, Science **264**, p. 413 (1994)
12. C. W. Kwon, M. C. Robson, K.-C. Kim, J. Y. Gu, S. E. Lofland, S. M. Bhagat, Z. Trajanovic, M. Rajewari,. T. Venkatesan, A. R. Kratz, R. D. Gomez, and R. Ramesh, J. Magn. Mag. Mat. **172** p229 (1997)
13. A. J. Millis, T. Darling, and A. Migliori, J. Appl. Phys. **83** p. 1,588 (1988)
14. R. Praus, B. Leibold, G.M. Gross and H.-U. Habermeier, Appl. Surf. Science **138-139** , p. 40 (1999)
15. T. Roch, F. S. Razavi, B. Leibold, R. Praus, and H.-U. Habermeier, Appl. Phys. A **67**, p.1 (1998)
16. G. M. Gross, R. Praus, B. Leibold, and H.-U. Habermeier, Appl. Surf. Science **138-139** p. 117 (1999)
17. H.-U. Habermeier and H. Kronmüller, Angew. Phys. **30** , p. 13 (1970)
18. O.I. Lebedev, G. van Tendeloo, S. Amelinckx, B. Leibold ans H.-U. Habermeier, Phys. Rev. B. **58** , p. 8,065 (1998)

THE EFFECTS OF SUBSTRATE-INDUCED STRAINS ON THE CHARGE-ORDERING TRANSITION IN $Nd_{0.5}Sr_{0.5}MnO_3$ THIN FILMS

W. PRELLIER [a,b,*], AMLAN BISWAS [a], M. RAJESWARI [a], T. VENKATESAN [a] AND R.L. GREENE [a]

[a] Center for Superconductivity Research and Department of Physics, University of Maryland, College Park, MD 20472-4111, USA.

[b] Current address: Laboratoire CRISMAT-ISMRA, CNRS UMR 6508, 6 Bd. du Maréchal Juin, 14050 Caen, FRANCE.

[*] Author to whom correspondence should be addressed : prellier@ismra.fr.

ABSTRACT

We report the growth and characterization of $Nd_{0.5}Sr_{0.5}MnO_3$ thin films deposited by the Pulsed Laser Deposition (PLD) technique on [100]-oriented $LaAlO_3$ substrates. X-ray diffraction (XRD) studies show that the films are [101]-oriented, with a strained and quasi-relaxed component, the latter increasing with film thickness. A post-annealing under oxygen leads to a quasi-relaxed film with a metallic behavior. We also observe that transport properties are strongly dependent on the thickness of the films. Variable temperature XRD down to 100 K suggests that this is caused by substrate-induced strain on the films.

INTRODUCTION

Perovskite manganites of the type $RE_{1-x}A_xMnO_3$ (RE=rare earth, A=alkaline earth) have received a renewed interest due to their colossal magnetoresistance (CMR) properties. They also exhibit a rich phase diagram as a function of the doping concentration x and the temperature [1-4]. For certain values of x and the average A-site cation radius ($<r_A>$), the metallic state below a certain temperature, becomes unstable and the material goes to an insulating state. This is due to the real space ordering of the Mn^{3+} and Mn^{4+} ions in different sublattices [5-7]. Such a charge ordering (CO) transition (T_{CO}) is associated with large lattice distortions. This phenomenon has been observed in $Nd_{0.5}Sr_{0.5}MnO_3$ single crystals [9] where $r_A=1.236$ Å [10]. However, most of the work published up to now has been on single crystals or ceramic samples [9,11,12]. One study on $Nd_{0.5}Sr_{0.5}MnO_3$ thin films has been reported [13] in which no CO behavior was observed. Epitaxial thin films have properties similar to those of single crystals and are also important for potential device applications [14]. For this reason, we have investigated the growth of thin films of $Nd_{0.5}Sr_{0.5}MnO_3$ and the effect of a post-annealing under oxygen. Their structural and physical properties have been analyzed. The transports properties of the as-grown films are significantly different from those observed in single crystals of $Nd_{0.5}Sr_{0.5}MnO_3$. We propose a model to explain these differences.

EXPERIMENTAL

Thin films of $Nd_{0.5}Sr_{0.5}MnO_3$ were grown using the Pulsed Laser Deposition technique. The target, made by the classical solid state method, had the nominal composition of $Nd_{0.5}Sr_{0.5}MnO_3$. The substrates were [100] $LaAlO_3$, which has a pseudocubic crystallographic structure with a=3.79 Å. The laser energy density on the target was about 1.5 J/cm², and the deposition rate was 10 Hz. The LaAlO₃ substrate was kept at a constant temperature of 820 °C during the deposition which was carried out at a pressure of 400 mTorr of flowing oxygen. After deposition, the samples were slowly cooled to room temperature at a pressure of 400 Torr of oxygen. Further details of the target preparation and the deposition procedure are given elsewhere [5]. The structural study was done, at room temperature, by X-ray diffraction (XRD) using a Rigaku diffractometer. Low temperature XRD experiments were performed with a Siemens Kristalloflex X-ray diffractometer. DC resistivity was measured by a four-probe method and magnetization was measured using a Quantum Design MPMS SQUID magnetometer. The composition analysis, checked by Rutherford Backscattering Spectroscopy (RBS), indicate a stochiometric composition within error limits.

RESULTS AND DISCUSSION

Fig.1a shows the room temperature Θ-2Θ scan of the films for different thicknesses (in the region 45°-50°). A peak is observed at 46.5° for all thicknesses which corresponds to an out-of-plane parameter of 1.955 Å. Another diffraction peak appears gradually for thicknesses above 1000 Å. This latter peak corresponds to a lattice parameter of 1.92 Å for a 2000 Å film.

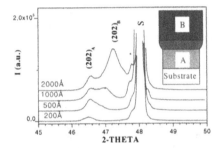

Fig.1a : Room temperature XRD for as-grown films. Peaks labeled S and * are due to the substrate and the sample holder.

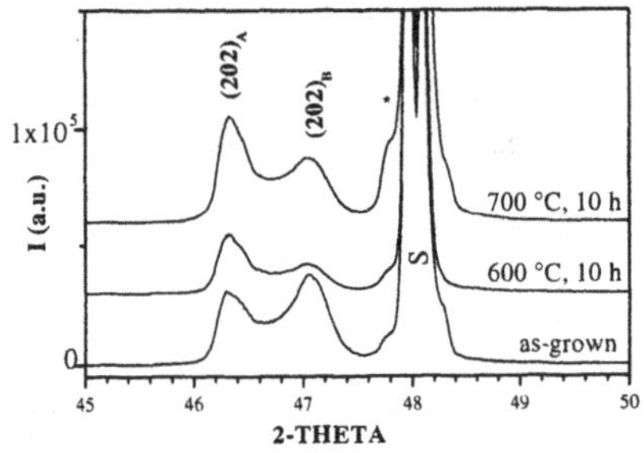

Fig.1b : Room temperature XRD for annealed films. Peaks labeled S and * are due to the substrate and the sample holder.

These two peaks correspond to two phases which we will call phases A and B (where A indicates the phase with the larger lattice parameter). We also plot the effects of a post-annealing under O₂ for a 1500 Å film in Fig.1b made at different temperatures during 10 hours. The structure of bulk Nd₀.₅Sr₀.₅MnO₃ is orthorhombic (*Pnma*) with a=5.43153 Å, b=7.63347 Å and c=5.47596 Å [9]. The out-of-plane parameter of the B-phase is equal to the d_{202} of bulk of Nd₀.₅Sr₀.₅MnO₃. Since this phase, which corresponds to the bulk value, appears for larger thicknesses, we assume that it is the relaxed part of the film with less influence from the substrate. The A-phase is always present, even in the thinnest film, which strongly suggests that it represents the strained phase of the initial layers of the film. The B-phase appears as the thickness is increased, which implies that the film is relaxed after a critical thickness of about 500 Å (see inset of Fig.1a). Under the oxygen annealing, the peak corresponding to the B-phase is increasing indicate that the film is relaxing along the out-of-plane direction. The effect has already be seen in previous studies of manganite thin films [15]. Moreover, the film is [101]-oriented, i.e. with the [101] axis perpendicular to the substrate plane, which is similar to an a-axis orientation according to the cubic perovskite cell. Note, that this orientation has already been reported in $(Pr_{0.7}Ca_{0.3})_{1-x}Sr_xMnO_3$ thin films grown by magnetron sputtering [16]. Detailed studies by Transmission Electron Microscopy are in progress to confirm these results. The resistivities of three films with different thicknesses are shown in Fig.2a. The 200 Å film is insulating at all temperatures. This layer is formed due to non-uniform distribution of the strain over the film and oxygen defects [15]. On increasing the thickness of the film, the resistivity shows the metal-insulator transition (T_{MI}) near 200 K, which is significantly lower than the T_{MI} observed in the bulk compound. This is an effect which has been observed in as-grown thin films of other CMR materials and is attributed to the substrate induced strain on the film [4].

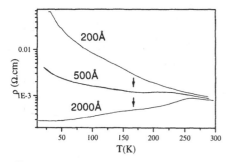

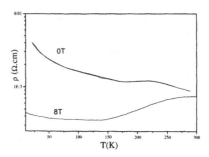

Fig.2a: Resistivity vs temperature for different thicknesses.

Fig.2b: MR vs temperature for a 500 Å film.

Also, around 170 K, the resistivity starts to increase when the temperature is decreasing. This is a typical signature of a charge-ordering transition, but the rise in the resistivity is not as sharp as observed in single crystals of $Nd_{0.5}Sr_{0.5}MnO_3$. The effect of a magnetic field on the resistivity is shown in Fig.2b. A field of 8 T shifts the T_{CO} to 140 K and also shifts the T_{MI} to higher temperatures. There is a large reduction of the resistivity for the entire temperature range. These features are qualitatively similar to the behavior observed in single crystals of of $Nd_{0.5}Sr_{0.5}MnO_3$. When the thickness of the film is increased to 2000 Å , the T_{MI} shifts to a higher temperature of 240 K as found in bulk. This is a signature of a reduction of the strain on the film [16]. But now the CO transition is suppressed and the resistivity shows a metallic behavior down to the lowest temperatures. This is unexpected since, as the strain of the substrate reduces, the properties should have approached those of the bulk whereas we find film sample is semiconductor below T_{CO}. We also saw that under the O_2 annealing the film is becoming metallic (not shown). In Fig.3, we indicate the magnetization versus temperature taken under a magnetic field of 2000 Oe for a 500 Å film. We clearly seen a transition from an antiferromagnetic state to a paramagnetic state at a temperature around 170 K, close to the T_{CO} transition. This behavior is the same as reported from the bulk compound [9]. However, a ferromagnetic component is still present at low temperature. The small deviation of composition between the film and the ideal composition of $Nd_{0.5}Sr_{0.5}MnO_3$, is not enough to explain the difference between the thin film and the bulk compounds. Two main features need to be clarified :

(1) Why does the 500 Å film show a CO-like behavior whereas the thicker 2000 Å film is metallic at low temperatures?

(2) Why is the transition seen in the 500 Å film not as sharp as seen in bulk?

The thickness dependence of the properties suggests that strain plays an important role in determining these properties. The substrate-induced strain can be expected to play an important role especially for the x=0.5 composition. In single crystals of $Nd_{0.5}Sr_{0.5}MnO_3$, there is an abrupt change of the lattice parameters in the orthorhombic structure which accompanies the CO transition [8]. Is this abrupt change seen in a thin film? To answer these questions, we performed variable temperature XRD down to 100 K in order to check the behavior of the out-of-plane parameter as the temperature is lowered below the bulk T_{CO}. Fig.4 shows the evolution of the d_{202} between 110 K and 260 K for a 2000 Å film. We also indicate in this graph the evolution of the 202 reflection of a single crystal at low temperatures, according to Ref.9 (at room temperature the 202 reflection of the bulk coincides with that of phase B).

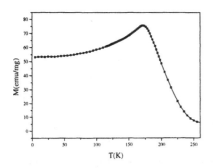
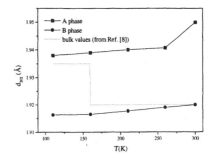

Fig.3 : Magnetization vs temperature recorded under 2000 Oe.

Fig.4 : Low temperature XRD for a 500 Å film.

From these data, we can see that neither of the film phases has the bulk value of the lattice parameters, below T_{CO} of the bulk. Thus, it seems impossible to have a sharp CO-transition in the films since the lattice parameters show no sharp change at T_{CO} nor do they have the same values as the bulk. In other words, the lattice is constrained by the substrate and is unable to change to the low temperature structure observed in single crystals. The distortion of the lattice is essential for the CO transition, as this decreases the Mn-O-Mn angle and thus the hopping probability, which results in the insulating behavior. The strain does not allow the Mn-O-Mn angle to change significantly in a thin film. However, the A-phase should exhibit a resistivity behavior close to that of a single crystal at low temperatures because the d_{202} is close to the value of the single crystal below T_{CO}. It can be seen in Fig.2a that the 500 Å film does show a behavior resembling a charge-ordering transition, with the low temperature resistivity similar to the resistivity behavior of a charge-ordering single crystal under a hydrostatic pressure of about 1.5 GPa [16]. It is known that the pressure induced by the substrate is of this order of magnitude. Indeed, it is possible to stabilize in thin film form, compounds similar to those synthesized by high-pressure methods [18]. Such effects of substrate-induced strain have also been observed in thin films of other charge-ordered compounds grown by nebulized spray pyrolysis on different substrates [19]. Even these films did not show the abrupt rise in resistivity when cooled below T_{CO}. The 2000 Å film is metallic because of the phase B which forms the top layer of the thicker films. The phase B is the quasi-relaxed phase at room temperature and has an out-of-plane parameter close to that of the bulk at room temperature. This is reflected in the fact that the T_{MI} of the 2000 Å film is close to that of the single crystal. But when the temperature is lowered the lattice parameters of the phase

B do not change as in a single crystal as seen (see Fig.4). This keeps the film in a metallic state. The overall resistivity behavior of the film is due to the parallel combination of the phase A and phase B. When the thickness of phase B exceeds a certain critical value, it changes the current distribution in the film, effectively shorting out the underlying phase A. The effect of phase A can still be seen in the resistivity curve of the 2000 Å film, as marked by the arrow.

CONCLUSION

In conclusion, we have grown $Nd_{0.5}Sr_{0.5}MnO_3$ thin films by PLD on [100] $LaAlO_3$ substrates. We found that the transport properties depend on the thickness of the film. We show that for intermediate thickness, it is possible to obtain a charge-ordering like behavior. However, thicker films lead to a metallic behavior. We also show, that under oxygen annealing, a film which is insulating and show a CO-like behavior, is becoming metallic. This anomalous behavior is due substrate induced strain on the film in the whole temperature range. Further investigations are in progress, in particular, growing films on different types of substrates, to confirm these results.

ACKNOWLEDGMENTS

We acknowledge Prof. R. Ramesh and Y. Zheng for help in low temperature XRD measurements. This work was partly supported by the MRSEC program of the NSF (Grant # DMR 96-32521).

REFERENCES

[1]   S. Jin, T.H. Tiefel, M. McCormack, R.A. Fastnacht, R. Ramesh and L.H.Chen, Science **264**, 413 (1994).

[2]   G.C. Xiong, Q. Li, H.L. Ju, S.N. Mao, L. Senapati, X.X. Xi, R.L. Greene and T. Venkatesan, Appl. Phys. Lett. **66**, 1427 (1995).

[3]   C.N.R. Rao, A.K. Cheetham and R. Mahesh, Chem. Mater. **8**, 2421 (1996).

[4]   W. Prellier, M. Rajeswari, T. Venkatesan and R.L.Greene, App. Phys. Lett **75**, (1999), P. Schriffer, A.P. Ramirez, W. Bao and S.W Cheong, Phys. Rev. Lett. **75**, 3336 (1995).

[5]   H.L. Ju, C. Kwon, Q. Li, R.L. Greene and T. Venkatesan, App. Phys. Lett. **65**, 2104 (1994).

[6]   Z. Jirak, S. Krupicka, Z. Simsa, M. Dlouha, S. Vratislav, J. Magn. Magn. Mater **53**, 153 (1985).

[7]   C.N.R. Rao, A. Arulraj, P.N. Santosh and A.K. Cheetham, Chem. Mater. **10**, 2714 (1998).

[8]   C.H. Chen and S.W. Cheong, Phys. Rev. Lett. **76**, 4042 (1996).

[9]   H. Kuwahara, Y. Tomioka, A. Asamitsu, Y. Moritomo, Y.Tokura, Science **270**, 961 (1995).

[10]   R.D. Shannon, Acta. Cryst. A **32**, 751 (1976).

[11]   A. Biswas, A.K. Ràychaudhuri, R. Mahendiran, A. Guha, R. Mahesh and C.N.R. Rao, J. Phys. Cond. Mater. **9**, L355 (1997).

[12]   H. Kawano, R. Kajimoto, H. Yoshizawa, Y. Tomioka, H. Kuwahara and Y. Tokura, Phys. Rev. Lett. **78**, 4253 (1997).

[13] P. Wagner, I. Gordon, A. Vantomme, D. Dierickx, M.J. Van Bael, V.V. Moshchalkov and Y. Bruynseraede, Europhys. Lett. **41**, 49 (1998).

[14] A. Goyal, M. Rajeswari, S.E Lofland, SM Bhagat, R. Shreekala, T. Boettcher, C. Kwon, R. Ramesh and T. Venkatesan, App. Phys. Lett. **71**, 2535 (1997).

[15] J.Q.Guo, H. Takeda, N.S. Kazama, K. Fukamichi, M. Tachini, J. App. Phys. **81**, 7445 (1997).

[16] B. Mercey, P. Lecoeur, M. Hervieu, J. Wolfman, Ch. Simon, H. Murray and B. Raveau, Chem. Mater. **9**, 1177 (1997).

[15] J.Z. Sun, D.W. Abraham, R.A. Roa, C.B. Eom, App. Phys. Lett. **74**, 3017 (1997).

[16] A.J. Millis, A. Goyal, M. Rajeswari, K. Ghosh, R. Shreekala, R.L. Greene, R. Ramesh and T. Venkatesan (1998) preprint.

[17] Y. Moritomo, H. Kuwahara, Y. Tomioka and Y. Tokura, Phys. Rev. B **55**, 7549 (1997).

[18] J.L. Allen, B. Mercey, W. Prellier, J.F. Hamet, M. Hervieu and B. Raveau, Physica C **241**, 158 (1995).

[19] V. Ponnambalam, S. Parashar, A.R. Raju and C.N.R. Rao, Appl. Phys. Lett. **74**, 206 (1999).

Poster Session II

Conductivity and NMR Study of LaGa$_{1-x}$Mn$_x$O$_3$ Single Crystals

N. Noginova, G. Loutts, L. Mattix, K. Babalola*, and E. Arthur
Center for Materials Research, Norfolk State University, Norfolk, VA, 23504
natalia@vger.nsu.edu

ABSTRACT

Electrical conductivity and magnetic relaxation has been studied in single crystals of Mn doped LaGaO$_3$ with doping concentrations of 0%, 0.5%, 2%, 10% and 50%. The resistivity of the crystals strongly depends on temperature and can be described with the hopping model. Both activation energy and resistivity coefficient grow significantly with decrease in Mn ion concentrations. Dramatic changes are observed in ^{71}Ga NMR spin-lattice relaxation times.

INTRODUCTION

In past few years, perovskite manganites (such as La$_{1-x}$Sr$_x$MnO$_3$, La$_{1-x}$Ca$_x$ MnO$_3$, *etc.*) attracted much scientific attention due to the discovery of colossal magnetoresistance (CMR) and a rich complex diagram of phase transitions in these materials [1]. Below the transition temperature, the materials demonstrate ferromagnetism and metal behavior. At high temperatures they are insulators with conductivity determined by hopping mechanism. According to theoretical models, the magnetic and transport properties of these materials are determined by magnetic interactions and exchange of electron between Mn ions. In CMR materials, Mn ions are ions of the crystal lattice, strongly interacting which each other. In our work we study similar type of materials, lanthanide perovskites, but with different relative concentration of Mn ions. In our materials, the Mn ions are diluted by nonmagnetic Ga ions, which enter the same place in the crystal structure as Mn ions. The goal of our work is to monitor changes in resistivity and magnetic properties of materials starting from materials with low manganese concentrations, where Mn ions are separate from each other, and finishing with materials where Mn ions form pairs and clusters.

Single crystals LaGaO$_3$, undoped and doped with 0.5%, 2%, 10%, 50% Mn, were grown by Czochralski technique. The undoped and 0.5% crystals have brownish color, the 2% crystal was dark-violet, the other crystals were practically black. The optical spectra of the materials are discussed elsewhere.

CONDUCTIVITY MEASUREMENTS

Due to extremely high resistance of the 10% and 2% doped crystals, in the experiments with these materials, and also in some experiments with 50% doped crystals, we used very thin plates (0.2- 0.5 mm of thickness). Two electrical contacts (with the area of approximately 1 mm^2 each) were mounted by silver paint on the opposite sides of the sample. In some experiments with 50% doped sample, a conventional four-point contact technique was used.

The voltage-current characteristics are linear at low currents. For currents higher than some critical current, J$_c$, the resistance started to drop, see Fig. 1. The value of J$_c$ depended on the Mn doping concentration. For 50 % doped LaGaO$_3$ J$_c$ $\approx$ 8-10 μA, for 10% and 2 % doped samples J$_c$ was much lower, about 0.1-0.2 μA.

To study the temperature dependence of the resistance, the current was chosen in the range J<J$_c$. The linearity of the voltage-current characteristics has been checked in the different temperatures.

---

* Summer CMR student

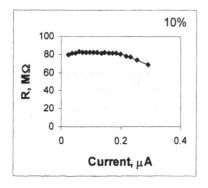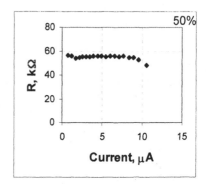

Fig. 1. Resistance as a function of current in Mn doped LaGaO₃ single crystals

Temperature dependences of the resistance for 2%, 10% and 50% doped crystals are shown in Fig. 2. As one can see, the dependences can be well described by the function

$$\rho(T) = B \, T \, \exp (E_g/kT),$$

where B is the coefficient, T is the temperature, k is the Boltzman constant, E_q is the activation energy. Assuming that the area of contacts was approximately 1 mm² we calculated the resistivity coefficients as shown in Table 1

Mn/Ga concentration ratio, %	E, meV	B, Ω•cm/K
50 : 50	336	$1.6*10^{-4}$
10 :90	491	$1.2*10^{-3}$
2: 98	570	1.5

Table 1. The activation energy and resistivity coefficient in Mn:LaGaO₃

As one can see from the Table 1, the value of activation energy, E_g increases dramatically from 336 meV to 491 meV when the concentration decreases in five times from 50% to 10%. E_g continues to increase with further decrease of the Mn doping concentration. For 2% doped sample $E_g \approx 570$ meV.

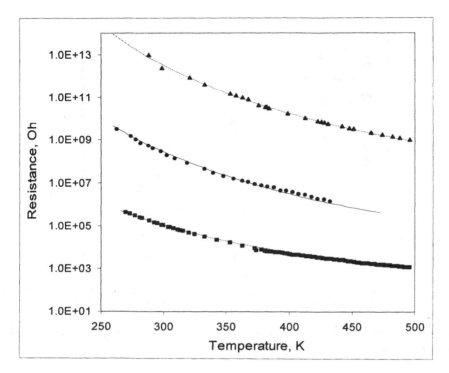

Fig. 2. Resistance as a function of temperature in 2% doped (triangles), 10% doped (circles), and 50% doped (squares) Mn:LaGaO$_3$ crystals. Solid curves are fits of the small polaron model.

Let us compare our results with experimental data for La$_{0.77}$Ca$_{0.33}$Ga$_x$Mn$_{1-x}$O$_3$ materials with low Ga concentration, obtained in Ref. [2]. Our data and data of Ref. [2] for activation energy, E$_g$, are shown together in Fig. 3. As one can see, both data fit approximately the same dependence, decreasing with increase of Mn concentration.

Strong dependence on Mn concentration is observed for the coefficient B as well. The values of the resistivity coefficient, B, measured in [2], were about 10^{-6} Ω•cm/K. In our materials, coefficients have much higher values, rapidly growing with decrease of Mn doped concentration, and changing in approximately 10^4 times, from 1.6*10^{-4} in 50 % doped crystal to 1.5 Ω•cm/K in 2 % doped crystal.

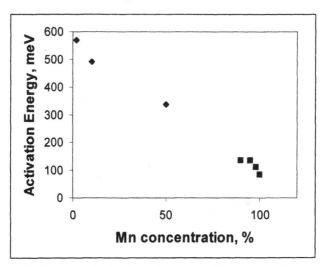

Fig. 3. Activation energy in the dependence on the Mn concentration. Diamonds: our data for Mn:LaGaO$_3$. Squares: data for Ga doped LaSrMnO$_3$ materials [2]

According to Emin-Holstein's small polaron hopping model, the resistivity of the materials with palaron hopping conductivity, may be described as

$$\rho = BT \exp(\frac{E_g}{kT})$$

where B is the resistivity coefficient

$$B = \frac{g_g k}{ne^2 a^2 v}$$

e is the electron charge, n is the carriers concentration, a is the hopping distance, $g_g \approx 1$ is the geometric factor depending on the lattice geometry, and v is the characteristic frequency of the optical phonon that carriers polaron through lattice.

The ratio of the coefficients in 10% and 50% doped samples is equal to 7.5, close to the ratio of the Mn ions concentrations in these samples. However, the resistivity coefficient in 2% doped material is too high to be explained just with a change in Mn doping concentration.

MAGNETIC RESONANCE STUDY

To characterize our materials magnetically, we perform NMR and relaxation measurements of ^{71}Ga. CXP-300 Bruker Pulse NMR Spectrometer (300 MHz) was used in the experiments. A spectrum of the crystal with low Mn concentration consists of one central line and two satellite lines corresponding to the quadropole splitting (the spin of ^{71}Ga is 3/2). With increase in Mn concentration, the central line becomes wider due to the inhomogeneous broadening. The NMR spectrum for the 50% doped material has width higher than 200 kHz, the spectral band of our spectrometer, and could not be recorded simultaneously.

To study spin-lattice relaxation, the crystal was oriented in such a way that the satellite lines were as close as possible to the central line. In this case, all the lines would be excited simultaneously, and relaxation can be described with a single exponential, see Fig. 4. Standard saturation-recovery sequence was used to study spin-lattice relaxation. The spin-spin relaxation time, T_2, in samples with 0%, 0.5%, 2% and 10% concentration was measured by echo technique.

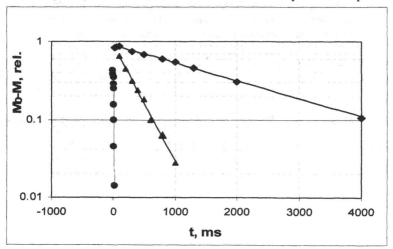

Fig.4. ^{71}Ga NMR spin-lattice relaxation in 0.5% doped (diamonds), 2% doped (triangles) and 10% doped (circles) Mn:LaGaO$_3$ single crystals. Solid lines are theoretical fits with T_1=1900 ms, 290 ms and 6 ms correspondingly.

Values of T_1 and T_2 for our samples at room temparature are shown in Table 2.

Mn doping concentration, %	T_1, ms	T_2, ms
0	1950	0.75
0.5	1900	0.7
2	290	0.65
10	6	0.48

Table 2. ^{71}Ga NMR spin-lattice and spin-spin relaxation times in Mn: LaGaO$_3$

As one can see from Table. 2, the time T_2 does not change significantly with increase of Mn concentration. In opposite, the spin-lattice relaxation time, T_1 strongly depends on Mn concentration. Especially prominent change observed in the value of T_1 when the Mn concentration changes from 2 to 10 %.

As known, the paramagnetic impurity mechanism of the relaxation implies a linear dependence of the relaxation rate with increase of the doping concentration. Our data point that starting with 10% and higher Mn doping concentration, another channel appears in the magnetic relaxation.

CONCLUSION

The resistivity and NMR relaxation were studied in Mn doped single crystals of $LaGaO_3$ The resistivity of the materials in study can be described with the hopping model, where resistivity coefficient and spin-lattice relaxation rates strongly depend on Mn concentration. The activation energy grows linearly with decrease of Mn concentration. Most prominent change in coefficient B is observed when concentration of Mn ions increases from 2 to 10%. At the same concentration, the dramatic acceleration of spin-lattice relaxation rate is observed.

Theoretical consideration of the results obtained will be published elsewhere.

ACKNOWLEDGMENTS

This work was supported by the U.S. Department of Energy, Grant No. DE-FG01-94EW11493.

REFERENCES.

1. T.H.Tiefel, M.mcCormack, R.A.Fastnacht, R.Ramesh, and L.H.Chen, Science 264, 413 (1994);
2. Y. Sun, X. Xu, L. Zheng, and U. Zhang. Phys. Rev. B, 60, 12317 (1999)

CRITICAL PHENOMENA AT THE ANTIFERROMAGNETIC TRANSITION IN MnO.

B.F. WOODFIELD,[*] J.L. SHAPIRO,[*] R. STEVENS,[*] J. BOERIO-GOATES,[*] M.L. WILSON[**]
[*]Department of Chemistry and Biochemistry, Brigham Young University, Provo, UT 84602
Brian_Woodfield@byu.edu
[**]University of Tulsa, Department of Physics, Tulsa, OK 74104

ABSTRACT

The specific heat of a polycrystalline sample of MnO was measured from $T \approx 1$ K to $T \approx 400$ K using two different experimental apparatuses at zero applied pressure. Features revealed by the data include a hyperfine contribution due to the Mn nuclei, a T^2 temperature dependence at low temperatures due to the type-II antiferromagnetic magnon contribution, and a sharp but well defined antiferromagnetic transition ($T_N = 117.7095$ K) that is clearly second order in nature. The critical exponent, α, deduced from the transition is consistent with a two dimensional Ising model. The specific heat of MnO is also compared with recent results on the type-A antiferromagnet $LaMnO_3$.

INTRODUCTION

Recent specific-heat measurements[1] on the colossal magnetoresistance materials $La_{1-x}Sr_xMnO_3$, where x = 0.0, 0.1, 0.2, and 0.3, revealed a low-temperature T^2 term in the undoped parent compound, $LaMnO_3$, due to spin waves associated with a type-A antiferromagnetic transition at 110 K. Contrary to most magnetic transitions, however, the transition in $LaMnO_3$ showed only a small anomaly with no excess specific heat. In order to better understand the magnetic properties of $LaMnO_3$ we decided to measure the specific heat of MnO, a material with a similar magnetic structure and ordering temperature.

MnO has long been considered one of the classic antiferromagnets and, consequently, many of its properties have been studied extensively. At room temperature, MnO is paramagnetic with a NaCl crystal structure and $Fm3m$ space group but, below 118 K, it transforms to an antiferromagnetic phase. The full magnetic structure of the antiferromagnetic state in MnO was determined by Shull, Strausser, and Wollan[2] and by Li.[3] These studies showed that the magnetic Mn spins align ferromagnetically within a given (111) plane and these planes are stacked antiferromagnetically in the direction normal to the given (111) plane; thus, MnO is classified as a type-II antiferromagnet.

Some twenty years later, interest in MnO was renewed by numerous theoretical and experimental investigations of tricritical phase transitions. In 1973 Bloch and Maury[4] used thermal expansion measurements on a single crystal of MnO to show that while the paramagnetic to antiferromagnetic transition is second order at ambient pressure it becomes first order when a uniaxial stress greater than ~200 bar is applied along the (111) direction. The application of the uniaxial stress removed the three T-domains not aligned with (111) thereby revealing the first order character intrinsic to the transition.

EXPERIMENTAL

The MnO sample used for all measurements was purchased from Alfa Aesar, lot number K13G04, with a stated purity of better than 99.5%. The lot assay reported MnO_2 as the only significant impurity at a concentration of 0.1% by weight. All sample masses have been

Mat. Res. Soc. Symp. Proc. Vol. 602 © 2000 Materials Research Society

corrected for the MnO_2 impurity, and the specific heat has been corrected for the MnO_2 contribution (at most 0.05% of the total specific heat) using data from Robie and Hemingway.[5] MnO crystals examined under a microscope were found to be cubic in shape and to range in size from 0.3 mm to 1.3 mm on a side. The median size was roughly 0.8 mm on a side.

The specific heat was measured using two different experimental apparatuses in three separate runs which we label as run A, run B, and run C, respectively. Run A used an adiabatic calorimeter and temperature scale described previously.[6] Run B used the same cryostat, instrumentation, and software as in run A but with a new, smaller calorimetric vessel having a volume of 10 cm³ and using a 25 Ω Pt resistance thermometer. The specific heat from run C used a new apparatus designed around a commercial ³He insert capable of attaining temperatures as low as 0.4 K. This apparatus employs a semi-adiabatic pulse technique from the lowest temperature up to 10 K and an isothermal technique to 100 K. This is the same apparatus used in prior measurements of the specific heat of $La_{1-x}Sr_xMnO_3$.[1]

RESULTS

The experimental specific heat for run A, run B, and run C, corrected for the addenda, curvature, and the MnO_2 impurity are shown in Figure 1 with symbols representing the separate runs. The agreement among the three sets of data is better than ±0.1%. Inset (a) shows the full height of the antiferromagnetic transition and demonstrates the magnitude of the transition relative to the rest of the data. Inset (b) of Figure 1 shows the low-temperature data on an expanded scale where the Mn-hyperfine contribution is clearly evident.

Low-Temperature Specific Heat

The low-temperature specific heat was fit to the equation

$$C = A_{-2}T^{-2} + B_n T^n + B_3 T^3 \tag{1}$$

from $T \approx 1$ K to $T \approx 10$ K where the T^{-2} term represents the high-temperature side of the Mn-hyperfine contribution (C_{hyp}), the T^3 represents the lattice (C_{lat}), and the T^n term represents the magnon contribution (C_{mag}) of unknown type. When n is an adjustable parameter in the fit, the optimal value for n is very nearly 2. With n fixed at 2, the low-temperature data were refit with no appreciable change in the %RMS deviation. This fit results in the values for the parameters A_{-2}, B_2 and B_3 given in Table I. The line through the data in inset (b) in Figure 1 was generated using this fit. Also given in Table I are the corresponding values from the fit of $LaMnO_3$.[1]

The Mn-hyperfine contribution to the specific heat is caused by the large local magnetic field at the Mn nucleus (H_{hyp}) due to the net spin of the Mn ion. H_{hyp} listed in Table I can be extracted from A_{-2} and is in

Table I: Fitting coefficients and derived parameters for MnO and $LaMnO_3$.

	MnO	LaMnO₃
A_{-2} (mJ)	21.94	8.03
B_2 (mJ)	0.00953	3.80
B_3 (mJ)	0.00285	0.00706
%RMS	0.53	1.0
H_{hyp} (T)	61	36
$D_\rho D_z$ (meV²·Å²)	800	55
θ_D (K)	515	516

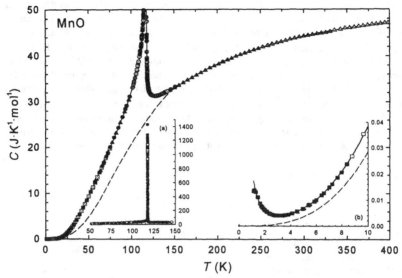

Figure 1:   Specific heat of MnO. Inset (a) shows the transition at full scale, and inset (b) shows the low-temperature data on an expanded scale.

agreement with a local field of 58.8 T at 0 K determined by zero-field ^{55}Mn NMR.[7] MnO has a
larger H_{hyp} than LaMnO$_3$ consistent with the larger spin of Mn^{2+} in MnO ($S = 5/2$) than of Mn^{3+}
in LaMnO$_3$ ($S = 2$).

The Debye temperature, θ_D, derived from B_3, is 515 K. The value of θ_D obtained for
LaMnO$_3$ from specific-heat measurements is nearly identical to that of MnO (see Table I) and
suggests comparable lattice dynamics at low temperatures.

Previous theoretical analyses have proposed that a T^2 dependence observed in the specific
heat is the result of reduced dimensionality in the system being thermally excited.[8] In the
present case, we posit that the T^2 term is the result of thermally excited long-wavelength spin
excitations in a layered antiferromagnetic structure.[1] The spin-wave stiffness product, $D_\rho D_z$,
extracted from B_2 for MnO, is given in Table I. It is considerably larger than $D_\rho D_z$ for LaMnO$_3$
implying a much stiffer magnetic lattice in MnO. This stiffness is due both to an increase in the
number of neighboring magnetic atoms and to an increase in the magnitude of the magnetic
coupling constants for MnO[9] over those found in LaMnO$_3$.[10]

Magnetic Entropy

Another useful quantity which can be extracted from the specific heat is the entropy, ΔS_{mag},
associated with the ordering and spin-wave excitations of the local magnetic moments and is
calculated using $\Delta S_{mag} = \int \dfrac{C_{mag}}{T} dT$. First, however, a careful estimate for C_{lat} must be calculated
and subtracted from the measured data on the interval $0 \text{ K} \leq T \leq 150 \text{ K}$ to yield C_{mag}. Below
10 K, the lattice and hyperfine contributions are easily calculated using the T^3 and T^{-2} terms
from the low-temperature fit leaving C_{mag}. Between 10 K and 150 K the lattice contribution was
estimated by interpolating the lattice specific heats of MgO, CaO, SrO, and BaO, all of which

have the same NaCl crystal structure. The onset of the transition at 150 K was determined by noting the deviation of the measured MnO specific heat from the systematic trend of the scaled MgO, CaO, SrO, and BaO specific heats. The lattice estimation is shown in Figure 1.

The calculated magnetic entropy, ΔS_{mag}, is 11.62 J·K^{-1}·mol^{-1} or $R\ln(4.0)$. This value is significantly smaller than $R\ln(6.0)$ expected from a freezing out of spin 5/2 Mn ions. One at first may argue that the Mn ions are spin 3/2 ions (consistent with a $R\ln(4.0)$ entropy) rather then the expected 5/2; however, the analysis of H_{hyp} reported above determined that the Mn ions are $S = 5/2$ at low temperatures. Hermsmeier et al.[11] have noted an abrupt shift in spin asymmetry near 530 K. Their observations, utilizing spin-polarized photoelectron diffraction, were interpreted as a transition involving a local spin reorientation which persisted as the sample was further cooled. Such a weak transition may account for the entropy missing below the Néel transition.

Antiferromagnetic Transition

The full antiferromagnetic transition as shown in inset (a) of Figure 1 could, at first glance, be misrepresented as a first order transition. However, when the low- and high-temperature sides of the transition are presented on an expanded scale (Figure 1), the transition has the characteristic lambda shape of a second order magnetic transition. That the specific heat distinctly shows a second order transition is not necessarily a contradiction to the conclusions of Bloch[4] and others[12-14] that the antiferromagnetic transition is intrinsically first order. Bloch[4] clearly demonstrates that MnO will undergo a second order transition if the crystal size is large enough for multiple T domains to exist within a single crystal. The application of sufficient uniaxial or hydrostatic pressure can remove all but a single T domain, however, the pressure required is highly sample dependent.[12-15] Our crystals are ~ (0.8 mm)^3 in size, hence, the presence of multiple T domains within each crystal is to be expected at the < 1 bar pressure in the calorimeter vessel.

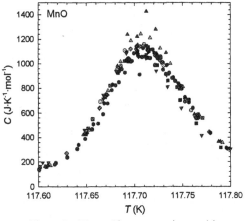

Figure 2: The antiferromagnetic transition on an expanded scale.

To determine the specific nature of the transition at T_N, C_{mag} has been analyzed for scaling behavior in the critical regime. Shown in Figure 2 is the antiferromagnetic transition on an expanded temperature scale. Near the peak of the transition, specific-heat values were measured using temperature intervals, ΔT, of 5 mK $\leq \Delta T \leq$ 10 mK. The Néel temperature, T_N, of 117.7095 K was determined as the intersection of linear extrapolations of the low- and high-temperature sides of the transition. Equilibrium times in the outer portion of the transition region were < 15 min but increased to ~3 hr very close to T_N. This dramatic increase in equilibration times as one approaches T_c (or T_N in the case of MnO) is characteristic of critical phenomena.[16,17] The rounding observed in C_{mag} at the peak of the transition (as seen in Figure 2) is a product of increased uncertainties in ΔT caused by the long equilibrium times and the limited thermometer resolution caused by the small ΔTs.

The high quality specific-heat data in the vicinity of the Néel temperature of MnO was fit to scaling approximations for the specific heat derived for different types of magnetic interactions. Typically, in a 3D system the critical exponent for the specific heat α is defined by[16,17]

$$C = \begin{cases} At^{-\alpha} & t > 0 \\ A'(-t)^{-\alpha} & t < 0 \end{cases} \tag{2}$$

where $t = \dfrac{T - T_c}{T_c}$, $A \neq A'$ and $\alpha > 0$.

For a 3D Ising model, theory and experiments have confirmed that $\alpha = 0.110$ and $A \neq A'$.[17] Shown in Figure 3 is a plot of C_{mag} versus $|t|$ on a log-log scale in the transition region from which A, A', and α are extracted. Data taken above and below the transition ($\blacktriangle$ and $\bullet$, respectively) can be seen to fall on the same curve. The flat portion at $|t| < 10^{-4}$ is caused by the rounding of the transition very near T_N.[17]

There are several features to this plot which are inconsistent with the 3D Ising predictions. First, $\alpha = 1.64$, as determined from a fit to the data in the temperature interval $5 \times 10^{-4} < |t| < 2 \times 10^{-3}$. This is more than ten times larger than that predicted by the 3D Ising model or measured in 3D magnetic systems.[16,17] Secondly, data above and below the transition clearly follow the same line, hence, $A = A'$ again in disagreement with 3D scaling models.[16,17]

In a 2D Ising model, the transition undergoes a logarithmic singularity ($\alpha = 0$),[18] and the specific heat diverges as

$$C = \begin{cases} -A\log|t| + B & t > 0 \\ -A'\log|t| + B' & t < 0 \end{cases} \tag{3}$$

where $A = A'$ and $B = B'$. Figure 4 shows a plot of C_{mag} versus $\log |t|$ consistent with this scaling

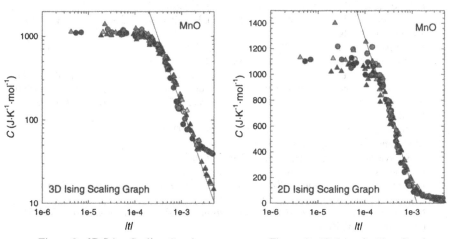

Figure 3: 3D Ising Scaling Graph.   Figure 4: 2D Ising Scaling Graph.

249

form. The same flat region is observed at $|t| < 10^{-4}$, but the linear portion of the scaling plot extends over a broader region than was discussed above, from $10^{-4} < |t| < 10^{-3}$. The data above and below the transition also fall on the same curve indicating that $A = A'$ and $B = B'$, as predicted by this 2D model.[18] The observance of scaling behavior spanning more than a decade in reduced temperature and the equivalencies, $A = A'$ and $B = B'$, provide support for characterizing the magnetic transition in MnO as similar to that of a 2D Ising magnet. This implies that the strongest contribution to the cooperative ordering may be from within the planes, with the antiferromagnetic coupling being much weaker.

CONCLUSIONS

Specific-heat measurements below 10 K have shown that MnO has a magnon contribution whose temperature dependence is consistent with a layered antiferromagnet and similar to that observed for $LaMnO_3$. The magnitude of the MnO spin-wave stiffness product is more than ten times larger than in the lanthanum compound. The Mn hyperfine field is also significantly larger in MnO than in $LaMnO_3$ which we attribute to the larger Mn spin in MnO. A calculation of the magnetic entropy associated with the antiferromagnetic transition yields an observed value of $R\ln(4.0)$ rather than the predicted $R\ln(6.0)$ suggesting that the short range ordering transition at 530 K observed by Hermsmeier et $al.$[11] is a real effect and does remove some of the magnetic entropy of the system. Measurements very close to T_N support the assignment of the magnetic transition as a second order transition with a critical exponent, $\alpha = 0$, and other factors compatible with a 2D Ising model. Additional measurements under pressure are needed to verify an intrinsic first-order character of the antiferromagnetic transition.

REFERENCES

1. B.F. Woodfield, M.L. Wilson, and J.M. Byers, *Phys. Rev. Lett.* **78,** 3201-3204 (1997).
2. C.G. Shull, W.A. Strausser, and E.O. Wollan, *Phys. Rev.* **83,** 333 (1951).
3. Y.-Y. Li, *Phys. Rev.* **100,** 627 (1955).
4. D. Bloch and R. Maury, *Phys.Rev. B* **7,** 4883 (1973).
5. R.A. Robie and B.S. Hemingway, *J. Chem. Thermodyn.* **17,** 165-181 (1985).
6. B.F. Woodfield, J. Boerio-Goates, J.L. Shapiro, R.L. Putnam, and A. Navrotsky, *J. Chem. Thermodyn.* **31,** 245-253 (1999).
7. M.E. Lines and E.D. Jones, *Phys. Rev.* **139,** 1313-1327 (1965).
8. R. Hotz and R. Siems, *Superlat. Microstruct.* **3,** 445-454 (1987).
9. M. Kohgi, Y. Ishikawa, I. Harada, and K. Motizuki, *J. Phys. Soc. Japan* **36,** 112 (1974).
10. K. Hirota, N. Kaneko, A. Nishizawa, and Y. Endoh, *J. Phys. Soc. Japan* **65,** 3736 (1996).
11. B. Hermsmeier, J. Osterwalder, D.J. Friedman, and C.S. Fadley, *Phys. Rev. Lett.* **62,** 478 (1988).
12 R. Boire and M.F. Collins, *Can. J. Phys.* **55,** 688 (1977).
13. M.S. Jagadeesh and M.S. Seehra, *Phys.Rev. B* **23,** 1185 (1981).
14. G. Srinivasan and M.S. Seehra, *Phys.Rev. B* **28,** 6542 (1983).
15. D. Bloch, C. Vettier, and P. Burlet, *Phys. Lett.* **75A,** 301 (1980).
16. H.E. Stanely, *Introduction to Phase Transitions and Critical Phenomena* (Oxford University Press, New York, 1971).
17. N. Goldenfeld, *Lectures on Phase Transitions and the Renormalization Group*, Vol. 85 (Addison Wesley, Reading, 1992).
18. L. Onsager, *Phys. Rev.* **65,** 117 (1944).

Tunnel Type Magnetoresistance in Electron Beam Deposited Films of Compositions $(Co_{0.5}Fe_{0.5})_x(Al_2O_3)_{(100-x)}$ $(7 \leq x \leq 52)$

H.R. Khan*, A. Ya Vovk**, A.F. Kravets**, O.V. Shipil** and A.N. Pogoriliy**
*FEM, Materials Physics Department, 73525 Schwäbisch Gmünd and Department
of Physics, University of Tennessee, Knoxville TN. U.S.A.
**Institute of Magnetism, 36-b Vernadskystr. 252142 Kiev, Ukraine

ABSTRACT

A series of 400 nm thick metal-insulator films of compositions $(Co_{50}Fe_{50})_x(Al_2O_{3(100-x)}$ $(7 \leq x \leq 52$; x is in vol.%) are deposited on glass substrates using dual electron beam evaporation technique. The films are nanocrystalline with crystallite sizes of 1-3 nm. Resistivity of the films varies as a function of $(1/T)^{0.5}$ showing a tunneling type behaviour. The films show isotropic and negative magnetoresistance (GMR). A film of composition $(Co_{50}Fe_{50})_{82.5}(Al_2O_3)_{17.5}$ show maximum tunneling magnetoresistance (TMR) of 7.2% at room temperature and in a magnetic field of 8.2 kOe.

INTRODUCTION

Tunnel type magnetoresistance (TMR) in ferromagnetic metal-insulator systems has attracted much attention due to their potential applications as magnetic sensors and memory devices. The mechanism of spin dependent tunneling in the metal-insulator granular films is similar to the mechanism in ferromagnetic/insulator/ferromagnetic junctions [1]. Fujimore [2] reported the giant magnetoresistance (GMR) of about 8% in a magnetic field of 12 kOe and at room temperature in the reactive r.f. sputtered granular films of Co-Al-O. The spin dependent tunneling of the electrons between the ferromagnetic grains across the insulating barrier is the cause of negative GMR. The GMR magnitude is sensitive to the microstructure of reactive r.f. sputtered granular Ferromagnet-Oxide films. In this paper we report the resistivity measurements as a function of temperature and magnetic field of nanocrystalline metal-insulator granular films of ferromagnetic alloy $Co_{50}Fe_{50}$ and insulator Al_2O_3 prepared by electron beam evaporation technique.

EXPERIMENT

The Granular films $(Co_{50}Fe_{50})_x(Al_2O_3)_{(100-x)}$ of compositions $(7 \leq x \leq 52$; x is in vol.%) of about 400 nm thickness were deposited on glass substrates using dual electron beam evaporation technique using $Co_{50}Fe_{50}$ and Al_2O_3-sources. The crystalline structure was investigated by X-ray diffraction and transmission electron microscopic techniques. The electrical resistance of the films was measured by a four point a.c. technique in a temperature range of 77 and 300 K. The room temperature magnetoresistance was measured in magnetic fields (H) upto 8.2 kOe in current in plane (CIP) configuration and three different geometries: H in the film plane and parallel to current I (II-Parallel); H in the film plane but perpendicular I (T-Transverse); H perpendicular to the film plane and I (Perpendicular).

RESULTS AND DISCUSSION

The X-ray diffractograms of the films show broad peaks at a Bragg angle corresponding to the <110> diffraction line of bcc $Co_{50}Fe_{50}$ indicating the nanocrystalline nature of the films. The

251

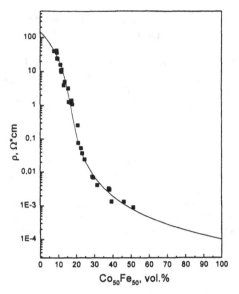

Fig. 1: Electrical resistivity ρ as a function of $Co_{50}Fe_{50}$ concentration of granular $(Co_{50}Fe_{50})_x(Al_2O_3)_{100-x}$ films.

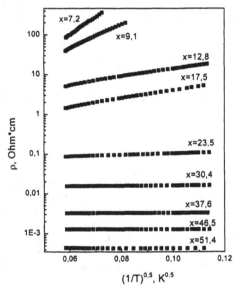

Fig. 2: Resistivity ρ as a function of inverse temperature $(1/T)^{0.5}$ for the $(Co_{50}Fe_{50})_x(Al_2O_3)_{100-x}$ $(7.2 \leq x \leq 51.4)$ films.

TEM investigations confirm the nanocrystallinity of the films and the average granular size varies between 1-3 nm which is typical of metal-insulator granular films [3-8].

The electrical resistivity of the films as a function of $Co_{50}Fe_{50}$ concentration is shown in Fig. 1 and is typical of cermet films [5, 9, 10]. The resistivity decreases with increasing concentration of the metallic component $Co_{50}Fe_{50}$. The resistivity vs. concentration curve can be fitted using generalized effective media equation [9].

$$(1-\phi)*(R_m^{1/t}-R_h^{1/t})/(R_m^{1/t}+A*R_h^{1/t})+ \phi *(R_m^{1/t}-R_l^{1/t})/(R_m^{1/t}+A*R_l^{1/t}) = 0,$$

where R_h, R_l, R_m are the resistances of the insulator phase, the metallic phase and the film respectively, ϕ is the volume fraction of metal phase, t is an exponent and $A=(\phi_c/(1-\phi_c)$, where ϕ_c is the critical percolation volume fraction of the metallic phase. The critical percolation parameters of the investigated films are determined and the values are: $\phi_c = 0.174 \pm 0.005$ (17.4 ± 0.5 vol.% of $Co_{50}Fe_{50}$) and t = 2.31 $\pm$ 0.04. The value of ϕ_c is in good agreement with the theoretical values predicted for 3-d system of spherical particles ($\phi_c = 0.16 \pm 0.02$ and t = 1.5 $\pm$ 0.1). The temperature dependent resistivity (ρ vs. $(1/T)^{0.5}$) of different compositions are shown in Fig. 2. These $\rho(T)$ curves are well fitted to the exponential relation $R \cong R_0 \exp (2*(E/kT)^{0.5})$ where E is a constant related to the charge energy of the particles [9]. The value of E is estimated from the $\rho(T)$ curves as a function of $(Co_{50}Fe_{50})$ concentration and the values are 4.27 x 10^{-2}, 1.96 x 10^{-2} and 1.8 x 10^{-3} eV for the films of compositions x = 7.2, 9.1 and 12.8 respectively and the E value decreases with increasing concentration of $Co_{50}Fe_{50}$. The temperature dependent resistivity of the films of higher metallic concentration is more complex. The magnetoresistance percent $(\Delta R/R\%) = (R(H)-R(0) / R(0) * 100$ (where R(H) and R(0) are the resistances in magnetic field H and without field) of the films of composition (x $\leq$ 50) are measured in three different geometries. All these films show negative and isotropic magnetoresistances of different amplitudes as a function of magnetic field. A typical example of $\Delta R/R\%$ versus H of a film of composition $(Co_{50}Fe_{50})_{17.5}(Al_2O_3)_{82.5}$ is shown in Fig. 3. This indicates that main contribution to magnetoresistance originates from the spin polarized electrons tunneling [2, 11]. The magnitude of MR% increases with increasing concentration of $Co_{50}Fe_{50}$ and a maximum tunneling magnetoresistance (TMR) is observed for the composition x = 17.5 (7.2% in a magnetic field of 8.2 kOe and at room temperature). After this composition, MR% decreases (Fig. 4). We are in the process of investigating the detailed magnetic and magnetoresistive properties of these films.

CONCLUSIONS

The electron beam evaporated films of compositions $(Co_{50}Fe_{50})_x(Al_2O_3)_{(100-x)}$ ($7 \leq x \leq 52$) are nanocrystalline with granular size of few nm. The resistivity vs. metallic concentration curve is fitted to effective media equation gives a critical percolation volume fraction value of 17.5 for the metallic component. Films of composition x $\leq$ 18 show spin dependent tunneling and a maximum GMR value of 7.2% for the composition $(Co_{50}Fe_{50})_{17.5}(Al_2O_3)_{82.5}$ in a magnetic field of 8.2 kOe and at room temperature.

ACKNOWLEDGMENTS

The work of A.Ya. Vovk was partly supported by the Fellowship of the President of Ukraine for young scientists and INTAS Program, Brussels Belgium. The technical assistance of A. Weiler is gratefully acknowledged.

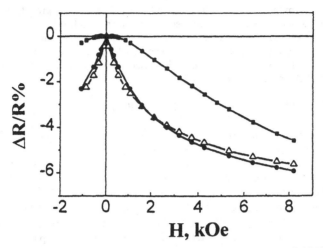

Fig. 3: Magnetoresistance $\Delta R/R\%$ as a function of magnetic field H in three different geometries: II (▲), T (●) and P (■) of a film of composition x = 17.5.

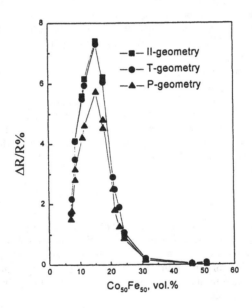

Fig. 4: Magnetoresistance $\Delta R/R\%$ as a function of $Co_{50}Fe_{50}$ concentration in $(Co_{50}Fe_{50})_x(Al_2O_3)_{100-x}$ films measured in a field of 8.2 kOe and at room temperature.

REFERENCES

1. S. Maekawa and U. Gafvert, IEEE Trans. Magn. **18**, 707 (1982).
2. H. Fujimori, S. Mitani, S. Ohnuma, Mater. Sci. Eng. B **31**, 219 (1995).
3. S. Mitani, H. Fujimori, S. Ohnuma, J. Magn. Magn. Mater. **165**, 141 (1997).
4. S. Honda, T. Ikada, M. Nawate, M. Tokumoto, Phys. Rev. B **56**, 14566 (1997).
5. B. Abeles, P. Sheng, M.D. Coutts, Y. Arie, Adv. Phys. **24**, 401 (1975).
6. W.D. Westwood, Reactive sputtering in: M.H. Francombe, J.L. Vossen (Eds.), Physics of Thin Films, Academic Press, New York, 1989, vol. **14**, Page 1.
7. A.Y. Vovk, A.V. Filatov, Met. Phys. Adv. Tech. **15** (7).
8. A.K. Butilenko, A.Ya. Vovk, H.R. Khan, Surf. and Coating Techn. **107**, 197 (1998).
9. D.S. McLachlan, M. Blaszkiewicz, R.E. Neownham, J. Am. Ceram. Soc. **73** (8), 2187 (1990).
10. A.Ya. Vovk, V.A. Zrazhevskiy,l Ukr. Fiz. Zhurn. **39**, 728 (1994).
11. T. Furubayashi and I. Nakatani, J. Appl. Phys. **79**, 6258 (1996).

COLOSSAL MAGNETORESISTANCE IN NEW MANGANITES

C. H. Shen [1], R. S. Liu [1], S. F. Hu [2], J. G. Lin [3], C. Y. Huang [4]
[1] Department of Chemistry, National Taiwan University, Taipei, TAIWAN
[2] National Nano Device Laboratories, Hsinchu, TAIWAN
[3] Center for Condensed Matter Sciences, National Taiwan University, Taipei, TAIWAN
[4] Center for Condensed Matter Sciences, Department of Physics and Department of Electrical Engineering, National Taiwan University, Taipei, TAIWAN

ABSTRACT

The evolution of structural, electrical and magnetic properties with the isovalent chemical substitution of Ca^{2+} into the Sr^{2+} sites in new series of two-dimensional $La_{1.2}(Sr_{1.8-x}Ca_x)Mn_2O_7$ compounds ($x = 0 \sim 1.8$) and three-dimensional $La_{0.6}(Sr_{0.4-x}Ca_x)MnO_3$ compounds ($x = 0 \sim 0.4$) are investigated. The highest magnetoresistance (MR) ratios $[\rho(0) - \rho(H) / \rho(0)]$ of 52 % ($H = 1.5$ T) at 102 K and 13 % ($H = 1.5$ T) at 210 K were observed for the $x = 0.4$ samples in $La_{1.2}(Sr_{1.8-x}Ca_x)Mn_2O_7$ and $La_{0.6}(Sr_{0.4-x}Ca_x)MnO_3$, respectively. The Curie temperatures (T_C) decreased from 135 K to 102 K and 370 K to 270 K for $x = 0$ to 0.4 in $La_{1.2}(Sr_{1.8-x}Ca_x)Mn_2O_7$ and $La_{0.6}(Sr_{0.4-x}Ca_x)MnO_3$, respectively. The compositional dependence of the structural variation has been found in $La_{0.6}(Sr_{0.4-x}Ca_x)MnO_3$. Our results confirm that the dimensionality as well as ionic size plays an important role in controlling the colossal magnetoresistance in manganites.

INTRODUCTION

Since the discovery of high temperature superconductivity in perovskite copper oxides, there has been revived interest in mixed valence manganese perovskites. The ABO_3-type manganites $R_{n+1}Mn_nO_{3n+1}$ (R = rare earth and n = ∞), exhibit a colossal magnetoresistance (CMR) in a relatively small temperature range around the Curie temperature (T_C) [1]. The colossal magnetoresistance is defined as a huge decrease of resistance under an applied magnetic field. It also corresponds to a paramagnetic insulator-like phase at high-temperature and a ferromagnetic metallic phase at low temperature. A suitable substitution of A^{2+} ions for R^{3+} ions results in Mn^{3+}/Mn^{4+} mixed valency, and hence the strong coupling between the magnetic ordering and the electrical conductivity demonstrates a strong relationship between the electrical resistivity and the spin alignment, which has been explained by the double-exchange mechanism. This mechanism describes the transfer of an electron between the Mn^{3+} and Mn^{4+} ions. The electrons in the e_g orbital of Mn^{3+} ($t_{2g}^3 e_g^1$) ions are the charge carriers that move in a background of Mn^{4+} (t_{2g}^3) ions. The alignment of the Mn^{4+} localized spins favors the delocalization of the e_g electrons and reduces the total energy of the system [2,3].

In the Ruddlesden-Popper $La_{n-nx}(Sr,Ca)_{1+nx}Mn_nO_{3n+1}$ system, the dimensionality can be varied by increasing the number of perovskite layers and by controlling the arrangement of MnO_6 octahedra. The compounds with the n = 2 member in $La_{n-nx}(Sr,Ca)_{1+nx}Mn_nO_{3n+1}$ exhibited a conducting ferromagnet with a MR ratio ($[\rho(0) - \rho(H) / \rho(0)]$) higher than those of the n = 3 and n = ∞ members, where $\rho(0)$ is the resistivity at the absence field and $\rho(H)$ is the resistivity at 1.5 T magnetic field. This implies that a lower dimensionality is more favorable for CMR. Many interesting studies have been carried out on the manganites with the chemical compositional $La_{n-nx}Sr_{1+nx}Mn_nO_{3n+1}$ with $x = 0.4$ and $n = 2$. The $La_{1.2}Sr_{1.8}Mn_2O_7$ compound is a conducting ferromagnet with a T_C of 130 K and with a high MR ratio, this composition is a paramagnetic insulator above T_C [4]. Moreover, a detailed study of the manganite $La_{n-nx}Ca_{1+nx}Mn_nO_{3n+1}$ has been

257

performed by Asano *et al.* [5]. In the $La_{1.2}Ca_{1.8}Mn_2O_7$ layered compound, they observed a ferromagnetic transition at $T_C = 240$ K and a higher MR ratio than those of $n = 3$ and $n = \infty$ materials. However, the properties of the solid solution of Sr^{2+} and Ca^{2+} resulting in $La_{1.2}(Sr_{1.8-x}Ca_x)Mn_2O_7$ have not been systematically studied.

In the n = ∞ compounds, the gradual replacement of Ca^{2+} by Sr^{2+} in $La_{0.75}(Ca_{0.25-x}Sr_x)MnO_3$ results in an increase of T_C from ~ 225 K (x = 0) to ~ 340 K (x = 0.25) [6]. The phase transformation from orthorhombic to rhombohedral has been observed in the $La_{0.75}(Ca_{0.25-x}Sr_x)MnO_3$ system with fixed Mn valence, which is difficult to be explained as the effect of increasing the Mn valence. Therefore, a detailed study on the phase transformation is very informative.

In this paper, we demonstrate the evolution of the structural, electrical and magnetic properties with the isovalent chemical substitution of Ca^{2+} into the Sr^{2+} sites in series of two-dimensional $La_{1.2}(Sr_{1.8-x}Ca_x)Mn_2O_7$ compounds (x = 0 ~ 1.8) and three-dimensional $La_{0.6}(Sr_{0.4-x}Ca_x)MnO_3$ compounds (x = 0 ~ 0.4).

EXPERIMENT

High purity powders of La_2O_3, $SrCO_3$, $CaCO_3$ and MnO_2 were weighted in appropriate proportions to obtain the nominal compositions of $La_{1.2}(Sr_{1.8-x}Ca_x)Mn_2O_7$ ($0 \leq x \leq 1.8$) [7, 8] and $La_{0.6}(Sr_{0.4-x}Ca_x)MnO_3$ (x = 0 ~ 0.4). The mixtures were calcined in air for 24 h at 1200 °C and 900°C, respectively. Then, the samples were sintered in air at 1400 ~ 1500 °C for 24 h with intermediate grinding after each heating step. X-ray powder diffraction (XRD) measurements were carried out with a SCINTAG (X1) diffractometer (Cu Kα radiation, $\lambda = 1.5406$ Å) at 40 kV and 30 mA. The program of GSAS [9] was used for the Rietveld refinement in order to obtain the information of crystal structures of $La_{n-nx}(Sr,Ca)_{1+nx}Mn_nO_{3n+1}$ ($n = 2$ and $n = \infty$). The high-resolution transmission electron microscopy (HRTEM) was carried out using a JEOL 4000EX electron microscope operated at 400 kV. Magnetization data were taken from a superconducting quantum interference device (SQUID) magnetometer (Quantum Design).

RESULTS

In Fig. 1 we show the XRD patterns of the series samples of $La_{1.2}(Sr_{1.8-x}Ca_x)Mn_2O_7$ (x = 0 ~ 1.8). All of the samples are single phase. The samples can be indexed to the $Sr_3Ti_2O_7$-type structure with tetragonal unit cell (space group: I4/mmm). Base on our XRD refinements, both lattice constants (*a* and *c*) and cell volume (V) decrease as the Ca content increases. The structural changes are simply due to a manifestation of the size effect between Ca^{2+} [1.18 Å for C.N. (coordination number) = 9] and Sr^{2+} (1.31 Å for C.N. = 9) [10].

The inset of Fig. 2 shows the ideal crystal structure of $La_{1.2}(Sr_{1.4}Ca_{0.4})Mn_2O_7$ (n = 2) which consists of double perovskite layers, and each layer is made up of a two dimensional network of MnO_6 octahedra. The unit cell is shown with a solid line. The arrangement of MnO_6 octahedra is shaded. This model is also supported by HRTEM observations. Fig. 2 shows HRTEM image with the incident electron beam along the [110] direction and the corresponding selected-area electron diffraction pattern. The lattice image clearly shows layers with uniform spacing of about 10 Å, which corresponds to a nearly half the c-axis unit length. There are no stacking faults in this observation, indicating a high homogeneity in this compound. The electron diffraction pattern can be indexed with respect to the $h + k + l = 2n$ reciprocal lattice sections of the tetragonal cell with $a = b$ ~ 3.9 Å and c ~ 20 Å.

The powder XRD patterns of the $La_{0.6}(Sr_{0.4-x}Ca_x)MnO_3$ (x = 0 ~ 0.4) samples are shown in

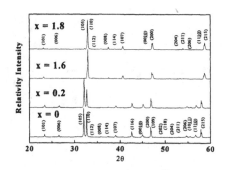

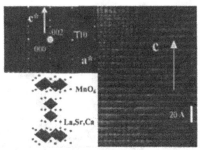

Fig. 1. Powder XRD spectra of the $La_{1.2}(Sr_{1.8-x}Ca_x)Mn_2O_7$ samples.

Fig. 2. HRTEM image along the *c*-axis for $La_{1.2}(Sr_{1.4}Ca_{0.4})Mn_2O_7$ sample. The inset of figure 2 shows the electron diffraction pattern along [110] of the tetragonal cell.

Fig. 3(a). The series samples are all single phase. For the samples with x = 0 and 0.4, all the peaks in each pattern can be indexed with rhombohedral unit cell (space group: *R-3c*) and orthorhombic (space group: *Pbnm*), respectively. As x increases to x = 0.3, some reflection planes merge together, which indicates that in the x < 0.3 region the structure is rhombohedral and in the x ≥ 0.3 region the structure becomes orthorhombic. An increase in the Ca content in $La_{0.6}(Sr_{0.4-x}Ca_x)MnO_3$ leads to an increase in the distortion of the MnO_6 octahedra. The basic perovskite lattice constants (a_p and b_p) also decrease as x increases [as shown in Fig. 3(b)].

The crystal structure and the results of HRTEM of $La_{0.6}(Sr_{0.4-x}Ca_x)MnO_3$ are show in Fig. 4 The HRTEM lattice image along [111] zone-axis direction of $La_{0.6}(Sr_{0.4-x}Ca_x)MnO_3$ (x = 0.1) is shown in Fig. 4(a). Inset of Fig. 4(a) shows the structure model along [111] of the rhombohedral cell (space group: *R-3c*) of $La_{0.6}(Sr_{0.4-x}Ca_x)MnO_3$. The corresponding lattice image is in aggrement with the model. Figure 4 (b) shows the HRTEM lattice image along the [010] zone-axis irection of $La_{0.6}(Sr_{0.4-x}Ca_x)MnO_3$ (x = 0.3). The structural model along [010] of the orthorhombic cell (space group: *Pbnm*) is shown in the inset of Fig. 4(b). The perfect order along the c-axis is confirmed by the HRTEM image. No superstructures were found in the $La_{0.6}(Sr_{0.4-x}Ca_x)MnO_3$ (x = 0 ~ 0.4) series samples.

The highest magnetoresistance (MR) ratios of 52 % (*H* = 1.5 T) at the temperature of 102 K

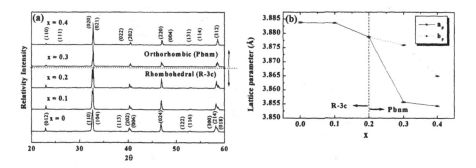

Fig. 3. (a) XRD spectra of the series $La_{0.6}(Sr_{0.4-x}Ca_x)MnO_3$ samples. (b) Basic perovskite cell parameters (a_p and b_p) as a function of x in $La_{0.6}(Sr_{0.4-x}Ca_x)MnO_3$.

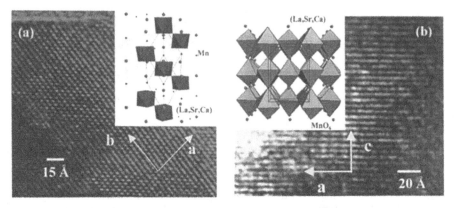

Fig. 4. HRTEM lattice images taken along the (a) [001] zone-axis directions of the La$_{0.6}$(Sr$_{0.4-x}$Ca$_x$)MnO$_3$ sample with x = 0.1. The corresponding rhombohedral structure is shown in the insets of Figures 4(a). (b) [010] zone-axis directions of the La$_{0.6}$(Sr$_{0.4-x}$Ca$_x$)MnO$_3$ sample with x = 0.3. The corresponding orthorhombic is shown in the insets of Figures 4(b).

and 13 % (*H* = 1.5 T) at the temperature of 210 K were observed for x = 0.4 in La$_{1.2}$(Sr$_{1.8-x}$Ca$_x$)Mn$_2$O$_7$ and La$_{0.6}$(Sr$_{0.4-x}$Ca$_x$)MnO$_3$, respectively, as measured by resistivity [7].The temperature dependence of the magnetization (emu/g vs. T) in a magnetic field of 0.1 T of La$_{1.2}$(Sr$_{1.8-x}$Ca$_x$)Mn$_2$O$_7$ (x = 0 ~ 1.8) and La$_{0.6}$(Sr$_{0.4-x}$Ca$_x$)MnO$_3$ (x = 0 ~ 0.4) is shown in Figs. 5(a) and 5(b), respectively. A systematic analysis of the phase diagram of the transition temperature (T$_C$) versus the concentration of Ca^{2+} for La$_{1.2}$(Sr$_{1.8-x}$Ca$_x$)Mn$_2$O$_7$ (x = 0 ~ 1.8) indicates that the T$_C$'s decreases at both regions (x < 1.0 and x >1.0) from 135 K to 71 K of x = 0 to 0.8 and from 355 K to 260 K of *x* = 1.2 to 1.8, respectively [8,11]. In Fig. 5(b), within the temperature range of 250 ~ 350 K, it indicates a sharp paramagnetic to ferromagnetic transition in the series of samples. A decrease in the T$_C$'s from > 350 K of x = 0 to 270 K of x = 0.4 was observed. These results lead to an understanding that the T$_C$'s is very sensitive to chemical pressure (i.e., structural distortions induced by changing the average radius of the cations).

In Fig. 6, we show the T$_C$'s of (Ln,A)MnO$_3$ as a function of the average radius (assuming

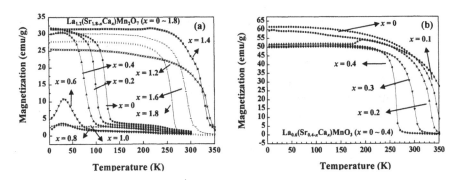

Fig. 5. Temperature dependence of magnetization at a magnetic field of 0.1 T for the series of (a) La$_{1.2}$(Sr$_{1.8-x}$Ca$_x$)Mn$_2$O$_7$ (x = 0 ~ 1.8) and (b) La$_{0.6}$(Sr$_{0.4-x}$Ca$_x$)MnO$_3$ (x = 0 ~ 0.4).

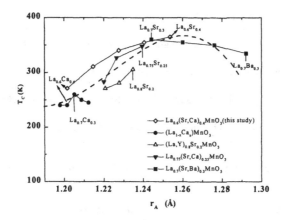

Fig. 6. Phase diagram of T_C versus $\langle r_A \rangle$ for (Ln,A)MnO$_3$, where the average radius $\langle r_A \rangle$ (assuming ninefold coordination) of the Ln/A (Ln= La or Y; A = Ca, Sr or Ba) cations is $[(1-x) r_{Ln} + x r_A]$.

ninefold coordination) of the Ln/A (Ln= La or Y; A = Ca, Sr or Ba) cations, $\langle r_A \rangle = [(1-x) r_{Ln} + x r_A]$. The data include our series La$_{0.6}$(Sr$_{0.4-x}$Ca$_x$)MnO$_3$ compounds and previous results [12-15]. An increase in $\langle r_A \rangle$ gives rise to an increase in T_C. Therefore, the isovalent chemical substitution of the smaller Ca^{2+} ions into the bigger Sr^{2+} sites in the series of La$_{n-nx}$(Sr,Ca)$_{1+nx}$Mn$_n$O$_{3n+1}$ causes the decrease of the T_C's. Detailed studies on the mechanism of CMR and their correlation to the crystal structure are still underway.

CONCLUSIONS

An investigation of the new series of two-dimensional La$_{1.2}$(Sr$_{1.8-x}$Ca$_x$)Mn$_2$O$_7$ ($x = 0 \sim 1.8$) and three-dimensional La$_{0.6}$(Sr$_{0.4-x}$Ca$_x$)MnO$_3$ ($x = 0 \sim 0.4$) mangnites has allowed us to establish the relationship between the ionic size and the Curie temperature. This study has demonstrated that the size of the interpolated cation by tuning the concentration between Ca^{2+} and Sr^{2+} plays a crucial role in controlling the magnetrotranport properities of the colossal magnetroresistance in La$_{n-nx}$(Sr,Ca)$_{1+nx}$Mn$_n$O$_{3n+1}$ materials.

ACKNOWLEDGMENTS

This research is financially supported by National Science Council of the Republic of China under Grant numbers NSC 89-2113-M-002-004 and NSC 89-2112-M-002-030.

REFERENCES

1. S. Jin, T. H. Tiefel, M. McCormack, R. A. Fastnacht, R. Ramesh, and L. H. Chen, Science **264**, 413 (1994).

2. C. Zener, Phys. Rev. **82**, 403 (1951).

3. P. -G. de Gennes, Phys. Rev. **118**, 141 (1960).

4. Y. Moritomo, A. Asamitsu, H. Kuwahara, and Y. Tokura, Nature **380**, 141 (1996).

5. H. Asano, J. Hayakawa, and M. Matsui, Phys. Rev. B **57**, 1052 (1998).

6. Z. B. Guo, W. Yang, Y. T. Shen, and Y. W. Du, Solid State Commun. **105**, 89 (1998).

7. R. S. Liu, C. H. Shen, J. G. Lin, C. Y. Huang, J. M. Chen, and R. G. Liu, J. Chem. Soc. Dalton Trans. 923 (1999).

8. R. S. Liu, C. H. Shen, J. G. Lin, J. M. Chen, R. G. Liu, and C. Y. Huang, J. Inorg. Mater.**1**, 61 (1999).

9. A. C. Larson and R. B. Von Dreele, *Generalized Structure Analysis System* (Los Alamos National Laboratory Los Alamos NM 1994).

10. R. D. Shannon, Acta. Cryst. Sect. A **32**, 751 (1976).

11. C. H. Shen, R. S. Liu, S. F. Hu, J. G. Lin, C. Y. Huang, and S. H. Sheu, J. Appl. Phys. **86**, 2178 (1999).

12. R. Mahendiran, S. K. Tiwary, A. K. Raychaudhuri, T. V. Ramakrishnan, R. Mahesh, N. Rangavittal, and C. N. R. Rao, J. Solid State Chem. **114**, 297 (1995).

13. Z. B. Guo, Y. W. Du, J. S. Zhu, H. Huang, W. P. Ding, and D. Feng, Phys. Rev. Lett. **78**, 1142 (1997).

14. H. Y. Huang, S. -W. Cheong, P. G. Radaelli, M. Marezio, and B. Batlogg: Phys. Rev. Lett. **75**, 914 (1995).

15. P. G. Radaelli, G. Iannone, M. Marezio, H. Y. Hwang, S. –W. Cheong, J. D. Jorgensen and D. N. Argyriou, Phys. Rev. B **56**, 8265 (1997).

B-SITE DOPED LANTHANUM STRONTIUM MANGANITES BY THE DAAS TECHNIQUE

S. YANG*, M.R. KOLODY*, C.-T. LIN*, P.M. ADAMS**, AND D.M. SPECKMAN**
*Department of Chemistry, Northern Illinois University, DeKalb, IL 60115
**The Aerospace Corporation, P.O. Box 92957, Los Angeles, CA 90009

ABSTRACT

Magnetic perovskites of the general form $La_{0.7}Sr_{0.3}Mn_{1-y}Fe_yO_3$ (y = 0, 0.05, 0.10, and 0.15) have successfully been synthesized using deposition by aqueous acetate solution (DAAS). Crystalline, iron-doped, lanthanum strontium manganite (Fe-doped LSMO) powders are obtained by preparing an aqueous solution of metal acetate precursors in the proper stoichiometry, drying the solution to generate a glassy gel, consolidating the gel, and then firing the gel for short periods of time (<2hrs). This novel technique has the potential for depositing large area thin films with high throughput and low cost. Powder samples of $La_{0.7}Sr_{0.3}Mn_{1-y}Fe_yO_3$ prepared by DAAS and annealed for 100 minutes at 1200°C are of high purity, are single phase, and exhibit excellent electrical and magnetic characteristics. Powders annealed at 1200°C or greater exhibit sharp metal-insulator transitions. Increasing the iron dopant concentration in these powders from 0% to 15% decreases the metal-insulator transition temperature of these samples from ~360K to about 140K. The resistivity of these powders also increases with increasing substitution of the lattice B-site with iron, as does the unit cell volume of the lattice. Preparation of an iron doped LSMO powder that exhibits a maximum magnetoresistance at 305K was successfully carried out via a careful selection of iron content and anneal temperature. This compound, $La_{0.7}Sr_{0.3}Mn_{0.93}Fe_{0.07}O_3$, exhibits a magnetoresistance of 40% at 305K and an applied field of 5 Tesla.

INTRODUCTION

The recent discovery of large magnetoresistive effects in doped rare-earth manganites, $Ln_{1-x}A_xMnO_3$ (Ln = rare earth metal, A = divalent alkaline earth cation such as Ca, Sr, Ba), has sparked a renewed interest in the study of these materials over the past few years [1]. Due to their unusual magnetic and electronic properties, rare-earth manganites have a variety of potential applications, including magnetic storage cells for magnetoresistive random access memories (MRAM), solid electrolytes for fuel cells, and infrared bolometers. Much of the research to date on magnetic perovskites has focused on studying the effects of alkaline earth substitution into the "A"-site of the ABO_3 perovskite structure. However, there have been relatively few investigations on the effects of "B"-site doping in magnetic perovskites. Since colossal magnetoresistance in rare earth manganites is believed to involve conduction pathways through the B-site of the lattice, such studies may lead to a greater understanding of the mechanisms responsible for colossal magnetoresistivity, and might provide some insight on how to best enhance these properties.

Strontium-doped lanthanum manganites, $La_{1-x}Sr_xMnO_3$, appear to be particularly promising materials for use advanced solid state magnetic memories, since these compounds exhibit magnetoresistive behavior at practical device operating temperatures of 300K or greater [2]. Because of this potential application, we decided to study the effects of B-site doping in lanthanum strontium manganites, and in particular, to examine the effect of iron doping on the physical and electronic properties of $La_{0.7}Sr_{0.3}MnO_3$. The methodology we used to produce the iron-doped LSMO materials for this investigation is a process we previously used to deposit powders and thin films of both ferromagnetic and magnetoresistive materials [3]. The DAAS, or "deposition by aqueous acetate solution", technique is a simple solution-based process that is capable of producing thin films over large area substrates, with reasonable throughput and low cost. The DAAS method offers the advantages of good stoichiometric control, flexibility in powder versus thin film deposition, stable precursor solutions, and low anneal times (< 2hr). We describe here the DAAS process for preparing iron-doped lanthanum strontium manganites, and discuss the properties of these B-site doped perovskites as determined by their physical, electrical, and magnetic characteristics.

Mat. Res. Soc. Symp. Proc. Vol. 602 © 2000 Materials Research Society

EXPERIMENTAL

The DAAS technique used to prepare powders of the general formula $La_{0.7}Sr_{0.3}Mn_{1-y}Fe_yO_3$ is described below. Lanthanum acetate hydrate, strontium acetate, iron acetate, and manganese acetate were dissolved in a deionized water/ acetic acid mixture in the same metal ratios as desired in the final Fe-doped LSMO product, and the resultant combination was sonicated to produce a clear, stable solution. For the preparation of bulk powders, the solution was then dried under an air purge to generate a hard, glassy gel, the gel was consolidated for 6 hours at 600°C to generate a crude solid, and the solid was subsequently annealed at either 600, 900, or 1200°C for 100 minutes. Product powders were analyzed by x-ray diffraction (XRD), energy dispersive x-ray spectroscopy (EDX), and by temperature-dependent electrical and magnetic measurements.

For x-ray diffraction measurements, samples were analyzed with Cu Kα and/or Cr Kα radiation using a personal computer controlled Philips Electronics Instruments APD 3720 vertical powder diffractometer equipped with a θ compensating slit, diffracted-beam graphite crystal monochromator and a scintillation detector. Bulk samples were initially scanned in the range from 20-130 degrees with 0.01 degree steps. In order to ensure accuracy, National Institute of Standards and Technology Standard Reference Material #640b silicon powder was added to all samples.

EDX measurements were carried out at 20 keV using a Cambridge Instruments scanning electron microscope equipped with an Oxford Instruments ISIS Energy Dispersive X-ray Analyzer. A pure cobalt sample was used as the calibration source for all of the quantitative measurements.

Resistivity was measured using a standard four-point probe in the temperature range 10-350K, and electrical contacts were established using silver paint. Magnetoresistance (R_H) was measured with the Quantum Design Physical Properties Measurement System equipped with a 7 Tesla superconducting magnet. The R_H measurements were performed as a function of temperature at constant dc bias magnetic field (H) for both zero field cooled (ZFC) and field cooled (FC) samples, and as a function of magnetic field at constant temperature.

RESULTS AND DISCUSSION

In order to assess the suitability of the DAAS technique for preparing uniform, iron-doped LSMO powders of desired stoichiometry, compounds of the general form $La_{0.7}Sr_{0.3}Mn_{1-y}Fe_yO_3$ were prepared (y = 0, 0.05, 0.10, and 0.15), annealed at 1200°C for 100 minutes, and then subjected to visual inspection and composition analyses. The general morphology of each sample was examined using scanning electron microscopy. The powders were all found to exhibit the same general physical characteristics, that is, they consisted of two types of crystalline microstructures: dark-colored masses of very small, densely packed particles, and white colored, medium-sized crystallites (Figure 1). The overall stoichiometry of each powder, as well as the individual compositions of white and dark crystallites within each powder, was determined using EDX. For the bulk powder EDX analysis, a scan area of 425 μm x 475 μm was measured for the analysis, whereas a spot mode analysis was used for determining the composition of individual crystallite regions.

The bulk EDX analysis for each iron-doped LSMO powder compared very well with the target composition of that sample. The mole fractions of A-site atoms (lanthanum and strontium) were typically within 5% of the predicted values, with the exception of those samples with high iron content (y = 0.15), which exhibited greater deviation. The mole fractions of B-site atoms (manganese and iron) were typically correct to within 1% of the anticipated values. Overall, the powders were found to be somewhat B-site deficient however. The ratio of B-site atoms to A-site atoms was always less that unity, with a deficiency that ranged from 3% to 10%. For example, the overall composition of a powder targeted to have a stoichiometry of $La_{0.7}Sr_{0.3}Mn_{0.90}Fe_{0.10}O_3$ was found to have an actual overall stoichiometry of $La_{0.74}Sr_{0.26}(Mn_{0.90}Fe_{0.10})_{0.91}O_{3-\delta}$ as measured by EDX.

Figure 1. Scanning electron micrograph of a $La_{0.7}Sr_{0.3}Mn_{0.95}Fe_{0.05}O_3$ powder annealed at 1200°C

There was more variability in the composition of the individual crystallites within a given sample, however. The medium sized, white crystallites typically had compositions very close to that of the target stoichiometry, but the dark crystalline regions exhibited lanthanum and strontium mole ratios that deviated from the targeted composition by sometimes as much as 15%. Interestingly, the relative manganese and iron mole fractions were typically correct to within 1% of their anticipated stoichiometries, even in these dark regions. The darker crystallites also tended to be more B-site atom deficient than the white crystallites.

The influence of annealing temperature, T_a, on the structure and electrical characteristics of iron-doped LSMO powders was also examined. Powders with a stoichiometry of $La_{0.7}Sr_{0.3}Mn_{0.95}Fe_{0.05}O_3$ were prepared by DAAS, subjected to annealing temperatures of 600, 900, or 1200°C, and then analyzed by x-ray diffraction and by electrical measurements. For each of the annealing temperatures, the annealing time was kept constant at 100 minutes. At all anneal temperatures, the powders were found to be crystalline, and Figure 2a shows a portion of the XRD spectra for each of these powders. For powders annealed at 600°C, the XRD spectrum consists of only broad, single 2θ peaks, but by increasing the annealing temperature, the peaks narrow and sharpen. At an annealing temperature of 1200°C, sharp doublet peaks appear in the XRD spectrum. The monoclinic lattice parameters "a" and "b" shown in Figure 2b for the sample annealed at 600°C show that this sample is pseudo-tetragonal ("a" ~ "b"). However, with increasing anneal temperature, the lattice becomes increasingly distorted, and at T_a = 1200°C, the powders can be indexed as rhombohedral.

Resistivity-temperature curves for the $La_{0.7}Sr_{0.3}Mn_{0.95}Fe_{0.05}O_3$ powders, plotted as a function of anneal temperature, are shown in Figure 3. The resistivities of each powder sample were found to decrease by an order of magnitude for each increase in anneal temperature, and the peak resistivity temperature for each sample was found to increase with increasing T_a. The resistivity peaks were observed to sharpen as T_a increased as well, with a sharp metal-insulator transition at ~ T_c observed for T_a = 1200°C. At the 1200°C anneal temperature, a second broad magnetic transition is also observed below T_c.

These variations in structure and electrical properties with anneal temperature may be primarily due to differences in sample homogeneity and crystallite size. For the sample annealed at 600°C, the broad XRD peaks may be due to the presence of many, slightly different phases or compositions in the material and/or may be due to the presence of small crystallites. Also, low temperature growths are known to produce LSMO materials with significant lanthanum and strontium vacancies [4], and these materials often contain more oxygen than samples grown or annealed at high temperatures. A more heavily oxidized powder will have a higher Mn^{4+}/Mn^{3+} ratio, which causes a reduction in the amount of Jahn-Teller (J-T) distortion in the lattice. The pseudo-tetragonal structure of the Fe-doped LSMO powders annealed at 600°C is consistent with a

265

reduced J-T distortion. Furthermore, the high resistivity of these samples is consistent with the presence of small crystallites with a large grain boundary area, since conduction is impeded at these interfaces.

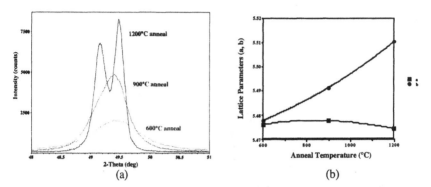

Figure 2. a) Powder XRD patterns for La$_{0.7}$Sr$_{0.3}$Mn$_{0.95}$Fe$_{0.05}$O$_3$ annealed at 600, 900, and 1200°C, b) monoclinic cell lattice parameters 'a' and 'b' for La$_{0.7}$Sr$_{0.3}$Mn$_{0.95}$Fe$_{0.05}$O$_3$ powders annealed at 600, 900, and 1200°C.

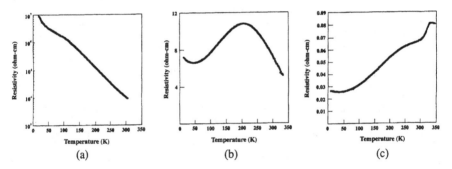

Figure 3. Resistivity-temperature curves for La$_{0.7}$Sr$_{0.3}$Mn$_{0.95}$Fe$_{0.05}$O$_3$ as a function of anneal temperature. a) 600°C anneal, b) 900°C anneal, and c) 1200°C

As T$_a$ increases, the changes in the XRD and in the resistivity-temperature curves are consistent with the formation of larger, more homogeneous crystallites. The growing distortion in the lattice with increasing T$_a$ is consistent with a decrease in the Mn^{4+}/Mn^{3+} ratio in these powders, and the sharp doublets in the XRD pattern at T$_a$ = 1200°C suggest that this powder is probably high purity and single phase. The 1200°C annealed sample also exhibits a very low resistivity indicative of bulk lattice conduction pathways, and the sharp metal-insulator transition at ~ T$_c$, indicates that this powder exhibits intrinsic bulk lattice properties. The broad transition observed below T$_c$ suggests some magnetic inhomogeneities exist in the sample, probably at the crystallite surfaces [5].

The influence of varying the amount of iron dopant on the structural and electrical characteristics of La$_{0.7}$Sr$_{0.3}$MnO$_3$ has also been examined. Powders of the form La$_{0.7}$Sr$_{0.3}$Mn$_{1-y}$Fe$_y$O$_3$ (y = 0, 0.05, 0.10, and 0.15) were annealed at 1200°C for 100 minutes, and then analyzed by XRD and electrical measurements. For each of the iron dopant concentrations studied, the powders were found to be highly distorted and could be indexed as either rhombohedral or monoclinic. The monoclinic lattice parameters a, b, c, beta, and unit cell volume were all found to increase as a function of increasing iron content in the films until the mole fraction of iron reached the high value of 15%. At 15% iron content, parameters c, beta, and cell volume decreased.

Previous literature has reported that iron is incorporated into LSMO materials only as the Fe^{3+} ion [6]. This ion has Pauling ionic radius of 78.5pm in its high spin (HS) state, and ionic radius of 69pm in its low spin (LS) state. The increase in lattice parameters with increasing iron content suggests that Fe^{3+} must be in its high spin configuration, and may be substituting for the smaller Mn^{4+} (67pm) in the lattice. If the Fe^{3+} ion substitutes for Mn^{3+} (78.5pm), the lattice size may still increase if the lattice contains fewer manganese vacancies as a result of this substitution. The HS Fe^{3+} ion has an electron in each of its e_g orbitals, and therefore interacts with the lattice much like Mn^{3+}, which is consistent with the observed increased lattice distortion with increasing iron dopant concentration. At an iron dopant concentration of 15%, the cell volume contracts, which may be due to Fe^{3+} substitution on a different lattice site, or may be due to changes in the A-site stoichiometry (EDX indicates that for Fe=0.15, the samples are strontium-rich).

Figure 4 shows the resistivity-temperature curves for $La_{0.7}Sr_{0.3}Mn_{1-y}Fe_yO_3$ powders as a function of increasing iron content. The resistivities of each powder sample were found to increase with increasing iron concentration, whereas the metal-insulator transition temperature was found to decrease with increasing iron content.

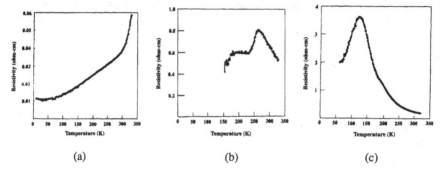

Figure 4. Resistivity-temperature curves for $La_{0.7}Sr_{0.3}Mn_{1-y}Fe_yO_3$ as a function of iron content. a) y = 0.05, b) y = 0.10, and c) y = 0.15

The increased resistivity with increased iron content suggests that the added dopant disrupts the long range Mn-O-Mn order in the B-site of the lattice, resulting in smaller polarons and fewer conduction pathways. Similarly, the decrease in metal-insulator transition temperature with added iron dopant could be attributed to the formation of localized antiferromagnetic regions in the B-site of the lattice (similar to Mn^{3+}-O-Mn^{3+} interactions), which disrupt long range ferromagnetic coupling through Mn^{3+}-O-Mn^{4+} bonds.

Through proper control of the iron dopant concentration and anneal temperature, we proposed that an iron-doped LSMO powder could be prepared that would exhibit a maximum magnetoresistive effect at room temperature. Using DAAS, a sample with the stoichiometry $La_{0.7}Sr_{0.3}Mn_{0.93}Fe_{0.07}O_3$ was prepared and annealed at 1300°C for 100 minutes. The resistivity as a function of temperature and applied magnetic field for this sample is illustrated in Figure 5, which demonstrates a magnetoresistance of 40% at 305K and an applied field of 5 Tesla.

CONCLUSIONS

It has been demonstrated that high purity, single phase, rhombohedral, iron-doped LSMO powders of the general form $La_{0.7}Sr_{0.3}Mn_{1-y}Fe_yO_3$ can be prepared using the novel DAAS

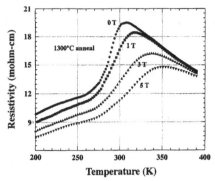

Figure 5. Resistivity as a function of temperature and applied magnetic field for
La₀.₇Sr₀.₃Mn₀.₉₃Fe₀.₀₇O₃ powder sample annealed at 1300°C

technique. Using the proper concentration of iron dopant and anneal temperatures of at least 1200°C, Fe-doped LSMO powders can be produced that exhibit a significant magnetoresistive effect at room temperature. The highly flexible DAAS process offers advantages over conventional preparations of rare-earth manganites in that the solution-based chemistry allows for intimate mixing of the precursors, the La:Sr:Mn:Fe ratio in the precursor solution can be well controlled, the metal acetate solutions are very stable, reaction conditions are relatively mild, and the required annealing times are very short. Furthermore, this technique can be used for the production of both powders and thin films. This technique has the potential for depositing excellent quality rare-earth manganites, with high throughput and low cost, for a wide range of applications.

ACKNOWLEDGMENTS

We wish to thank Mr. Michael Tueling for carrying out the EDX analyses of these samples, and we thank the Aerospace Independent Research and Development program and ONR/ARPA for financial support of this project.

REFERENCES

1. C.N. R. Rao, A.K. Cheetham, and R. Mahesh, Chem. Mater. **8**, 2421 (1996)

2. A. Urushibara, Y. Moritomo, T. Arima, A. Asamitsu, G. Kido, Y. Tokura, Phys. Rev. B **51**, 14103 (1995)

3. S. Yang, C.T. Lin, K. Rogacki, B.M. Dabrowski, P.M. Adams, and D.M. Speckman, Mat. Res. Soc. Symp. Proc. **474**, 241 (1997); C.T. Lin, "Production of PT/PZT/PLZT Thin Films, Powders, and Laser 'Direct Write' Patterns", U.S. patent 5,188,902, February 23, 1993; U.S. patent 5,348,775, September 20, 1994

4. S. Otoshi, H. Sasaki, H. Ohnishi, M. Hase, K. Ishimaru, M. Ippommatsu, T. Higuchi, M. Miyayama, and H. Yanagida, J. Electrochem. Soc. **138**, 1519 (1991)

5. N. Zhang, W. Zhong, and W. Ding, J. Electron. Mater. **27**, 1 (1998)

6. C.H. Kim, S.I. Park, and S.Y. Bae, unpublished

TWO PEAK EFFECT IN GMR : A CHEMICAL EFFECT ?

B. Vertruyen[1], A. Rulmont[1], S. Dorbolo[2], H. Bougrine[3], Ph. Vanderbemden[3], M. Ausloos[2] and R. Cloots[1]
[1] SUPRAS and LCIS, Institute of Chemistry B6, University of Liège, B-4000 Liège, Belgium
[2] SUPRAS, Institute of Physics B5, University of Liège, B-4000 Liège, Belgium
[3] SUPRAS, Montefiore Electricity Institute B28, University of Liège, B-4000 Liège, Belgium

ABSTRACT

We show that the synthesis conditions have a dramatic influence on the resistivity behavior of calcium and sodium doped LaMnO$_3$. Several samples prepared by low-temperature techniques exhibit a double-peaked curve of resistance versus temperature. The model of spin-polarized intergranular tunneling provides a good approach to discuss our experimental results. Grain size and crystallinity are proved to be essential parameters, which are strongly influenced by the preparation process (e.g. the precursors nature and the thermal treatment).

INTRODUCTION

Since the discovery of colossal magnetoresistance in perovskite manganates LaMnO$_3$, these compounds have attracted great interest. It is well established that the magnetic and electrical transport properties are very sensitive to the Mn^{4+}/Mn^{3+} ratio as well as to the lattice distorsions. For example, a considerable change in magnetoresistance can be induced by doping the lanthanum site by another trivalent metal ion with a different ionic radius [1].

The physical properties depend also on the synthesis conditions, which influence the material microstructure. Mono- and polycrystalline samples of the same chemical composition exhibit significantly different properties [2]. Those differences are generally attributed to the presence of grain boundaries in the granular sample, the interface between each grain being considered as a barrier for the exchange interaction between Mn^{4+} and Mn^{3+} ions. The quality of the interface thus plays a significant role in the understanding of the physical properties.

In this paper, we report an anomalous temperature behavior of the resistivity for polycrystalline samples prepared by low-temperature methods. Amorphisation or degradation of the crystallinity at the grain surfaces has to be taken into account, depending on the preparation process. Different mechanisms can be considered : unconventional substitution-induced strengthening of the potential barrier for intergranular tunneling [3], or magnetic disorder at the grain surfaces reducing the probability of electrons tunneling across the intergranular barrier [4], or both. The model of spin-polarized intergranular tunneling proposed by Huang et al. [5] and extensively reviewed by Zhang et al. [4] will provide a good approach to discuss our results.

EXPERIMENTAL

La$_{0.7}$Ca$_{0.3-x}$Na$_x$MnO$_{3+\delta}$ samples were prepared by three different methods, starting from an aqueous solution of the metallic cations : sol-gel method with urea as gelifiant agent (x = 0.1), combustion of a nitrate-urea mixture (x = 0 and 0.1) and combustion by decomposition of ammonium nitrate (x = 0.1). Two different thermal treatments were applied to the as-formed precursors. After a first treatment at 773 K for 15 h, the powders were calcined at powders were next pressed into pellets and sintered at 1523 K for 10 h. In order to make the discussion easier, labels for the different samples are indicated in table I.

The homogeneity of the samples was checked by X-ray diffraction and EDX analysis. A small amount of Mn_3O_4 was found in the combustion samples. The average grain size was estimated by scanning electron microscopy (SEM). The electrical resistance as a function of temperature was measured using the conventional four-probe method.

TABLE I : Summary of labels, synthesis conditions and average grain sizes

Composition	label	synthesis method	thermal treatment	average grain size
$La_{0.7}Ca_{0.2}Na_{0.1}MnO_3$	LCNSG1	sol-gel	short[a]	4 μm
	LCNCU1	combustion with urea	short[a]	4 μm
	LCNCU2	combustion with urea	long[b]	15 μm
	LCNC1	combustion without urea	short[a]	5 μm
$La_{0.7}Ca_{0.3}MnO_3$	LCCU1	combustion with urea	short[a]	1.5 μm

[a] : 773 K (15 h) – 1523 K (3 h) – 1523 K (10 h)   [b] : 773 K (15 h) – 1573 K (24 h) – 1523 K (10 h)

RESULTS

Transport measurements

Figure 1 shows the temperature dependence of the resistance for the LCNCU1 combustion sample. Contrary to the "canonical" behavior (a sharp single-peak curve as in a $La_{0.7}Ca_{0.3}MnO_3$ sample prepared by the conventional solid state reaction [6]), two peaks can be seen in the resistance as a function of temperature [R(T)] curve : a broad peak at low temperature (ca. 240 K) and a sharper one at high temperature (ca. 300 K). Further experiments were performed to determine the origin of this unusual behavior.

The R(T) curve of the LCCU1 combustion sample is shown in figure 2. The presence of a double peak indicates that the presence of two different doping cations on the La site is not responsible for the double-peak behavior seen in figures 1 and 2.

A double peak is also observed in the R(T) curve of the LCNSG1 sol-gel sample (Figure 3). Contrary to the combustion samples, this sol-gel sample contains no Mn_3O_4. We can conclude that neither a combustion synthesis nor the presence of Mn_3O_4 are necessary conditions for the appearance of the double-peak behavior.

The R(T) curve of the LCNC1 combustion sample shows only one peak (Figure 4). This peak is situated at about the same temperature as the high-temperature peak in the LCNCU1 sample. The LCNCU1 and LCNC1 samples have been respectively prepared in presence and in absence of urea : thus, synthesis conditions seem to play a significant role in the appearance of a double peak in the R(T) curve.

Given that the LCNC1 sample contains a small amount of Mn_3O_4, it turns out that the presence of Mn_3O_4 does not induce the double-peak behavior.

Figure 5 shows the R(T) curve of the LCNCU2 combustion sample. Only one peak can be seen in this sample, which only differs from the LCNCU1 sample by the long high-temperature treatment time. As the other "double-peak samples" undergo the same high-temperature treatment, a decrease of the broad peak and an increase of the sharp peak are systematically observed.

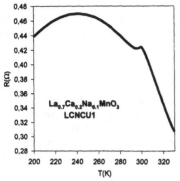

Fig. 1 : Zero-field resistance as a function of temperature for the LCNCU1 sample

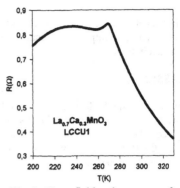

Fig. 2 : Zero-field resistance as a function of temperature for the LCCU1 sample

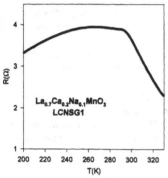

Fig. 3 : Zero-field resistance as a function of temperature for the LCNSG1 sample

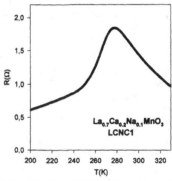

Fig. 4 : Zero-field resistance as a function of temperature for the LCNC1 sample

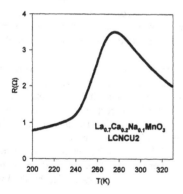

Fig. 5 : Zero-field resistance as a function of temperature for the LCNCU2 sample

Grain size

The transition width has been proposed to depend on the grain size [7]. The grain sizes of our polycrystalline samples were estimated by scanning electron microscopy. The results are summarized in table 1.

DISCUSSION

By confronting the R(T) curves with the average grain size values for the many different samples that we prepared (several of them are not mentioned in this paper), we have observed a rough correlation between the grain size and the shape of the R(T) curve. As the grain size increases, the R(T) curve shows a shape evolution ; from a very broad low-temperature peak to a sharp high-temperature peak, with intermediate situations where a double-peak behavior is observed.

A similar trend was reported by Zhang *et al.* [4] for samples of the $La_{1-x}Sr_xMnO_3$ system prepared by citrate sol-gel method. (Their samples with different grain sizes are obtained by varying the sintering temperature.) These authors have proposed a model based on spin-dependent tunnelling. They assume that a grain can be divided into a core phase and a surface phase, with different T_C. The sharp high-temperature peak is supposed to reflect the "intrinsic" properties of the material, whereas the broad low-temperature peak would result from a spin-dependent interfacial tunnelling, which would stem from the existence of a different grain surface state.

Experimental results recently reported by Hossain *et al.* [8] support the hypothesis of a difference between core and surface states being at the origin of the double-peak behavior. In that study, samples that initially exhibited a single-peak R(T) curve were submitted to a treatment in reducing atmosphere (very low O_2 partial pressure) : the appearance of a second peak at low temperature was observed. Given that the grain boundary diffusion coefficient for oxygen is higher than the bulk diffusion coefficient [8], it can be concluded that the performed treatment induced the appearance of less oxygenated regions at the grain surface.

In the present work, the different shapes of the R(T) curves would thus be due to the different grain sizes. When the grain size increases, the sharp peak (which reflects the intrinsic properties of the material) becomes predominant, since the ratio between the surface and the volume of the grain varies with 1/r where r is the radius of the grain (assumed to be spherical).

However, if we compare the R(T) curves of the LCNCU1 and LCNC1 samples, we can note that they are quite different (double-peaked and single-peaked, respectively), though the difference in grain size is not very high. Moreover, by comparing our results with data from the literature (for other compositions) [9], we can find a few contradictions with respect to the expected dependence.

It is useful to examine more carefully the results of Hossain *et al* [8]. In the experiments they performed, the annealing temperature for the reducing treatment was only 900°C. Moreover, this treatment lasted only a few hours. As a consequence, it is well-founded to consider that the grain size was left unchanged after the reducing treatment. Thus, different shapes of R(T) curves occur for identical grain sizes and the grain size is not the only parameter that has to be taken into account to explain the double-peak behavior !

However, the zone in which the material is "different" from the core material has a non-zero thickness. We will use henceforth the term "shell" rather than "surface" in the discussion. The grain shell has not necessary the same thickness in all samples. (For example, the shell is grain size, the volume of the "intrinsic" material increases and the high-temperature peak becomes predominant when the shell thickness decreases.

Our LCNC1 sample shows a single-peak R(T) curve, whereas the LCNCU1 sample exhibits a double peak. Both samples have similar grain sizes. The shell of the LCNCU1 sample is thus thicker. The only difference in the synthesis scheme of those samples is the presence of urea in the case of the LCNCU1 sample. This urea is not totally decomposed during the combustion. Carbonaceous residues remain unburnt (at least at the beginning of the thermal treatment) and probably interfere with the crystallization of the perovskite phase. It is worth emphasizing that double-peak R(T) curves are generally observed for samples prepared by low-temperature synthesis techniques with carbonaceous precursors and short thermal treatment [10].

We conjecture that the grain shell corresponds to a less crystallized, more disordered phase. A change in cell parameters (possibly due to a disorder in cations or oxygen vacancies distribution) induces a modification of the Mn-O distance. The physical properties of manganate perovskites have been proved to depend strongly on crystallographic distorsions [11]. The disordered material has thus a lower T_C than the well-crystallized compound, whose "intrinsic" properties are reflected in the sharp high-temperature peak. The considerable width of the low-temperature peak is attributed to a distribution of T_C's : there is no sharp transition between a crystallized material and a disordered phase, but more probably a smooth evolution of the cell parameters as the grain radius increases during the grain growth.

We have seen that when a sample is maintained at high temperature for a long time, the broad peak decreases or even disappears. This is consistent with our model : since the ionic diffusion in the solid state is easier at high temperature, a sintering process promotes grain growth and improves the crystallization. Therefore, the thickness of the disordered shell decreases and the sharp peak becomes predominant.

Before concluding, some assumptions have to be justified.

The little difference in grain size between the LCNCU1 and LCNC1 samples is not high enough to explain the huge difference between the shapes of the R(T) curves, but it is interesting to discuss its origin. Two mechanisms (possibly concomitant) can account for this result. First, the carbonaceous residues probably slow down the crystallization. Besides, the temperature reached during the combustion is not necessary identical for both techniques (in absence or presence of urea). It has been shown that the combustion temperature varies with the fuel content and the composition of the sample [12]. The microstructure of the precursor will thus be influenced.

We did not discuss the results in terms of resistivity values. The connectivity of the grains possibly have an effect on the resistivity of a compound, and the porosity of our different samples was not constant. On the contrary, the porosity has no influence on the shape of the R(T) curve, since we have observed double-peak behavior as well for compact as for porous samples.

CONCLUSION

In this study we have shown that the synthesis conditions have a dramatic influence on the shape of the R(T) curve. The appearance of a second peak, very broad and at a lower temperature than the usual T_C, has been attributed to the existence of a more disordered phase in the shell of the grains. Many low-temperature synthesis methods require carbonaceous precursors, which seem to be responsible for the double-peak behavior. Materials obtained by those low-temperature techniques generally exhibit high MR values, which make them potentially useful for practical applications. The study of those phenomena is thus interesting not only for fundamental but also for technological reasons.

ACKNOWLEDGMENTS

BV and PhV are particularly grateful to FNRS for a research grant. SD thanks FRIA for a research grant.

REFERENCES

1. J. Blasco, J. Garcia, J. M. De Teresa, M. R. Ibarra, P. Algarabel and C. Marquina, J. Phys.: Condens. Matter **8**, 7427 (1996)

2. G. J. Snyder, R. Hiskes, S. DiCarolis, M. R. Beasley and T. H. Geballe, Phys.Rev.B **53**(21), 14434 (1996)

3. S. Sergeenkov, M. Ausloos, H. Bougrine, A. Rulmont and R. Cloots, JETP Lett. **70**(7), 481 (1999)

4. N. Zhang, F. Wang, W. Zhong and W. Ding, J. Phys. : Condens. Matter **11**, 2625 (1999)

5. H. Y. Hwang, S. W. Cheong, N. P. Ong and B. Batlogg, Phys. Rev. Lett. **77**(10), 2041 (1996)

6. A. P. Ramirez, S. W. Cheong and P. Schiffer, J. Appl. Phys. **81**(8), 5337 (1997)

7. R. Mahesh, R. Mahendiran, A. K. Raychaudhuri and C. N. R. Rao, Appl. Phys. Lett. **68**(16), 2291 (1996)

8. A. K. M. A. Hossain, L. F. Cohen, T. Kodenkandeth, J. MacManus-Driscoll and N. McN. Alford, J. Magn. Magn. Mater. **195**, 31 (1999)

9. N. Zhang, W. Ding, W. Zhong, D. Xing and Y. Du, Phys.Rev.B **56**(13), 8138 (1997)

10. J. Pierre, F. Robaut, S. Misat, P. Strobel, A. Nossov, V. Ustinov and V. Vassiliev, Physica B **225**, 214 (1996)

11. C. N. R. Rao, R. Manesh, A. K. Raychaudhuri and R. Mahendiran, J. Phys. Chem. Solids **59**(4), 487 (1998)

12. A. Chakraborty, P. S. Devi and H. S. Maiti, J. Mater. Res. **10**(4), 918 (1995)

Magnetic Oxide Thin Films and Heterostructures

A-SITE ORDERED, PEROVSKITE-LIKE MANGANITES GROWN BY PLD OR LASER-MBE : THEIR GROWTH, STRUCTURAL AND PHYSICAL CHARACTERIZATION

B. MERCEY*, A.M. HAGHIRI-GOSNET*, Ph. LECOEUR*, D. CHIPPAUX*, B. RAVEAU*, P.A. SALVADOR**
*Laboratoire CRISMAT-ISMRA, UMR CNRS 6508, CAEN FRANCE.
**Carnegie Mellon University, Department of Materials Science and Engineering PITTSBURGH, PA, USA.

ABSTRACT

Using both a "classical" pulsed laser deposition process and a laser-MBE process, superlattices in the $Pr_{0.5}Sr_{0.5}MnO_3$ system were grown from $PrMnO_3$ and $SrMnO_3$ targets. A-site ordering of perovskite-like manganite was achieved using both deposition processes. However the electric and magnetic properties of the superlattices are shown to be strongly dependent on the growth process.

INTRODUCTION

Colossal Magnetoresistance (CMR) properties make manganese perovskites very attractive for fundamental physicochemical studies and for potential applications such as magnetic sensors or bolometers [1,2]. Studies carried out on bulk materials have shown the importance, for the electronic and magnetic properties, of the manganese valence and of the average A-site cationic radius and its variance [3,4]. The possibility of preparing bulk material possessing A-site ordering has been shown for $La_{0.5}Ba_{0.5}MnO_3$ [5]. A higher metal to insulator transition as well as a higher Curie temperature are observed for this A-site ordered material as compared to that observed in the disordered phase. The preparation of such ordered structures is difficult to achieve using classical solid state methods, however, thin films methods offer the advantage to artificially induce this A-site ordering which is not commonly obtained by solid state reactions. Artificial ordering has already been obtained using pulsed laser deposition (PLD)[6,7], and in some cases has resulted in original CMR behavior [8]. However, this was not the first time that A-site ordered manganese oxides were grown by PLD. Previous studies undertaken by our group [9], have shown that this order can be obtained for the $La_{0.74}Sr_{0.26}MnO_3$ composition. Thin films with this composition were successfully grown using individual $LaMnO_3$ and $SrMnO_3$ targets. The overall La/Sr ratio is maintained constant but different sequences of similar adjacent layers were obtained. With regard to the bulk material with the equivalent La/Sr ratio it has been shown that the electric and magnetic properties of the superlattices were strongly dependent on the number of layers. This paper addresses the effect of the A-site order on the CMR properties of $Pr_{0.5}Sr_{0.5}MnO_3$ which, unlike other $Ln_{0.5}AE_{0.5}MnO_3$ materials, does not present a charge order (CO) phase transition [10]. Thin films are prepared from two targets, $PrMnO_3$ and $SrMnO_3$ and the superlattices are grown keeping the Pr/Sr ratio equal to one. Two different PLD methods were utilized : the "classical" one using molecular oxygen (0.3 mbar) and a laser-MBE system working at low pressure (10^{-3} mbar range) using a mixture of oxygen and ozone. It will be shown that the electric and magnetic properties of superlattices grown in these two systems are different.

EXPERIMENTAL

Dense ceramic targets of PrMnO$_3$ and SrMnO$_3$ were prepared using standard ceramic synthesis methods. Appropriate ratios of Pr$_6$O$_{11}$, SrCO$_3$ and MnO$_2$ powders were mixed and intimately ground using a semi-planetary ball mill. The powder was annealed twice for 12 h at 900°C and once for 12 h at 1100°C, with intermediate grindings. Pellets having a diameter of 2.5 cm were cold uniaxially pressed and sintered for 24 h at 1500°C.

Two different deposition systems were used, a "classical" one and a laser-MBE one, and an excimer laser (Lambda Physics) with λ = 248 nm (KrF) was utilized with both systems.

"Classical" PLD system : Optical quality single-crystal substrates of [001] SrTiO$_3$ were ultrasonically cleaned in acetone and then in alcohol, and were attached to the heater using silver paste. A thermocouple was mechanically attached to the heater block close to the substrate and the temperature measured is referred to as the deposition temperature. The background pressure of the chamber was 10^{-5} mbar and the distance between the substrate and target was 47 mm. During deposition, the substrate was held at a constant temperature of 580°C in a dynamic vacuum of 0.3 mbar. After deposition, the oxygen pressure was increased to a static value of 500 mbar and the films were cooled to room temperature at 20°C/min. Superlattices were deposited by alternating the number of laser pulses according to the deposition rate of each material. The deposition rate was observed to be relatively constant when identical deposition parameters were used. The deposition rate for each target is determined by depositing several superlattices and determining the thickness of each layer and the overall superperiod from the satellite peaks observed in the x-ray diffraction patterns. This gives a fairly accurate deposition rate under the conditions of superlattice deposition.

Laser-MBE : Optical quality single-crystal substrates of [001] SrTiO$_3$ were also used but after the cleaning in acetone and alcohol they were etched with a NH$_4$F/HF buffered solution to obtain a terraced surface [11]. The substrate was then attached to the heater block with silver paste. The temperature referred to as the deposition temperature is measured by a thermocouple attached to the heater block. The base pressure of the system was 10^{-8} mbar and the distance between target and substrate was 6 cm. To improve the quality of the surface, prior to the deposition of the superlattices, the substrate was buffered with about 25 layers of SrTiO$_3$ deposited at 720°C in a dynamic vacuum of 1.5x 10^{-4} mbar of molecular oxygen. During the deposition of the superlattices, the heater was held at constant temperature in a dynamic vacuum of 7.5 x10^{-4} mbar, while the gas mixture used during deposition was 94% O$_2$ and 6% O$_3$. A differentially pumped electron gun was used to monitor RHEED oscillations during the deposition. After deposition, the superlattices were cooled to 300°C in the same pressure and atmosphere as used during the deposition. Between 300°C and 200°C, the ozone flow was stopped but the oxygen pressure was kept constant. Below 200°C, films were cooled in 5x10^{-7} mbar and subsequently RHEED was used during cooling to check for the surface stability of the superlattice. Oscillations of the specular beam of the RHEED pattern were used to control the deposition rate of each target.

X-ray diffraction was carried out using a Seiffert XRD 3000 diffractometer with Cu-Kα_1 radiation. Temperature-dependent four-probe measurements were performed using a Quantum Design PPMS between 4 and 400K in applied fields up to 7T. The electrical leads were thermocompressed on thermally evaporated silver contacts. X-ray photoelectron spectroscopy (XPS) measurements were performed in an analytical chamber fitted with a Leybold hemispherical analyzer.

RESULTS AND DISCUSSION

Figure 1 shows the resistive behavior of films grown in the "classical" PLD system using the deposition conditions described above. Fig 1a shows the behavior of a film which does not present A-site ordering. This film is also grown with the two target system using a number of laser pulses on each target that is not large enough to grow a complete layer of

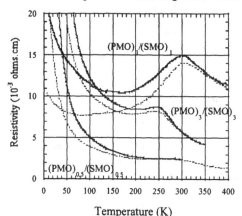

SrMnO₃ or PrMnO₃. However, the number of pulses is adjusted to keep the Pr/Sr ratio equal to one. A small bump is observed at 290K, which is the temperature of the paramagnetic (PM)-ferromagnetic (FM) transition determined by neutron diffraction on the bulk material. The FM antiferromagnetic (AFM) transition, observed on bulk material at 150K, is scarcely visible however the increase in resistivity of the film for temperatures below 150K suggests that it could possibly exist. A small decrease of the resistivity is observed in a magnetic field of 7 tesla. Figure 1b presents resistive behavior of the artificially ordered film $(PrMnO_3)_1(SrMnO_3)_1$. The PM-FM

Figure 1 : resistivity **1a** $(PMO)_{0.5}(SMO)_{0.5}$, **1b** $(PMO)_1(SMO)_1$, **1c** $(PMO)_3(SMO)_3$, dashed line 7T.

transition is now observed at 313K and the metallic behavior of the superlattice is observed in the 313-150K temperature range. For temperatures lower than 150K, a semiconducting behavior is observed as in the bulk material. Under an applied magnetic field of 7 tesla, the superlattice presents an important CMR effect, which is not observed in the bulk material. This effect is observed between 313 and 4K and exhibits a maximum value at 100K. For comparison, the $(PrMnO_3)_3(SrMnO_3)_3$ superlattice presents a different behavior (figure 1c) consisting of a small bump observed at 250K, but as for the disordered material (fig. 1a), a metallic domain is hardly visible. The semiconducting state seems to begin close to the bump and only a weak decrease of the resistivity is observed in a magnetic field of 7 tesla. It should be noted that this superlattice behaves like the $(LaMnO_3)_{14}(SrMnO_3)_6$ previously reported [9]. This type of transition could be related to a charge order transition observed for other $Ln_{0.5}AE_{0.5}MnO_3$ systems but not observed in bulk $Pr_{0.5}Sr_{0.5}MnO_3$. The growth of such an A-site ordered superlattice could favor the charge order, thus leading to a behavior different from that observed in the disordered bulk material.

Different films were grown in a mixture of oxygen and ozone in the RHEED-monitored laser-MBE system. In figure 2 are presented the intensity variations of the specular beam of the RHEED pattern during the growth of a disordered $Pr_{0.5}Sr_{0.5}MnO_3$ film. Since this film is grown from two targets, variations of the intensity are correlated to the growth. For example during the deposition of half a layer of SrMnO₃ the intensity decreases, then, when the growth is stopped, the intensity increases, this is due to the surface rebuilding. When the growth of half a layer of PrMnO₃ begins, the intensity decreases and then increases to reach a maximum value, then the

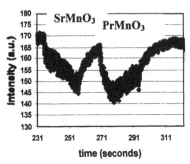

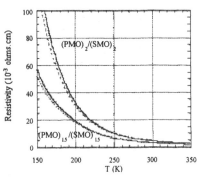

Figure 2 : Oscillations of the intensity of the specular beam observed during the deposition of 0.5 layer of $SrMnO_3$ and 0.5 layer of $PrMnO_3$.

Figure 3 : Resistivity versus temperature of 3a : $(PrMnO_3)_{1.5}(SrMnO_3)_{1.5}$ and 3b : $(PrMnO_3)_2(SrMnO_3)_2$ dashed line 7 T.

growth is stopped and again rebuilding of the surface is observed. When superlattices are grown variations in the intensity of the specular beam of the RHEED pattern are an indication for us to stop the growth after the completion of each layer. Figure 3 displays the resistivity of the films grown in the laser-MBE system. The resistivity of a disordered film, (not shown), exhibits a behavior is similar to that observed for the disordered film prepared in the "classical" PLD system (fig 1a). On the other hand, superlattices $(PrMnO_3)_{1.5}(SrMnO_3)_{1.5}$ and $(PrMnO_3)_2(SrMnO_3)_2$ whose resistivities are higher than that observed in the disordered film, and are presented in figure 3a and 3b, respectively do not have the same behavior as the

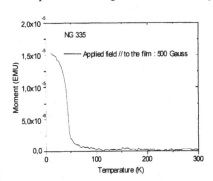

Figure 4 : Magnetization versus temperature of a $(PrMnO_3)_2(SrMnO_3)_2$ film deposited by laser-MBE. A small increase of the magnetization is observed below 50K

superlattices grown in the "classical" PLD. No bump is observed and a semiconducting behavior is observed in the entire temperature range. The effect of the magnetic field on such superlattices is rather weak. The magnetization measured for these superlattices is rather low (fig 4), however, to explain this, two hypotheses can be put forward: this small magnetization is either due to a PM behavior, or to an AFM behavior. To explain such a difference in behavior between superlattices grown in the "classical" PLD system as compared to the laser-MBE system, two important point must be taken into consideration : first, the oxidation state of the manganese in the $SrMnO_3$ films grown in the two systems and, secondly, the lattice parameter

of the superlattices in the direction perpendicular to the substrate plane. Figure 5 shows the XPS spectra of $PrMnO_3$ and $SrMnO_3$ films. While the energy of the Mn peaks of the $PrMnO_3$ films does not depend on the system used for the growth of the film and could be attributed to "pure" Mn^{3+}. The energy for manganese peaks of the $SrMnO_3$ films, which have been shown to be stabilized with the cubic structure [9], strongly depends on the system used for the growth.

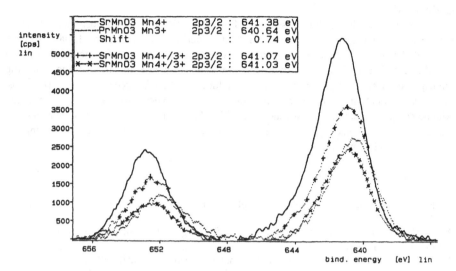

Figure 5 : XPS spectra (Mn 2p 3/2 range) of PrMnO₃ films deposited by laser-MBE, and of SrMnO₃ film deposited by laser-MBE (_____) or by "classical" PLD (..+...+...+) or (..x...x...x)

When grown in the laser-MBE system the films present a peak energy in good agreement with "pure" Mn^{+4}, while grown in the "classical" PLD system they present a peak energy corresponding to an average value between Mn^{3+} and Mn^{+4}. The X-ray diffraction studies of two 1/1 superlattices show that the lattice parameter of the superlattice grown in a mixture of oxygen and ozone is different from that of the superlattice grown in pure oxygen. For the laser-MBE grown the measured lattice parameter perpendicular to the substrate plane is 3.812Å in agreement with the theoretical value. While for "classical" PLD grown films this lattice parameter is somewhat shorter 3.765Å , not in agreement with the theoretical value.

Taking into consideration these two features, the difference between resistive behavior of superlattices grown in the two systems can be explained. For superlattices grown in the laser-MBE system, the strong oxidation potential of ozone enables the stabilization of a SrMnO₃ layer containing only Mn^{4+}, while for superlattices grown in the "classical"system, the SrMnO₃ layer contains both $Mn^{3+}$and Mn^{4+}. The superlattices grown in a mixture of oxygen and ozone are A-site ordered, and thus induces an ordering between Mn^{3+} and Mn^{4+}. The film is then stabilized with an AFM order, which might explain the absence of a magnetic transition and the semiconducting behavior with no CMR effect observed for such films. The high resistivity of these superlattices, as compared to that of the disordered film, must also be pointed out. When the superlattices are grown in oxygen, the A-site order does not induce a complete order between Mn^{3+} and Mn^{4+}, since the SrMnO₃ layer is stabilized with these two cations. This might explain the observation of a transition due to the ordering of the two manganese ions, which is not completely achieved by the deposition process.

CONCLUSION

The stabilization, using "classical" PLD or laser-MBE thin film deposition methods, of perovskite-like managanites presenting an A-site ordering has been shown. It has also been shown that magnetic and electric properties depend on the system used for the growth of such superlattices. When films are grown at a low deposition rate and in a strong oxidizing atmosphere, A-site ordering seems to also induce an order between Mn^{+3} and Mn^{+4} ions. Whereas films grown at a medium deposition rate under oxygen only present A-site ordering with no complete order of the manganese ions. The electric and magnetic study at temperatures higher than 400K of superlattices grown in the laser-MBE system should be very interesting and are currently under investigations.

REFERENCES

1. R. Von Helmolt, J. Wecker, R. Holzapfel, L. Schultz and K. Samwer, Phys. Rev. Lett. **71**, 2331, (1993)
2. T. Venkatesan, M. Rajeswari, Zi-W. Dong, S.B. Ogale, R. Ramesh, Phil. Trans. R. Soc. Lond. A, **356**, 1661, (1998).
3. C.N.R. Rao, A.K. Cheetham and R. Mahesh, Chem. Mater. **8**, 2421, (1996).
4. B. Raveau, A. Maignan, C. Martin, and M. Hervieu, Chem. Mater. **10**, 2641, (1998).
5. F. Millange, V. Caignaert, B. Domenges, B. Raveau and E. Suard, Chem. Mater. **10**, 1974, (1998).
6. C. Q. Gong, A. Gupta, G. Xiao, P. Lecoeur and T. R. McGuire, Phys. Rev. B, **54**, R3742, (1996).
7. A. Gupta, Curr. Opin. Solid. State Mater. Sci. **2**, 23, (1997).
8. B. Mercey, P. A. Salvador, W. Prellier, T. D. Doan, J. Wolfman, J.F. Hamet, M. Hervieu and B. Raveau, J. Mater. Chem. **9**, 233, (1999).
9. P. A. Salvador, A. M. Haghiri-Gosnet, B. Mercey, M. Hervieu and B. Raveau, Appl. Phys. Lett. **75**, 2638, (1999).
10. C. Martin, A. Maignan, M. Hervieu and B.Raveau, Phys. Rev. B, in press (1999).
11. M. Kawasaki, K. Takahashi, T. Maeda, R. Tsuchiya, M. Shinohara, O. Ishiyama, T. Yonezawa, M. Yoshimoto and H. Koinuma, *Science,* **256**, 1540, (1994)

STRUCTURE AND MAGNETISM OF NANOCRYSTALLINE K$_\delta$MnO$_2$

R.M. STROUD, E. CARPENTER, V.M. BROWNING, J.W. LONG, K.E. SWIDER,
D.R. ROLISON
Naval Research Laboratory, Washington, DC 20375

ABSTRACT

The structure and magnetic properties of sol-gel-synthesized, nanocrystalline K$_\delta$MnO$_2$ were investigated. The nanoparticles were determined by x-ray diffraction and high-resolution transmission electron microscopy to be single-crystal rods of the cryptomelane phase of MnO$_2$, with a typical particle size of 6 nm x 20 nm. The field and temperature dependence of the magnetization indicates superparamagnetic behavior, with a blocking temperature of 15K. The dependence of the magnetic properties on particle size, surface layers and mixed valency is discussed.

INTRODUCTION

Similarly to the perovskite-type manganites, MnO$_2$-based materials exhibit a complex interplay of structure, magnetism and transport. A century of research into these materials, as naturally occurring minerals, and for battery applications, has documented the extreme polymorphism of MnO$_2$ and related oxyhydroxides [1]. The polymorphs are built up from MnO$_6$ octahedra with different long-range order that depends on the balance of charge between the oxygen anions and mixed valent manganese cations (Mn^{3+} and Mn^{4+}), dopant cations, and incorporated OH$^-$ and H$_2$O. The sensitivity of the electrochemical properties of MnO$_2$ materials to structure is well documented, however relatively little is known about the magnetic properties.

The room temperature field-dependent magnetization for some commercially available, chemically synthesized MnO$_2$ phases has been surveyed [2]. In fields ranging from 2 to 8 kGauss, the magnetization increased linearly with field, indicating paramagnetic behavior, for MnO$_2$ samples with varying octahedral packings. In general, higher magnetic moments were observed for phases with more tightly packed MnO$_6$ octahedra. The particle size of the materials was not reported.

In order to better understand the complex structural and magnetic properties of MnO$_2$-based battery materials, we have chosen to thoroughly characterize a single system rather than survey a multitude of poorly determined systems. We present herein the results from the investigation of sol-gel synthesized cryptomelane-type K$_\delta$MnO$_2$.

EXPERIMENT

Nanocrystalline K$_\delta$MnO$_2$ samples were obtained by sol-gel synthesis. The synthesis procedure was adapted from an established method for producing MnO$_2$ aerogels [3]. Solid fumaric acid was added to 0.2 M KMnO$_4$ (0.579 g : 75 mL) and the solution was subjected to vacuum for 8 min before pouring the sol into polypropylene molds. Gelation occurred in ~ 1 h; the gels were aged for 24 h. The resultant gels were washed with H$_2$O, then 1 M H$_2$SO$_4$,

Mat. Res. Soc. Symp. Proc. Vol. 602 © 2000 Materials Research Society

followed by additional washes of H_2O. To control the pore volume and pore size distribution of the final gels, the method for pore liquid extraction and drying was varied. In order of increasing pore volume: xerogels were obtained by drying in ambient air for at least 24 h and then heating to 80 °C under ambient pressure; ambigels were obtained by replacement of the pore liquid with hexane and subsequent drying under vacuum at 60 °C for 24 h; and aerogels were obtained by replacement of the pore liquid with liquid CO_2 and supercritical drying. All dried gels were sintered at 300 °C for 2 h, producing rigid monoliths for the xerogel and ambigel, and a fragile monolith for the aerogel. The residual K content of the annealed gels was determined by energy dispersive x-ray spectroscopy to be between 1 and 2 atom %. A more complete chemical and physical characterization of these samples is described elsewhere [4].

Structural characterization of the samples was performed by conventional x-ray diffraction (XRD) and high-resolution transmission electron microscopy (HRTEM). For XRD measurements the powders were mounted on single crystal silicon wafers and the spectra accumulated using a Philips 1710 diffractometer. Samples investigated by HRTEM, using a 300 kV Hitachi H9000 TEM, were prepared by dipping lacey carbon grids in an acetone suspension of the nanoparticles.

Magnetic measurements were made using a superconducting quantum interference device (SQUID) magnetometer. The samples were contained in gelcaps, sealed with kapton tape. Hysteresis (M,H) loops were obtained at 300K, 30K and 5K, at fields up to 60 kOe. The temperature dependence of the magnetization, 5K to 350K, was measured in field-cooled and zero-field cooled states, at fields of 100 Oe, 1 kOe, and 10 kOe. The data were corrected for the diamagnetic contribution of the sample holder.

RESULTS

The XRD data (Fig 1.) reveal that the annealed xerogel, ambigel and aerogel nanoparticle samples all take the cryptomelane form of MnO_2. This monoclinic structure, 0.99 nm × 0.284 nm × 0.972 nm, 90.3°, is composed of parallel chains of edge sharing MnO_6, arranged to form a 2 × 2 tunnel structure, as shown in Fig 2. These large tunnel sites (0.46-nm wide) accommodate hydroxyl ions, water and K^+, or other interstitial cations, such as Li.

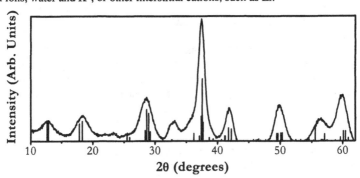

Figure 1. X-ray diffraction of K_8MnO_2 xerogel and PDF 44-1386, cryptomelane. The ambigel and aerogel XRD patterns also index to cryptomelane. The broad peaks and texturing result from nanocrystalline particles with a (010) growth direction.

Figure 2. Schematic of the cryptomelane structure of $K_\delta MnO_2$, with one occupied tunnel site. The structure is formed of parallel double chains of edge-sharing MnO_6 octahedra. The 0.46 nm 2 x 2 tunnels accommodate hydroxyl ions, water and K^+, or other ions.

The HRTEM images (Fig 3.) show a "rice-grain" morphology, consisting of randomly-oriented, single crystal rods, 5 to 7 nm in diameter and 10 to 30 nm long. The nanoparticles exhibit lattice fringes indicating a (010) growth direction along the axis of the rod. No evidence for the stacking fault defects commonly found in the pyrolusite and ramsdellite forms of MnO_2 is observed. However, features consistent with the collapse of the tunnel structure at the rod edges are seen, when the rods are imaged end-on, i.e., $\perp$ (010).

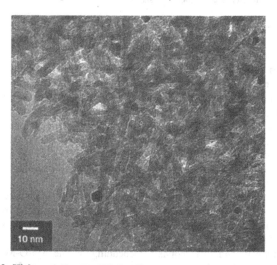

10 nm

Figure 3. High-resolution transmission electron micrograph of $K_\delta MnO_2$ xerogel.

The field dependence of the magnetization of the xerogel sample is shown in Fig. 4a. The 300K and 30K curves are linear, with zero remanence, indicating paramagnetic, or superparamagnetic behavior. The data taken at 5K show some hysteresis and a decreasing slope at high field, which indicate the presence of some ferromagnetic interactions. The temperature dependence of the susceptibility (Fig. 4b), shows a split between the field-cooled (FC) and zero-field cooled (ZFC) behavior for temperatures below the blocking temperature (T_b) of 15K. The ambigel and aerogel samples exhibit quantitatively similar magnetization properties.

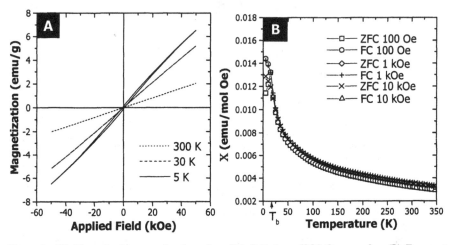

Figure 4. (A) Magnetization as a function of applied field for a K_8MnO_2 xerogel. (B) Temperature dependence of the susceptibility of a K_8MnO_2 xerogel.

DISCUSSION

The magnetic properties of nanocrystalline particles are frequently complicated. As the size of the particles decreases, the exchange interactions that dominate the bulk properties must compete with the increasingly important thermal fluctuations and surface effects. Particles smaller than 100 nm in diameter frequently exhibit superparamagnetism at room temperature, whereby the moments of the individual particles are decoupled, due to rapid thermal fluctuation of the magnetization directions. The superparamagnetic blocking temperature (T_b) depends on the particle size and the effective anisotropy constant (K): $k_BT_b = 25 \, V/K$, for uniaxial particles [5]. The effective magnetic particle size may also be smaller than the physical particle size, because differences in bond arrangements and stoichiometry at the surface can alter the exchange interactions, producing a "dead zone".

For the cryptomelane particles in this study, the full explanation is further complicated by considering the interplay of valence, structure and magnetism exhibited by manganese oxides, and the fact that the sol-gel synthesis route produces hydrous materials with OH$^-$-terminated surfaces. The exchange interactions between Mn^{3+} and Mn^{4+} cations in this material must be indirect, mediated by O and OH- bonds. Depending on the Mn-O-Mn bond lengths, the exchange interactions will be either paramagnetic, ferromagnetic, or antiferromagnetic. Based on the low temperature magnetization and susceptibility data, some ferromagnetic exchange interactions do occur in these particles. However, the ferromagnetic ordering is relatively weak, and there may be competing antiferromagnetic interactions.

The divergence of the FC and ZFC susceptibility at low temperature, could result from either size-induced superparamagnetism, or local competition of ferromagnetic and antiferromagnetic interactions, or a combination of both effects. The size of the particles, ~ 6 nm × 20 nm, is sufficiently small for thermal fluctuations of the particle magnetization direction to be important. In future experiments, this could be tested by increasing the particle size, using a slower gelling synthetic route. Assuming the divergence is due solely to size, and that the effective anisotropy constant is fixed, the blocking temperature of the particles could be raised to 300K from 15K, by increasing the particle dimensions by $(20)^{1/3}$ or a factor of 2.7.

Determining the spatial variation in Mn oxidation state is another factor that merits further study. Recent advances in energy-filtered TEM imaging have made it possible to map transition metal valence changes across grain boundaries in bulk and thin film samples [6]. The valence maps are obtained by imaging in energy windows corresponding to the L_3 and L_2 "white-lines" in the core-loss spectra of a transition metal element such as Mn or Co. We hope to extend this technique to the mapping of nanoparticle surfaces, in order to measure difference in bulk and surface valence, with a spatial resolution of ~ 2 nm. For the cryptomelane nanoparticles, this technique could reveal surface/core segregation of Mn^{3+} and Mn^{4+}, or Mn^{2+}, if present.

CONCLUSION

We have synthesized nanocrystalline K_8MnO_2 of the cryptomelane tunnel structure. At temperatures above 15 K, paramagnetism is observed. Below the blocking temperature of 15K, significant hysteresis in the M,H loops is observed, indicating the presence of weak ferromagnetic ordering. More work is necessary to determine the relative contributions of particle size, surface layers and charger localization on Mn sites to the reported magnetic properties. Future work will include the development of valence-mapping techniques based on energy-filtered TEM imaging.

ACKNOWLEDGEMENTS

This work was supported by the U.S. Office of Naval Research. J.W.L. is an NRC Postdoctoral Associate (1997-2000).

REFERENCES

1. Y. Chabre and J. Pannetier, *Prog. Solid St. Chem.* **23**, pp.1-130 (1995).

2. M.V. Ananth, V. Venkatesan, K. Dakshinamurthi, *J. of Power Sources* **72**, pp. 99-102 (1998).

3. S. Bach, M. Henry, N. Baffier, J. Livage, *J. Solid- State Chem.* **88**, 325-333 (1990).

4. J.W. Long, K. E. Swider, R. M. Stroud, D. R. Rolison, *Electrochem. Solid-State Lett.*, in preparation.

5. B. D. Cullity, *Introduction to Magnetic Materials*, Addison-Wesley, Reading, MA, 1972, p. 414.

6. Z. L. Wang, J. Bentley, and N. D. Evans, *Microscopy and Microanalysis*, vol. 5, supplement 2, Proceedings of Microscopy and Microanalysis '99, Portland, OR August 1-5, 1999, (Springer p. 102-103, 1999).

SPIN/ORBITAL MODULATION IN PEROVSKITE MANGANITE SUPERLATTICES

M. IZUMI *, T. MANAKO *, M. KAWASAKI *,**, and Y. TOKURA *,***

*Joint Research Center for Atom Technology (JRCAT), Tsukuba 305-0046, Japan
**Department of Innovative and Engineered Materials, Tokyo Institute of Technology, Yokohama 226-8502, Japan
***Department of Applied Physics, University of Tokyo, Tokyo 113-8656, Japan

ABSTRACT

A systematic study is presented for structural characterization and physical properties of two kinds of perovskite oxide superlattices composed of ferromagnetic (FM) and antiferromagnetic (AF) layers. Spin ordering structures is modulated in FM $La_{0.6}Sr_{0.4}MnO_3$/G-type AF $La_{0.6}Sr_{0.4}FeO_3$ superlattices, whereas, ordering structures both in spin and orbital are modulated in $La_{0.6}Sr_{0.4}MnO_3$/A-type AF $La_{0.45}Sr_{0.55}MnO_3$ along the growth directions. Large magnetoresistance subsists down to low temperature in $La_{0.6}Sr_{0.4}MnO_3$ /$La_{0.6}Sr_{0.4}FeO_3$ (F/G) superlattices as a result of recovery of ferromagnetism, which is once suppressed by spin frustration at the interface between FM and G-type AF layers. In contrast, the constituent layers in the $La_{0.6}Sr_{0.4}MnO_3$/$La_{0.45}Sr_{0.55}MnO_3$ (F/A) superlattices appear to keep their ground states due to the absence of spin frustration at the interface. Magnetoresistance is pronounced in this type of superlattices at low temperatures when the AF layer is very thin, indicating restoration of the electronic coupling between the neighboring FM layers which are otherwise decoupled by intervening A-type AF spin ordering and $d_{x^2-y^2}$ orbital ordering in $La_{0.45}Sr_{0.55}MnO_3$ layers.

INTRODUCTION

Perovskite type manganese oxides with chemical formula of $RE_{1-x}AE_xMnO_3$, where RE is trivalent rare earth element and AE is divalent alkaline earth element, show a variety of spin-charge coupled properties [1]. One of the most striking phenomena observed in doped manganites is the colossal magnetoresistance (CMR); a large drop of resistance by magnetic field is observed near the ferromagnetic (FM) transition temperature. The appearance of such an interesting phase can be explained in terms of subtle competition and compromise between the double-exchange interaction mediated with the conduction e_g electron and the super-exchange interaction between local t_{2g} electrons. In actual complex phase diagrams, not only the FM phase but also a variety of antiferromagnetic (AF) phases appear. This is partly due to the variety of coexistence of the AF superexchange interactions between the t_{2g} spins and partly due to the electronic anisotropy arising from the orbital degree of freedom in the conduction e_g electrons.

The purpose of this report is to understand the physical properties of FM/AF artificial oxide superlattices where FM layer is $La_{0.6}Sr_{0.4}MnO_3$. $La_{1-x}Sr_xMnO_3$ is a prototypical compound which shows CMR for the doping level of $0.15 \leq x \leq 0.5$ [2]. FM transition takes place above 300 K at $0.3 \leq x \leq 0.5$. $La_{1-x}Sr_xMnO_3$ is free from charge ordering phenomenon even when $x = 0.50$, unlike $Pr_{1-x}Sr_xMnO_3$ [3] or $Nd_{1-x}Sr_xMnO_3$ [4], but changes its ground state to an AF metal at $x > 0.5$ [5]. By combining with other perovskite compounds which have AF spin ordering in a form of superlattice, we can tune the physical properties by utilizing the competition between FM and AF magnetic ordering structures in the constituent layers. Since the physical properties of $La_{1-x}Sr_xMnO_3$ are mainly dominated by double-exchange interaction in this doping region, the charge dynamics is quite sensitive to the spin angle between the neighboring t_{2g} spins. Therefore, the perturbation in spin arrangement in FM $La_{0.6}Sr_{0.4}MnO_3$ layer due to the proximity effect of AF layer can be sensitively detected not only by magnetization, but also by transport measurements.

We chose here two different types of AF compounds, $La_{0.6}Sr_{0.4}FeO_3$ and $La_{0.45}Sr_{0.55}MnO_3$. Bulk crystal of $La_{0.6}Sr_{0.4}FeO_3$ has G-type spin ordering with Néel temperature (T_N) of 310

K [6]. $La_{0.45}Sr_{0.55}MnO_3$ was reported to have an anisotropic of AF state but the details are still unknown since the physical properties of the $La_{1-x}Sr_xMnO_3$ compounds with $x > 0.5$ have been studied only using polycrystalline samples because of difficulty in the fabrication of a bulk single crystal [5].

In this paper, the characteristics of $La_{0.6}Sr_{0.4}MnO_3$, $La_{0.6}Sr_{0.4}FeO_3$, and $La_{0.45}Sr_{0.55}MnO_3$ single-layer films grown on $SrTiO_3$ substrates are described so as to establish the magnetic, electronic, and structural basis of the superlattice research. Physical properties of superlattices are explained with an emphasis on the differences of physical properties between two type FM/AF superlattices.

EXPERIMENT

The superlattices of $La_{0.6}Sr_{0.4}MnO_3/La_{0.6}Sr_{0.4}FeO_3$ and $La_{0.6}Sr_{0.4}MnO_3/La_{0.45}Sr_{0.55}MnO_3$ as well as single-layer (100 nm thick) films of component compounds were fabricated by pulsed laser deposition method employing stoichiometric targets [7, 8]. For achieving atomically regulated epitaxy, $SrTiO_3$ (001) single crystal substrates treated with NH_4F-HF solution were used [9]. Prior to the deposition, the substrate was in-situ annealed at 900°C in 1 mTorr of oxygen for 20-30 minutes, resulting in the well defined TiO_2 terminated surface having straight and evenly aligned steps apart by about 150 nm (corresponding to a miscut angle of 0.15°). After the procedure, the film deposition was carried out at a substrate temperature of 800°C while keeping 1 mTorr oxygen pressure. KrF excimer laser pulses of 100 mJ were focused on a target at a fluence of 3 J/cm^2. During the deposition, reflection high energy electron diffraction (RHEED) pattern was monitored by a CCD-camera and real-time analyses were carried out by a computer. After the deposition, the film was cooled in 760 Torr of oxygen.

The surface morphology of the films were analyzed by an atomic force microscope (AFM). X-ray diffraction (XRD) was carried out by a four-circle diffractometer with $CuK\alpha$ source to obtain conventional 2θ-θ diffraction pattern and reciprocal space mapping. Magnetization was measured by a superconducting quantum interference device magnetometer. Magnetic field was applied parallel to the film plane to avoid the geometric demagnetization effect. Resistivity was measured by a four-probe method in a magnetic field up to 7 T applied along the film plane.

RESULTS AND DISCUSSION

Physical Properties of Single-Layer Films

Magnetization and resistivity of $La_{0.6}Sr_{0.4}MnO_3$, $La_{0.6}Sr_{0.4}FeO_3$, and $La_{0.45}Sr_{0.55}MnO_3$ single-layer films are shown in Fig. 1. The properties of the three films are quite different from each other. The $La_{0.6}Sr_{0.4}MnO_3$ film shows FM transition at $T_C = 330$-340 K and saturation magnetization of 3.5 μ_B/Mn at 5 K corresponding to full magnetic moment. Large negative magnetoresistance (MR) is observed around T_C which is well known as CMR, driven by the double-exchange mechanism [1]. The reduced T_C compared with that of 370 K for the bulk single crystals is due to the epitaxial strain caused by lattice mismatch as reported previously [7]. The lattice parameters determined from the reciprocal space mapping for single-layer films are listed in Table 1. The in-plane lattice parameter a for all the films is matched with that of the substrate. Consequently, the out-of-plane lattice parameter c is elastically deformed. Such a coherent strain can be observed even for 100 nm thick films when the substrate surface structure and the deposition conditions are well optimized as in the present study.

The $La_{0.6}Sr_{0.4}FeO_3$ film shows AF and insulating behavior. Resistivity at room temperature was 0.08 Ω cm and increased almost exponentially with lowering the temperature to 100 Ω cm at 210 K. There was no detectable MR in this temperature range. From the temperature dependence of magnetization, T_N was estimated to be 250 K. Since the crystal structure is not deformed because of the excellent lattice matching (see Table 1) and the properties are similar to those of bulk crystals [6], spin ordering in the ground state is G-type

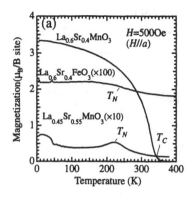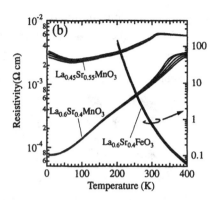

Figure 1: Physical properties of La$_{0.6}$Sr$_{0.4}$MnO$_3$, La$_{0.6}$Sr$_{0.4}$FeO$_3$, and La$_{0.45}$Sr$_{0.55}$MnO$_3$ single-layer (100 nm thick) films grown on SrTiO$_3$ (001) substrate. (a) Temperature dependence of magnetization measured during cooling in a magnetic field of 500 Oe (FC) applied along the [100] direction in the film plane. Ferromagnetic transition at T_C = 330-340 K and antiferromagnetic transitions T_N = 250 K and 220-230 K were observed for, La$_{0.6}$Sr$_{0.4}$MnO$_3$, La$_{0.6}$Sr$_{0.4}$FeO$_3$, and La$_{0.45}$Sr$_{0.55}$MnO$_3$, respectively. (b) Temperature dependence of resistivity measured during cooling at various magnetic fields of H = 0, 1, 3, 5, and 7 T. Magnetic fields were applied parallel to the current in the film plane.

AF as shown in Fig. 2 .

Although the ground state is AF, the La$_{0.45}$Sr$_{0.55}$MnO$_3$ films quite shows different transport behavior from that of the La$_{0.6}$Sr$_{0.4}$FeO$_3$ film. It is known that the ground state in La$_{1-x}$Sr$_x$MnO$_3$ is changed from FM to AF metal with increasing hole doping level from x = 0.4 to x = 0.55 [5]. AF transition was indeed observed at a T_N of 220-230 K, as seen as a broad cusp in Fig. 1. The transport property stays fairly metallic and negative MR remains at low temperatures. We attribute the ground state of La$_{0.45}$Sr$_{0.55}$MnO$_3$ to the A-type antiferromagnetism with $d_{x^2-y^2}$ orbital ordering as depicted in Fig. 2. In the A-type spin structure, magnetic moment is ordered ferromagnetically in the ab plane of MnO$_2$ layers and these layers are antiferromagnetically coupled along the c direction. Our assignment of the observed AF metallic state of the La$_{0.45}$Sr$_{0.55}$MnO$_3$ film to the A-type state with $d_{x^2-y^2}$ orbital ordering is based on the following features of the present epitaxial film [10]. First, the composition (x = 0.55) is in the regime of the overdoping that diminishes the double-

Table 1: Crystal symmetry and lattice parameters of the 100 nm thick single-layer films of La$_{0.6}$Sr$_{0.4}$MnO$_3$, La$_{0.6}$Sr$_{0.4}$FeO$_3$, and La$_{0.45}$Sr$_{0.55}$MnO$_3$ together with those of the SrTiO$_3$ substrate.

	symmetry	a (nm)	c (nm)	c/a
La$_{0.6}$Sr$_{0.4}$MnO$_3$	tetragonal	0.390$_5$	0.384$_0$	0.983
La$_{0.6}$Sr$_{0.4}$FeO$_3$	tetragonal	0.390$_5$	0.390	1.00
La$_{0.45}$Sr$_{0.55}$MnO$_3$	tetragonal	0.390$_5$	0.378$_6$	0.970
SrTiO$_3$	cubic	0.3905	0.3905	1.00

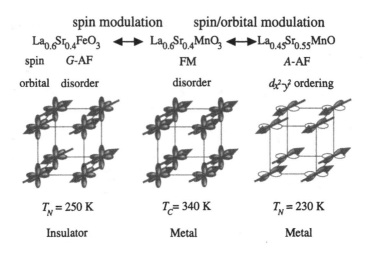

spin modulation   spin/orbital modulation

$La_{0.6}Sr_{0.4}FeO_3$ ⟷ $La_{0.6}Sr_{0.4}MnO_3$ ⟷ $La_{0.45}Sr_{0.55}MnO$

spin G-AF   FM A-AF

orbital disorder   disorder $d_{x^2-y^2}$ ordering

T_N = 250 K T_C = 340 K T_N = 230 K

Insulator   Metal   Metal

Figure 2: Schematic illustration of spin and orbital ordering structures in constituent compounds of superlattices. $La_{0.6}Sr_{0.4}MnO_3$ has ferromagnetic spin ordering with quantum disordered orbitals, $La_{0.6}Sr_{0.4}FeO_3$ has G-type antiferromagnetism with orbital disordering, and $La_{0.45}Sr_{0.55}MnO_3$ has A-type antiferromagnetism with $d_{x^2-y^2}$ orbital ordering.

exchange interaction and relatively stabilizes the super-exchange interaction as observed for the corresponding bulk crystals. More importantly, the presence of the macroscopic biaxial strain in the present $La_{0.45}Sr_{0.55}MnO_3$ thin film, expressed as a small c/a value (0.970) as shown in Table 1, favors such an orbital-ordered A-type state. The details are discussed elsewhere [10, 11].

It is known for the bulk crystals of $La_{1-x}Sr_xMnO_3$ and $Nd_{1-x}Sr_xMnO_3$ systems, for example, the compounds with $0.5 < x < 0.6$ show the AF and metallic ground state without charge ordering [5, 12]. The appearance of this interesting phase indicates that the $d_{x^2-y^2}$ type orbital ordering appears in ab plane to maximize the carrier kinetic energy via the double-exchange interaction, however, spins are coupled antiferromagnetically along the c axis to gain the super-exchange energy [13, 14]. Upon the phase transition to the layer-type (A-type) spin ordering associated with such an orbital ordering, the bulk crystal shows the expansion of the ab plane lattice parameters and the shrinkage of the c axis parameter, due to coupling of the ordered orbital with the cooperative Jahn-Teller distortion [12, 15]. Conversely, the external biaxial strain should stabilize the A-type AF spin ordered state. The $d_{x^2-y^2}$ orbital ordering in $La_{0.45}Sr_{0.55}MnO_3$ is in sharp contrast to the $d_{x^2-y^2}/d_{3z^2-r^2}$ quantum-disordered orbital structure in the FM-metallic state.

Fabrication and Structural Characterization of Superlattices

As discussed above, we have three perovskite films having quite different ground states as drown in Fig. 2, those are to be combined for making superlattices. These films can be made under the same deposition conditions and have similar lattice constants as shown in Table 1, which are advantages to make superlattices. Hereafter, we express the $La_{0.6}Sr_{0.4}MnO_3$, $La_{0.6}Sr_{0.4}FeO_3$, and $La_{0.45}Sr_{0.55}MnO_3$ layers as F, G, and A layers, respectively. The combination and the layer thicknesses of component layers in superlattices are expressed as $[F_m, G_n (A_n)]$, where m and n represent the thickness in unit cell numbers for F and G (A) layers, respectively. Where we combine the F and G layers, we can learn about spin frustration at the interfaces, because G-type antiferromagnetism has an staggered type AF ordering

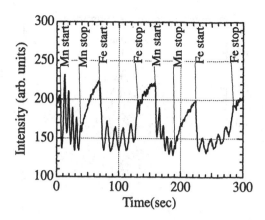

Figure 3: RHEED intensity oscillation observed during the deposition of F/G superlattice on SrTiO$_3$ (001) substrate. The thicknesses of the constituent layers are 5 unit cells.

in the (001) interface plane. Here, we will discuss the competition between spin ordered structures, ferromagnetism $vs.$ antiferromagnetism in these spin superlattices. When A layer is combined with F layer, both spin and orbital structures are modulated along growth direction. In this case, there can be no spin frustration at the interfaces because the A layer have FM ordering in the (001) interface plane. However, since electron cannot hop between adjacent F layers due to alternatingly stacked spin polarization in the intervening A layer, the F layers are decoupled. Magnetic field should give spin canting in A layer, resulting in the enhancement of the coupling between adjacent F layers.

During the deposition of the superlattices, we could routinely observe persisting oscillation of specular beam intensity in RHEED analysis. By counting the RHEED oscillation, the thicknesses of the constituent layers in superlattices were regulated on an atomic scale. An example during the growth of [F$_5$, G$_5$] superlattice is shown in Fig. 3, where completion of 5 unit cells (u.c.) of F and G layers were detected by the 5th peak. The period of the oscillation corresponds to the deposition of 0.4 nm thick layer ($i.e.$, unit cell of perovskite). Surface morphology of superlattices was also very smooth as examined by AFM images which showed 0.4 nm height steps, corresponding to the unit cell height of perovskite. The results of RHEED and AFM indicate that the film growth is an almost ideal two-dimensional layer-by-layer growth mode.

Figure 4 shows the reciprocal space mapping for an [F$_{10}$, G$_4$] superlattice. The substrate peak and fundamental diffraction peak of superlattice have the same Q_x values, indicating that the in-plane lattice constant of the superlattice is fit to that of the substrate. Since SrTiO$_3$ has a cubic crystal symmetry, the crystal structure of the superlattice is modified to the tetragonal one to keep the coherency at the interfaces in the superlattice. Although the satellite peaks originating from the superlattice periodicity could be observed between the fundamental perovskite peaks by conventional 2θ-θ XRD measurement, the intensities were very low. This is due to the small difference between the atomic scattering factor between Mn and Fe. However, by employing synchrotron radiation x-ray at the Photon Factory, KEK, not only satellite peaks but also clear Laue peaks due to finite thickness was clearly observed to be consistent with the designed structures. The simulation analyses make it clear that the interface is atomically flat and rigid [8].

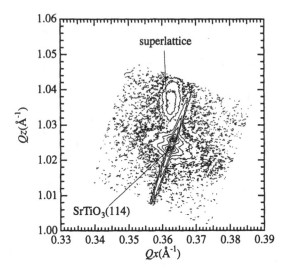

Figure 4: Logarithmic contour mapping around (114) diffraction of $[F_{10}, G_4]$ superlattice measured with Cu $K\alpha$ radiation. Q_x and Q_z correspond to [110] and [001] directions, respectively. In-plane lattice parameter for the superlattice estimated as $a = \sqrt{2}/Q_x$ is identical to that of the substrate, indicating coherent epitaxy.

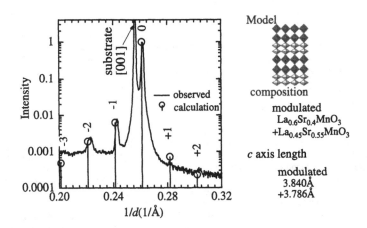

Figure 5: 2θ-θ scan of x-ray diffraction around the perovskite (002) fundamental peak for a $[F_{10}, A_3]$ superlattice. The circles given in the figure are the intensities of satellite peaks calculated with the one-dimensional step model by assuming that La/Sr composition and lattice parameter are ideally modulated as illustrated in the right panel (see text).

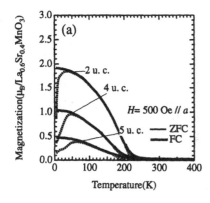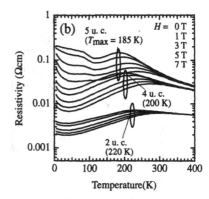

Figure 6: Temperature dependence of resistivity (left panels) and magnetization at 500 Oe (right panels) of $[F_{10}, G_n]$ ($n = 0, 2, 4,$ and 5) superlattices. Magnetization was measured during warming after zero-field cooling (ZFC) and during cooling in magnetic field of 500 Oe (FC). Resistivity was evaluated by assuming that only $La_{0.6}Sr_{0.4}MnO_3$ layers are conductive. Magnetic field was applied along the [100] in both cases.

The F/A superlattices also have the same in-plane lattice parameter as that of the substrate. To analyze the superstructure along the growth direction, 2θ-θ scan was performed for an $[F_{10}, A_3]$ superlattice. Although satellite peaks can be seen in the both sides as shown in Fig. 5, we note the fact that the intensity of satellite peak is asymmetric around the fundamental peak; -1 satellite peak intensity is 20 times as high as that of the +1 for instance. This large asymmetry can be explained only when we take account both of the c lattice parameter modulation and the compositional modulation between the constituent layers. We have carried out the simulation of the XRD profile based on the one-dimensional step model. The calculated peak position and relative intensity normalized by that for the fundamental perovskite peak are plotted in Fig. 5. In the model, we assumed that La/Sr concentration has perfect modulation as designed, and the lattice parameter c is perfectly modulated between 0.384 nm and 0.379 nm. The experimental result well agrees with the model. This fact is significant because the large tetragonal distortion c/a is kept the same as that of the thick A single-layer film even the A layer is as thin as 3 u.c. in the superlattice. Therefore, A-type AF spin ordering should be stabilized via $d_{x^2-y^2}$ orbital ordering. This result gives an important conjecture on the carrier dynamics in the superlattice; the carriers in the respective constituent layers will be well confined due to the persistent A-type AF spin ordering as well as to the persistent $d_{x^2-y^2}$ orbital ordering in the A layers, as discussed later.

Physical Properties of F/G Superlattices

The differences in ground state and associated physical properties between G layer $(La_{0.6}Sr_{0.4}FeO_3)$ and A layer $(La_{0.45}Sr_{0.55}MnO_3)$ give quite different effect on the physical properties of F layers $(La_{0.6}Sr_{0.4}MnO_3)$ in the superlattices. Key factor for the former superlattice is the spin frustration at the interfaces. In the latter superlattices, both layered AF spin ordering and $d_{x^2-y^2}$ orbital ordering in A layer play roles for the confinement of spin-polarized and orbital disordered electrons in F layers. Such spin interaction can be understood by the analog of spin valve compounds. Understanding the effect of modulation in orbital ordering structures at the interface may open a way to new device design as possible *"orbital valve"*.

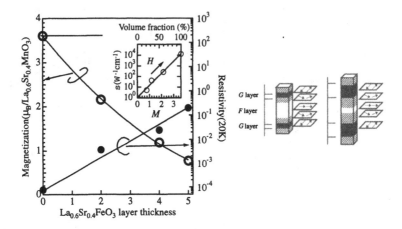

Figure 7: Magnetization and resistivity of the superlattices, $[F_{10}, G_n]$ ($n = 0$, 2, 4, and 5), as a function of $La_{0.6}Sr_{0.4}FeO_3$ layer thickness. Figures shown in right panel are schematic illustrations of the superlattices. G layers are shown as black areas. Hatched areas indicate spin-canting region due to the existence of antiferromagnetic interaction at the interfaces. The volume of the spin-canting region increases with increasing the G layer thickness. Inset: Plot of conductivity versus magnetization. Volume fraction in the upper abscissa corresponds to the magnetization of superlattices divided by $3.6\mu_B$ which is saturation magnetization of fully ferromagnetic F single layer film.

Here, we concentrate on three samples composed of F layers with the common thickness of 10 u.c. and G layers with various thicknesses of 2, 4, and 5 u.c. The temperature dependence of magnetization and resistivity in various magnetic fields is shown in Fig. 6. The T_C of the superlattices evaluated from the resistivity maximum decreases as 220 K, 200 K, and 185 K with increasing the G layer thicknesses as 2 u.c., 4 u.c., and 5 u.c., respectively. In addition, the resistivity value itself also increases significantly and large MR tends to subsist down to low temperatures with an increase of the G layer thickness. The results are obviously different from those of $La_{0.6}Sr_{0.4}MnO_3$ single layer film as shown in Fig. 1.

On the basis of these transport and magnetic properties, the properties of F layer at the interface is gradually modified from a FM metal to an AF insulator as increasing the thickness of the adjacent G layers. Strong frustration of spin arrangement as shown in Fig. 2 should be the main reason of such spin modification of F layer. To substantiate the above picture, we show in Fig. 7 the magnetization and resistivity of the superlattices as a function of G layer thickness. Since the observed magnetization is presented as per F layer, the decrease of magnetization is due to the penetration of AF order into F layers, resulting in spin canting from the original FM spin arrangement in F layer at the interface. The increase of G layer thickness makes the FM volume fraction small as represented by the change in the magnetization value (open circle) and also shown in the schematic illustration in the right panel of Fig. 7. Therefore, the robustness of the AF spin structures in the G layer and effective penetration depth of spin canting region of the F layer sensitively depend on the thickness of the G layers in the superlattice structure.

Such a change in magnetic coupling is reflected in charge transport in the F layers. The inset of Fig. 7 shows the relation between magnetization and conductivity σ at a low temperature (20 K) indicating σ increases exponentially with the effective thickness of the FM layers deduced from apparent magnetization. Magnetic field can increase the magnetization of superlattices gradually with reducing the volume of the AF region, and hence increase σ

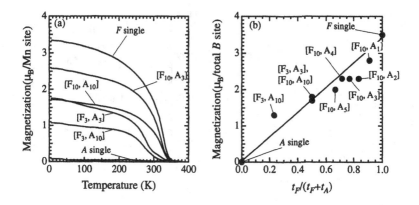

Figure 8: (a) Temperature dependence of magnetization for the F/A superlattices measured during cooling in a magnetic field of 500 Oe applied along [100] in the film plane and normalized by the total number of Mn ions in the superlattices. The respective layer thickness of the constituent layers is given by m and n in the notation of $[F_m, A_n]$, respectively. The data for the F and A single-layer films are also shown for comparison. (b) Remanent magnetization at 5 K normalized by the total layer thickness of superlattices is plotted as a function of volume fraction of the F layers, $t_F/(t_F + t_A)$.

via an increase of FM volume fraction of F layer. This can explain the huge MR observed at temperatures enough lower than T_C.

Physical Properties of F/A Superlattices

Figure 8(a) shows magnetization of the superlattices as a function of temperature. In this plot, the magnetization is normalized by the total layer thickness of the superlattices, instead of F layer thickness; all the superlattices show clear FM transition. T_C is scarcely modified by changing the F layer thickness unlike F/G Superlattices. Even if the F layer is as thin as 3 u.c. (1.2 nm), T_C is 280-290 K. Figure 8(b) shows variation of the remanent magnetization obtained from the hysteresis loop measured at 5 K after ZFC. The magnetization appears to be simply proportional to the volume fraction of the F layer, $t_F/(t_F + t_A)$, where t_F and t_A are layer thickness for the F and the A layers, respectively as shown in Fig. 8(b). In other words, it appears that the observed magnetization comes from the F layers alone, and that the A layers have the same AF state to that of the single-layer film. This assignment is consistent with the structural characterization results: c/a of $La_{0.45}Sr_{0.55}MnO_3$ layers is kept anisotropic even in the superlattices, perhaps reflecting the $d_{x^2-y^2}$ type orbital ordering and A-type AF ordering. Therefore, we conclude that spin and orbital structures are modulated in the F/A superlattices.

Figure 9 shows the resistivity in magnetic fields for superlattices. At a glance, it is noticeable that the resistivity and MR behaviors are similar to each other regardless of the combination: All the superlattices show metallic temperature dependence of resistivity. Although the magnitude depends on the combination, negative MR was observed over a whole temperature-region. The resistivity was calculated with taking the total superlattice thickness into account. In the ground state of the A-type antiferromagnet, conduction electrons can move only in the ab plane, and cannot hop between adjacent MnO_2 layers because of the interlayer AF coupling. The results of structural characterization and magnetization measurement indicate that the F and the A layers keep original FM and A-type AF properties, respectively. Therefore, the resistivity of the superlattices can be regarded to the first

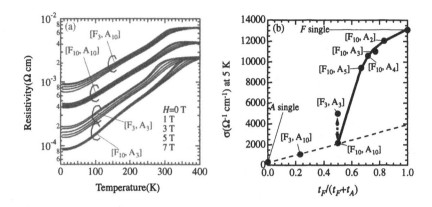

Figure 9: (a) Temperature dependence of resistivity for [F$_3$, A$_3$], [F$_3$, A$_{10}$], [F$_{10}$, A$_3$], and [F$_{10}$, A$_{10}$] super-lattices in magnetic fields. The resistivity was measured during cooling. (b) Conductivity σ at 5 K as a function of volume fraction $t_F/(t_F + t_A)$, where t_F and t_A represent the thicknesses of the constituent F and A layers, respectively. σ is normalized by the total thickness of the superlattice. The σ values for the 100 nm thick single-layer films (F and A) are shown for the comparison. A solid line represents a relationship between $t_F/(t_F + t_A)$ and σ of the F single-layer film or the superlattices whose F layer thickness is fixed to be 10 u.c. A broken line connects the data of the A single-layer film and the superlattices whose A layer thickness is fixed to be 10 u.c. The intersection between the broken line and the right ordinate may indicate σ of a hypothetical F single-layer where thickness is thin enough ($\leq$ 10 u.c.). Two superlattices of [F$_3$, A$_3$] and [F$_{10}$, A$_{10}$], having the same $t_F/(t_F + t_A)$ of 50%, show different σ as indicated by the broken arrow.

approximation as a parallel-circuit of the component layers.

However, if the constituent layers are very thin as the order of a few unit cells, we should carefully consider the effect of orbital mixing at the interface on the magnetotransport properties. Figure 9(b) shows σ at 5 K for F/A superlattices as a function of volume fraction $t_F/(t_F + t_A)$. The parallel-circuit model can explain, as in the case of magnetization results, overall transport properties qualitatively. For example, when the intervening A layer thickness is as large as 10 u.c., σ of the superlattice appears to be simply proportional to the volume fraction of the constituent F films, as represented by the broken line. However, when the thickness of the constituent A layers is small enough ($\leq$ 5 u.c.), the superlattice cannot be considered as a parallel-circuit. As indicated by a solid line in Fig. 9(b), the superlattice with [F$_{10}$, A$_n$] ($n \leq 5$) shows much higher σ than the extrapolated value (the broken line) of the $n = 10$ based parallel-circuit model. Figure 10 shows the temperature dependence of magnetoconductivity, $\Delta\sigma = \sigma(7\ \text{T}) - \sigma(1\ \text{T})$, for the superlattices and the single-layer films (note a logarithmic scale on the ordinate). In a simple parallel-circuit model, $\Delta\sigma$ in the superlattice should position between the $\Delta\sigma$ curves for the F and the A single-layer films, as indicated by a hatched area in Fig. 10. However, $\Delta\sigma$ of the superlattices at low temperatures is much larger than that for single-layer films and rather enhanced *with decreasing* the constituent A layer thickness.

Here we note again that the orbital ordering serves as a driving force for the A-type spin ordering in the A layer. If $d_{x^2-y^2}$ electron orbitals in the A layer are partially mixed with $d_{x^2-y^2}/d_{3z^2-r^2}$ quantum-disordered orbitals in the F layer to possess finite component of $d_{3z^2-r^2}$, the spin angle between neighboring MnO$_2$ atomic layers in the A layer may deviate from 180°. According to the double-exchange model, electron hopping is expressed as

$$t = t_0 \cdot \cos(\theta/2), \tag{1}$$

298

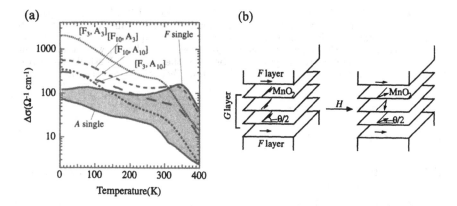

(a)

(b)

Figure 10: (a) Temperature dependence of magnetoconductivity $\Delta\sigma$, $[\sigma(7\ T)\text{-}\sigma(1\ T)]$, of the respective superlattices. $\Delta\sigma$ of the F and the A single-layer films are shown for the comparison. A hatched area indicates the region which is represented by the parallel-circuit model (see text). (b) Schematic illustration of the spin arrangement of the A layer in the superlattices. The spin angle between neighboring MnO_2 layers in the A layer is denoted as θ. θ is deviated from 180° as a result of orbital mixing and a further decrease of θ by field-induced spin-canting enhances the possibility of the electron hopping through the A layer.

where θ is the relative angle between the spins on adjacent sites [16]. Therefore, AF coupling ($\theta = 180°$) along the c axis in the A layers can confine the carrier motion within the FM layers. When a magnetic field is applied, the AF spin structure is canted to reduce θ from 180°. The reduced θ (and perhaps the associated mixing-in of $d_{3z^2-r^2}$ orbital component) induces the c axis hopping of electrons as expressed by Eq.(1). Further increase of electron hopping between the F layers through the A intervening layers in the superlattice can be caused by field-induced spin canting in AF layers as depicted in Fig. 10(b). This mechanism should give rise to the effective coupling between the F layers (i.e., dimensional crossover) and the large $\Delta\sigma$ increasing toward low temperature as observed.

CONCLUSIONS

Structure and magnetic and transport properties are characterized for oxide superlattices, $La_{0.6}Sr_{0.4}MnO_3/La_{0.6}Sr_{0.4}FeO_3$ and $La_{0.6}Sr_{0.4}MnO_3/La_{0.45}Sr_{0.55}MnO_3$. Although both $La_{0.6}Sr_{0.4}FeO_3$ and $La_{0.45}Sr_{0.55}MnO_3$ have AF spin ordering, the influences on the ferromagnetism in $La_{0.6}Sr_{0.4}MnO_3$ are quite different because of difference in spin ordering structures. In the former superlattices, the ferromagnetism in $La_{0.6}Sr_{0.4}MnO_3$ layers is strongly suppressed because of the presence of spin frustration at the interface between FM and G-type AF states. Large MR appears as a result of recovery of ferromagnetism in $La_{0.6}Sr_{0.4}MnO_3$ layer upon magnetic field. In the latter superlattices, magnetic ordering structures of constituent layers are kept since there is no spin frustration at the interface. Overall features of transport properties can be well understood by taking a simple parallel-circuit model into account. Excess conductivity in superlattices having very thin A layers can be understood by orbital mixing. Enhanced magnetoconductivity can be interpreted in terms of magnetic field induced spin canting in A layer, resulting in leakage of carriers across the AF layers and the enhanced coupling between the F layers.

ACKNOWLEDGMENTS

We thank Y. Murakami for his collaboration in structural characterization of the super-

299

lattices with use of synchrotron radiation, and N. Nagaosa and R. Maezono for stimulating discussions on the spin arrangements. We also thank Y. Konishi and Y. Ogimoto for fruitful discussions and technical support. This work, partly supported by New Energy and Industrial Technology Development Organization (NEDO) of Japan, was performed in the JRCAT under the joint research agreement between the National Institute for Advanced Interdisciplinary Research (NAIR) and the Angstrom Technology Partnership (ATP).

REFERENCES

[1] For late advance, see for example, M. Imada, A. Fujimori, and Y. Tokura, Rev. of Mod. Phys. **70**, 1218 (1998).

[2] A. Urushibara, Y. Moritomo, T. Arima, A. Asamitsu, G. Kido, and Y. Tokura, Phys. Rev. B **51**, 14103 (1995).

[3] Y. Tomioka, A. Asamitsu, Y. Moritomo, H. Kuwahara, and Y. Tokura, Phys. Rev. Lett. **74**, 5108 (1996).

[4] H. Kuwahara, Y. Moritomo, Y. Tomioka, A. Asamitsu, M. Kasai, R. Kumai, and Y. Tokura, Phys. Rev. B **56**, 9386 (1997).

[5] H. Fujishiro, M. Ikebe, and Y. Konno, J. Phys. Soc. Jpn. **67**, 1799 (1998).

[6] W. C. Koehler and E. O. Wollan, J. Phys. Chem. Solids **2**, 100 (1957): S. K. Park, T. Ishikawa, and Y. Tokura (unpublished).

[7] M. Izumi, Y. Konishi, T. Nishihara, S. Hayashi, M. Shinohara, M. Kawasaki, and Y. Tokura, Appl. Phys. Lett. **73**, 2497 (1998).

[8] M. Izumi, Y. Murakami, Y. Konishi, T. Manako, M. Kawasaki, and Y. Tokura, Phys. Rev. B **60**, 1211 (1999).

[9] M. Kawasaki, K. Takahashi, T. Maeda, R. Tsuchiya, M. Shinohara, T. Yonezawa, O. Ishihara, M. Yoshimoto, and H. Koinuma, Science **266**, 1540 (1994).

[10] Y. Konishi, Z. Fang, M. Izumi, T. Manako, M. Kasai, H. Kuwahara, M. Kawasaki, K. Terakura, and Y. Tokura, J. Phys. Soc. Jpn. *in press.*

[11] M. Izumi, T. Manako, Y. Konishi, M. Kawasaki, and Y. Tokura, submitted to Phys. Rev. B.

[12] H. Kuwahara, T. Okuda, Y. Tomioka, T. Kimura, A. Asamitsu, and Y. Tokura, Mat. Res. Soc. Symp. Proc. **494**, 83 (1998).

[13] R. Maezono, S. Ishihara, and N. Nagaosa, Phys. Rev. B **57**, R13993 (1998).

[14] S. Yunoki, A Moreo, and E. Dagotto, Phys. Rev. Lett. **81**, 5612 (1998).

[15] H. Kuwahara, T. Okuda, Y. Tomioka, A. Asamitsu, and Y. Tokura, Phys. Rev. Lett. **82**, 4316 (1999).

[16] P. W. Anderson and H. Hasegawa, Phys. Rev. **100**, 675 (1955).

IS MAGNETIC CIRCULAR DICHROISM SURFACE SENSITIVE IN THE MANGANESE PEROVSKITES ?

C.N. BORCA[1], R.H. CHENG[1], SHANE STADLER[2], Y.U. IDZERDA[3], JAEWU CHOI[1], D.N. MCILROY[1], Q.L. XU[1], S.H. LIOU[1], Z.C. ZHONG[1] AND P.A. DOWBEN[1]
[1]Department of Physics and Astronomy and the Center for Materials Research and Analysis, University of Nebraska, Lincoln, NE, 68588-0111, pdowben@unl.edu
[2]Naval Research Laboratory, NSLS, Upton, NY, 11973
[3]Naval Research Laboratory, Mater. Phys. Branch, Washington DC, 20375, Idzerda@nrl.navy.mil

ABSTRACT

We compare the magnetization versus temperature, determined by magnetic circular dichroism, with bulk magnetization determined by SQUID magnetometry. The results for two different manganese perovskites, $La_{0.65}A_{0.35}MnO_3$ systems (A=Pb, Sr), epitaxially grown on $LaAlO_3$ substrates compared. The results suggest that the magnetic circular dichroism results may not be representative of the bulk magnetization. Similar materials, but with a different secondary electron mean free path lengths or different surface compositions, may thus be expected to have different behavior in MCD or MXLD that depends on a photoyield for detection.

INTRODUCTION

Using total electron yield or photoemission as the signature of the magnetic circular dichroism (MCD) or magnetic x-ray linear dichroism (MXLD) signal, as is often used, means that the secondary electron mean free path does play a role in determining the probing depth of MCD. This probing depth of the secondary electrons or low kinetic energy can vary substantially, and in fact can be quite small. The mean free path of secondary electrons through Ni(110) was estimated [1] to be 30 Å for roughly 0 eV kinetic energy and 10 Å for 10 eV kinetic energy, while for 9 eV electrons through thin films of iron, the mean free path was estimated to be about 5 Å [2,3] and for gadolinium the value is even less [4]. These mean free path lengths are substantially less than the 100 to 200 Å one might infer from the "universal mean free path curve" for electrons in solids [5,6]. The use of the photoemission yield, at specific, but low kinetic energies is generally recognized as making MCD a surface sensitive or near surface sensitive magnetometer [7].

The electron mean free path is clearly a function of energy, and expected to increases at both low kinetic energies and high kinetic energies with a minimum mean free path at a kinetic energy of about 50 eV. Calculated models for the mean free path are typically based on the Bethe equation [8]. Thus at low kinetic energies, the mean free path should strongly depend upon the inverse of the free electron plasmon energy squared ($\propto E_p^{-2}$) or, alternatively, roughly proportional to the inverse of the number of valence electrons [9-11]. Thus we can obtain some evidence of surface sensitivity, albeit very indirectly, by comparing magnetic circular dichroism signals from two

similar doped with different metals and ostensibly different numbers of valence electrons. We compare MCD and SQUID magnetometry of $La_{0.65}Pb_{0.35}MnO_3$ with the previously reported $La_{0.7}Sr_{0.3}MnO_3$ where the surface magnetization has been compared with the bulk [12-13].

Mat. Res. Soc. Symp. Proc. Vol. 602 © 2000 Materials Research Society

While roughly similar materials, we anticipate that the Pb doped material has a larger number of valence electrons than the Sr doped material.

EXPERIMENTAL

The $La_{0.65}Pb_{0.35}MnO_3$ films, and $La_{0.65}A_{0.35}MnO_3$ (A=Ca,Sr,Ba) films with a nominal thickness of 1000 Å, were grown on a (100) LaAlO₃ substrates by RF sputtering in a 4:1 argon/oxygen atmosphere maintained at 15 mTorr. The Pb samples were subsequently annealed in two steps in order to prevent Pb evaporation: at a temperature of 650° C for 10 hours, at 850° C for 2 hours and then again for 10hrs at 650° C in an oxygen pressure of 1 atmosphere. The bulk chemical composition was determined from energy dispersive analysis of the X-ray emission (EDAX).

The XAS and MCD spectra were recorded by monitoring the sample current, which corresponds to the total electron yield. For magnetizing the sample, a 400 Oe pulse field was applied along the in-plane easy axis. The spectra for two different helicities were recorded by alternating the magnetization at every photon energy. In the total electron yield mode, the XAS spectra probes approximately 100 Å in depth, which is an intermediate length scale between the bulk and the surface boundary.

RESULTS

It is evident that the magnetization determined from MCD is not the same as that determined

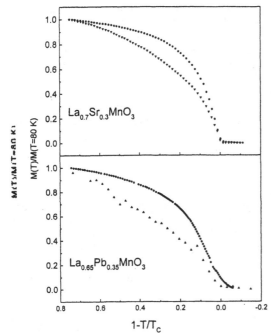

from SQUID magnetometry in the lanthanum manganates $La_{1-x}A_xMnO_3$ (A=Pb [14], Sr [12]), as seen in Figure 1. These results, with the knowledge that the electron mean free path of even very low kinetic electrons is limited, suggests that the MCD signal does have, at least, a limited surface contribution.

In Figure 2, we compare inverse photoemission $La_{1-x}A_xMnO_3$ (A = Pb, Ca, Sr, Ba) and the photoemission for $La_{0.65}A_{0.35}MnO_3$ (A = Ca, Sr, Ba). While the inverse spectra are similar in many respects, the position of the La/D/O inverse peak about 8 eV

Figure 1: Magnetic circular dichroism magnetization from the Mn 2p cores (▲) compared to the magnetization versus SQUID magnetometry (●). The comparison is shown for La₀.₆₅Pb₀.₃₅MnO₃ with the previously reported La₀.₇Sr₀.₃MnO₃ (data taken from ref. [12]).

above the Fermi level differs slightly for the different dopants, with the smallest energy above E_F for the Pb doped material. The same is true for the Mn-O occupied feature below the Fermi level. In addition, the Pb doped material has an apparent density of states, at E_F, 3 to 5 times larger than is observed for the Sr doped material. Both of these results support the reasonable premise that the Pb doped material has a larger number of valence electrons that the Sr doped material.

DISCUSSION

We need to address the surface sensitivity of the MCD signal. One method for determining mean free path of the secondary electrons are to grown uniform films of varying thickness on a well defined substrate of a different material and measure the attenuation of the signal with increasing thickness of the film [2]. This approach is really not yet really accessible experimentally for the manganese perovskites, though with the increasing sophistication of the *in situ* pulsed laser deposition process (PLD, as was used in [12]), this may ultimately be possible for the manganese perovskites. Another approach is to use mean field arguments and look at the magnetization with temperature [1].

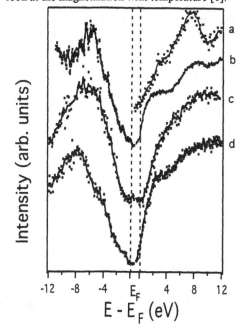

Figure 2: Inverse photoemission for $La_{0.65}Pb_{0.35}MnO_3$ (a) and combined photoemission and inverse photoemission for (b) $La_{0.65}Ca_{0.35}MnO_3$, (c) $La_{0.65}Sr_{0.35}MnO_3$, and (d) $La_{0.65}Ba_{0.35}MnO_3$. Photoemission was undertaken with helium I and the inverse photoemission in the isochromat mode ($\hbar\omega=9.4$ eV), in each case off the same sample.

The basic ideas behind this latter approach are that the surface magnetization is roughly linear in temperature, that is to say the surface magnetization $M_s(T)$ goes as:

$$M_s(T) = \left[M_b(T)/\xi(T) \right]$$

(1)

where $\xi(T)$ is the correlation length. Since, to a rough approximation, $\xi(T)$ varies as $(1-T/T_C)^{-1/2}$, and the bulk magnetization $M_b(T)$ varies as $(1-T/T_C)^{1/2}$, the surface magnetization should then vary as $(1-T/T_C)$ between 0.8 and 1.0 T/T_C [15]. Unfortunately, this type of surface magnetization can also be 'mimicked' by a paramagnetic surface or paramagnetic overlayer.

A paramagnetic overlayer on a ferromagnetic substrate, to a very good approximation, can be modeled by the Landau-Ginzburg mean field equation [16-17]:

$$\frac{1}{\kappa^2}\left[\frac{dM(z)}{dz}\right]^2 = [M(z)]^2 + b[M(z)]^4 + a \qquad (2)$$

where a is a constant of integration and κ^{-1} is the paramagnetic correlation length and M(z) is the magnetization induced at the surface of a paramagnetic overlayer of thickness z. This means that the induced magnetization goes as [16,18]:

$$M(z) = R \exp(-\kappa z) \qquad (3)$$

where R is a constant roughly proportional to the substrate magnetization. Thus, as the temperature decreases, the induced magnetization in the paramagnetic overlayer decreases not only because the substrate magnetization decreases but also because the paramagnetic correlation length $\chi(T) = \kappa^{-1}$ decreases. Thus, even though a surface may appear to be ferromagnetic in MCD, as in the case of Mn overlayers on ferromagnetic cobalt [19], the data may just as easily fit the mean field behavior expected for a paramagnetic overlayer [19], as seen in Figure 3. This induced surface magnetization should exhibit an effective Curie temperature T_c very close to the Curie temperature (T_o) of the ferromagnetic bulk if the paramagnetic overlayer is thin enough.

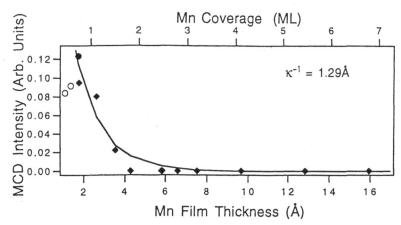

Figure 3: Fit of the paramagnetic mean field induced magnetization curve to MCD data for an Mn overlayer on cobalt. The data is taken from O'Brien and Tonner ref. [19]. Note that the derived paramagnetic correlation length is very short.

Thus a linear behavior of the magnetization could be due to the surface sensitivity or because of the presence of a paramagnetic surface layer (or paramagnetic overlayer). A distinction between these two extreme examples of surface magnetism based solely on the critical exponent of the MCD signal (or indeed most other surface sensitive magnetization spectroscopies) may not be compelling, in spite of current claims to the contrary [12-13]. This is important because there is now compelling evidence of surface segregation in $La_{0.65}A_{0.35}MnO_3$ (A = Ca [20-21], Sr [22], Pb [14]), so the surface may well be a different material from the bulk.

Indeed, even with enhanced or suppressed surface layer magnetization, the greater linearity in the magnetization derived from MCD for the lead doped perovskite (a functional exponent = 0.75) as compared to the strontium doped perovskite (a functional exponent = 0.56) suggests that there is greater surface sensitivity in the lead doped material in MCD, where the magnetization is related to the functional exponent β by $(1-T/T_C)^\beta$ [mean field arguments do not work at T_C so this is not the typical critical exponent of the phase transition, but the functional between 80% of T_C and T_C]. The value of β, from the SQUID measurements, rises from 0.36 for the Pb doped material and 0.39 for the Sr doped material towards 0.46, the greater the temperature range considered, as is expected [15], and approaching to the expected value of 1/2. This is consistent with the expectation of a greater number of valence electrons and a correspondingly smaller mean free path. Unfortunately, the data are not, as yet, compelling. We do not know if the thickness of the segregated surfaces layers are of similar thickness and identical structure with similar short range order parameters in the surface and bulk materials. The data is sufficient to raise the question of whether we can be sure that surface sensitivity is similar in the different manganese perovskites.

This question of surface sensitivity of MCD and MXLD has more general significance than just in the case of the complex oxides. For example, studies of the moment contributions in a range of iron nickel alloys from MXLD [23], based again upon electron yield, assumes the films are compositionally uniform [23-24]. This is an alloy that close $FeNi_3$ where there is some controversy about regarding enhanced surface magnetization with evidence for [25] and against [26] a surface Curie temperature higher than the bulk.

CONCLUSION

While our data thus far is not conclusive, the data suggest that there is some surface sensitivity of MCD and MXLD based upon electron yield detection in the manganese perovskites. This surface sensitivity may well depend on the kinetic energy of the (secondary) electrons and this may well be sample dependent. The MCD and MXLD is not a bulk measurement in this sense and the extent of surface sensitivity must be explored if MCD and MXLD is taken as representative of the bulk material in the various perovskite materials.

ACKNOWLEDGMENTS

This work was supported by NSF through grant # DMR-98-02126 and the Center for Materials Research and Analysis (CMRA) at the University of Nebraska.

REFERENCES

1. D.L. Abraham and H. Hopster, Phys. Rev. Lett. 58, 1352 (1987)
2. D.P. Pappas, K.-P. Kämper, B.P. Miller, H. Hopster, D.E. Fowler, C.R. Brundle, A.C. Luntz, and Z.-X. Shen, Phys. Rev. Lett. 66, 504 (1991)
3. F. Passek, M. Donath and K. Ertl, J. Magn. Magn. Mater. 159, 103 (1996)
4. M. Donath and B. Gubanka, in *Magnetism and Electronic Correlations in Local Moment Systems: Rare Earth Elements and Compounds*, World Scientific (1998) pp 217-234
5. M.P. Seah and W.A. Dench, Surf. Int. Anal., 1, 2 (1979)
6. I. Lindau and W.E. Spicer, J. Electron Spectrosc. Rel. Phenom., 3, 409 (1974)

7. K. Starke, E. Navas, E. Arenholz and G. Kaindl, in: *Spin-Orbit-Influenced Spectroscopies of Magnetic Solids*, edited by H. Ebert and G. Schütz, Springer 1996, Proceedings of the International Workshop at Herrsching, April 20-23 (1995)
8. H. Bethe, Ann. Phys. 5, 325 (1930)
9. S. Tamura, C.J. Powell, and D.R. Penn, Surf. Interface Anal. 17, 911 (1991)
10. S. Tamura, C.J. Powell, and D.R. Penn, Surf. Interface Anal. 21, 165 (1994)
11. C.J. Powell, A. Jablonski, J. Vac. Sci. Technol. A17, 1122 (1999)
12. J.-H. Park, E. Vescovo, H.-J. Kim, C. Kwon, R. Ramesh and T. Venkatesan, Phys. Rev. Lett. 81, 1953 (1998)
13. J.-H. Park, et al., Nature (London) 392, 794 (1998)
14. C.N.Borca, R.H. Cheng, Q.L. Xu, S.H. Liou, Shane Stadler, Y.U. Idzerda and P.A. Dowben, Journ. Appl. Phys. (2000) in press
15. Takahito Kaneyoshi, *Introduction to Surface Magnetization*, CRC Press, Boca Raton (1991)
16. P.A. Dowben, D. LaGraffe, D. Li, A. Miller, L. Zhang, L. Dottl, and M. Onellion, Phys. Rev. B43, 3171 (1991)
17. D.N. McIlroy and P.A. Dowben, Mat. Res. Soc. Symp. Proc. 375, 81 (1995)
18. K. Binder and P.C. Hohenberg, Phys. Rev. B6, 3461 (1972)
19. W.L. O'Brien and B.P. Tonner, Phys. Rev. B51, 2963 (1994)
20. Jaewu Choi, C. Waldfried, S.-H. Liou, and P.A. Dowben, J. Vac. Sci. Technol. A16, 2950 (1998)
21. Jaewu Choi, Jiandi Zhang, S.-H. Liou, P.A. Dowben, and E.W. Plummer, Phys. Rev. B59, 13453 (1999)
22. Hani Dulli, P. A. Dowben, Jaewu Choi, S.-H. Liou, and E.W. Plummer, submitted to Science; Hani Dulli, C.N. Borca, Jaewu Choi, Q.-L. Xu, S.-H. Liou, P. A. Dowben, and E.W. Plummer, in preparation
23. F.O. Schumann, R.F. Willis, K.G. Goodman and J.G. Tobin, Phys. Rev. Lett. 79, 5166 (1977)
24. F.O. Schumann, S.Z. Wu, G.J. Mankey and R.F. Willis, Phys. Rev. B56, 2668 (1997)
25. Yu. A. Mamaev, V.N. Petrov and S.A. Starovoitov, Sov. Tech. Phys. Lett. 13, 642 (1987)
26. J. Reinmuth, M. Donath, F. Passek, V.N. Petrov, J. Phys. Condens. Matter 10, 4027 (1998)

DISCOVERY OF A HIGH SPIN FREEZING TEMPERATURE AND CONTROL OF A SPIN-GLASS STATE IN $Mg_{1.5}FeTi_{0.5}O_4$ SPINEL FILMS

Y. Muraoka, H. Tabata, T. Kawai

ISIR-Sanken, Osaka University, 8-1 Mihogaoka, Ibaraki, Osaka 567-0047, Japan

ABSTRACT

A spin-glass state up to 210 K has been found in $(Mg,Fe)\{Mg,Fe,Ti\}_2O_4$ spinel ferrite thin films formed on α-Al_2O_3(0001) substrates. The long-time relaxation of the magnetization in zero-field-cooled operation, which is characteristic feature of the spin-glass state, has been observed below 210 K. We have also achieved the change of magnetic state in the film from spin-glass to ferrimagnet over a wide temperature range below 160 K by means of light-irradiation. The amount of spin-melt on light-irradiation is calculated to be 41 % at 10 K and 26 % at 100 K.

INTRODUCTION

A spin-glass occurs due to the combination of 'randomness' and 'frustration' in the spin ordering caused by the random mixing state of the ferromagnetic (spin parallel) and antiferomagnetic (spin antiparallel) interaction[1] in materials. Such an unique spin state has been reported in spinel[2], perovskite[3] oxides, dilute magnetic alloys[4,5] and amorphous[6-8]. In many cases, spin-glass state is observed at temperatures below 50 K (spin freezing temperature $T_f \leq 50$ K). Exceptionally, a spinel cobalt-ferrite is known to exhibit a high T_f around room temperature (T_f=284 or 320 K)[9,10].

The spin-glass system consists of various metastable states depending on the degree of 'spin freezing'[11,12]. Accordingly, applying appropriate external fields such as light perturbation to the spin-glass state is able to cause a 'melt' of the frozen spins and accelerate the magnetic relaxation, giving rise to an increase of the magnetization up to a new steady state value. This type of photo-induced magnetization has been demonstrated for amorphous spin-glass in oxides[13,14], although the operating temperature is too low (≤ 2 K) due to the low T_f (≤ 6 K) of the material. However, if such spin-glass control can be obtained at sufficiently high temperatures, then it could be applied for many magneto-optical devices.

According to the three-dimensional Heisenberg theory[15], the spin freezing temperature T_f depends heavily on the absolute value of the exchange interaction J. Hence, the spinel oxides, especially spinel ferrites can be one of the most suitable materials for achieving a high

Mat. Res. Soc. Symp. Proc. Vol. 602 © 2000 Materials Research Society

T_f because of their large J.

In this work, we have focused on the spinel ferrites of $Mg_{1.5}FeTi_{0.5}O_4$ which exhibit spin-glass behavoir around 20 K in bulk[16]. However, considering the fact of a large J ($|J_{Fe-Fe}|/k_B$=20 K, k_B stands for Boltzmann constant)[17,18] in this material, it should have a great possibility to show a high T_f. We have prepared the spinel ferrite in a form of thin film to realize the high T_f. We have also tried to control the spin-glass state of the ferrite film by means of light-irradiation. Since the ferrite film is expected to exhibit a high T_f, the spin-glass state can be controlled over a wide temperature range below the T_f.

EXPERIMENT

Target sample, $Mg_{1.5}FeTi_{0.5}O_4$ was prepared by normal solid state reaction. Films were prepared by a pulsed laser deposition (PLD) using an ArF excimer laser (wave length: 193 nm) (Lambda Physics, model COMPex 100). The substrate temperature and oxygen pressure was maintained at 500 °C and 1.5×10^{-5} Torr, respectively, during the film preparation. The typical deposition rate was about 10 Å/min. The thickness of the film was 6600 Å. The film was examined by XRD, X-ray pole figure (Rigaku rint-2000) and SQUID(Quantum design MPMS-5S). Magnetic fields were applied parallel to the substrate surface.

RESULTS AND DISCUSSION

Figure 1 shows the XRD pattern of $Mg_{1.5}FeTi_{0.5}O_4$ film. This film is clearly shown to be (111)-oriented single phase. The lattice parameter of the (111) out-of-plane d_{111} was determined to be 4.864(1) Å, which is smaller than that of bulk material (4.870(1) Å) with a cubic symmetry. By X-ray pole figure measurement, the film is found to be in well epitaxy

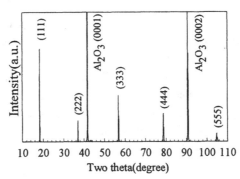

Figure 1: XRD pattern of (111)-oriented $Mg_{1.5}FeTi_{0.5}O_4$ film formed on $Al_2O_3(0001)$ substrate.

with the substrate.

Figure 2 depicts the temperature dependence of the magnetization for $Mg_{1.5}FeTi_{0.5}O_4$ film under the applied fields of H=50, 500 and 1000 Oe. In this figure, the zero-field-cooled (ZFC) and field-cooled (FC) data are shown. The magnetic measurement indicates that this film exhibits spin-glass behavior. On warming from 10 K to 350 K under an external field of 50 Oe, the cusp of zero field cool curves was observed at 210 K, but no cusp of field cool curves was observed. The cusp shape became broader and its temperature decreased with increasing the applied field. These are typical characteristics of a spin-glass. This bulk material is known to exhibit a spin-glass behavior and the freezing temperature (T_f) is reported as 20 K at an applied field of 50 Oe[16]. The T_f of the film (T_f=210 K) is much higher than that of the bulk material. This is presumably caused by the lattice mismatch between the substrates and the thin films. The stress of a (111) in-plane of films makes the bond angle close to 180° between cations on A and B-site bridged through oxygen anion in the lattice, which enhances the exchange interaction J and therefore increases the T_f of the ferrite film.

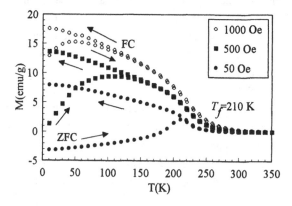

Figure 2: Temperature dependence of the magnetization of (111)-oriented $Mg_{1.5}FeTi_{0.5}O_4$ film on Al_2O_3(0001) under H=50, 500 and 1000 Oe.

The spin-glass character in the $Mg_{1.5}FeTi_{0.5}O_4$ film has been confirmed also by the measurement of the time dependence of the isothermal remanent magnetization. In this measurement, the film was first cooled from 300 to 50, 170 and 220 K in ZFC process and then the field of 100 Oe was applied. As soon as the field was applied, measurements were started. Figure 3 shows the magnetization M_{ZFC} vs log (t) plot at 50, 170 and 220 K. No change was observed with time at 220 K, but at 170 and 50 K the long-time relaxation, characteristic of a spin-glass was observed. The temperature of 170 K is below the cusp temperature T_f(=210 K) for DC susceptibility (SQUID, see Fig. 2); hence the long-time relaxation occurs below the T_f.

This result supports that our film is in a spin-glass state below 210 K in ZFC operation.

We have controlled the spin-glass state of $Mg_{1.5}FeTi_{0.5}O_4$ film by light-irradiation. White light from a Xe lamp was guided via an optical fiber (200-1060 nm) into a SQUID magnetometer for the illumination of the film. The light intensity was 10-100 $\mu W/mm^2$. Figure 4 shows the temperature dependence of the magnetization for $Mg_{1.5}FeTi_{0.5}O_4$ film without and with light-irradiation under the applied field of H=350 Oe. Without irradiation, the ZFC curves show a cusp at 140 K. On the light-irradiation, the magnetization in ZFC curves increases in the temperature range below 160 K, whereas no change is observed in the FC curves with light-irradiation. Furthermore, the cusp temperature in ZFC curves shifts toward low

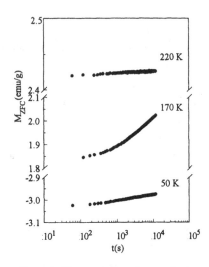

Figure 3: Time evolution of magnetization M_{ZFC} at 100 Oe after zero-field cooling from 300 K to 50, 170 and 220 K in $Mg_{1.5}FeTi_{0.5}O_4$ film.

temperatures from 140 to 130 K. These results suggest that the spin-glass state in this film changes, namely, the melting of the frozen spins in the spin-glass state is accelerated by the irradiation and the glass state approaches the more stable ferrimagnetic state (FC curves). The magnetization at 10 and 100 K in ZFC curves approach about 41 and 26 % to those in the FC curves, respectively, due to irradiation, which corresponds to the amount of melt of frozen spins.

A noteworthy point of above result is that the photo-induced magnetization is observed up to 160 K, indicating that the spin-glass state is controlled by the light below 160 K. This temperature region is much wider compared with previous reported value (up to 2 K)[13,14]. The reason for the increase in magnetization, that is, a melt of frozen spins could be due to two possible effects, photo- or thermal excitations of spins in the lattice. The detailed experimental studies for the elucidation of the mechanism by the light-irradiation are under investigation.

In conclusion, we have shown the high spin freezing temperature T_f of 210 K in $Mg_{1.5}FeTi_{0.5}O_4$ film formed on α-Al_2O_3(0001) substrate. The long-time relaxation of the magnetization has been observed below 210 K in the film. We have also demonstrated the control of the spin-glass state up to160 K in the ferrite film by light-irradiation. The amount

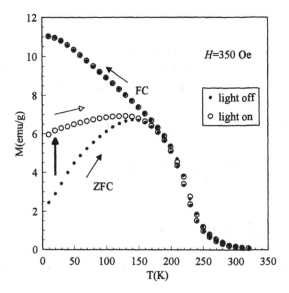

Figure 4: Temperature dependence of the magnetization without (closed circle) and with (opened circle) light-irradiation of (111)-oriented $Mg_{1.5}FeTi_{0.5}O_4$ film under H=350 Oe.

of melt of frozen spins on light-irradiation is 41 % at 10 K and 26 % at 100 K.

ACKNOWLEDGMENTS

The authors thank Mr. K. Ueda for helpful discussions. The authors also acknowledge Dr. S. Koshihara and Mr. Y. Ogawa for useful discussions about the results of SQUID measurements. This work has been performed under the Center of Excellence (COE) program supported by the Ministry of Education, Science, Sports, and Culture, Japan.

REFERENCES

1. S. F. Edwards and P. W. Anderson, *J. Phys. F* **5**, 965 (1975).

2. J. Hubsch and G. Gavoille, *Phys. Rev. Lett.* **26**, 3815 (1982).

3. J-W. Cai, C. Wang, B-G. Shen, J-G. Zhao, and W-S. Zhan, *Appl. Phys. Lett.* **71**, 1721 (1997).

4. V. Cannella and Mydosh, *Phys. Rev. B* **6**, 4220 (1972).

5. S. K. Burke and B. D. Rainford, *J. Phys. F* **13**, 451 (1983).

6. F. S. Huang, L. H. Bieman, A. M. De Graaf, and H. R. Rechenberg, *J. Phys. C* **11**, L271 (1978),

7. Y. Syono, A. Ito, and O. Horie, *J. Phys. Soc. Jpn.* **46**, 793 (1979).

8. Y. Yoshurun, M. B. Salamon, K. V. Rao, and H. S. Chen, *Phys. Rev. Lett.* **45**, 1366 (1981).

9. K. Muraleedharan, J. K. Srivastava, V. R. Marathe, and R. Vijayaraghavan, *J. Phys. C: Solid State Phys.* **18**, 5355 (1985).

10. S. N. Okuno, S. Hashimoto, K. Inomata, S. Morimoto, and A. Ito, *J. Appl. Phys.* **69**, 5072 (1991).

11. D. Sherrington and S. Kitkpatrick, *Phys. Rev. Lett.* **35**, 1792 (1975).

12. S. Kitkpatrick and D. Sherrington, *Phys. Rev.* **B17**, 4384 (1978).

13. M. Ayadi and J. Ferré, *Phys. Rev. Lett.* **50**, 274 (1983).

14. L. Rei, J.-S. Jung, J. Ferré, and C.-J. O'Connor, *J. Phys. Chem. Solids* **54**, 1 (1993).

15. B. W. Morris, S. G. Colborne, M. A. Moore, A. J. Bray, and J. Canisius, *J. Phys. C: Solid State Phys.* **19**, 1157 (1986).

16. R. A. Brand, H. G. Gibert, J. Hubsch, and J. A. Heller, *J. Phys. F: Met. Phys.* **15**, 1987 (1985).

17. A. Broese Van Groenou, P. F. Bongers, and A. L. Stuyts, *Mater. Sci. Eng.* **3**, 317 (1968/69)

18. E. De Grave, A. Govaert, D. Chambaere, and G. Robbrecht, *Physica* **96B**, 103 (1979).

Magnetic Oxide Heterostructures and Devices

STRUCTURAL AND MAGNETIC STATES IN LAYERED MANGANITES: AN EXPANDING VIEW OF THE PHASE DIAGRAM

J.F. MITCHELL, J.E. MILLBURN, C. LING, D.N. ARGYRIOU AND H. N. BORDALLO*
Materials Science Division, Argonne National Laboratory, Argonne, IL 60439
*Intense Pulsed Neutron Source, Argonne National Laboratory, Argonne, IL 60439

ABSTRACT

Colossal magnetoresistive (CMR) manganites display a spectacular range of structural, magnetic, and electronic phases as a function of hole concentration, temperature, magnetic field, etc. Although the bulk of research has concentrated on the 3-D perovskite manganites, the ability to study anisotropic magnetic and electronic interactions made available in reduced dimensions has accelerated interest in the layered Ruddlesden-Popper (R-P) phases of the manganite class. The quest for understanding the coupling among lattice, spin, and electronic degrees of freedom (and dimensionality) is driven by the availability of high quality materials. In this talk, we will present recent results on synthesis and magnetic properties of layered manganites from the $La_{2-2x}Sr_{1+2x}Mn_2O_7$ series in the Mn^{4+}-rich regime $x > 0.5$. This region of the composition diagram is populated by antiferromagnetic structures that evolve from the A-type layered order to G-type "rocksalt" order as x increases. Between these two regimes is a wide region ($0.7 < x < 0.9$) where an incommensurate magnetic structure is observed. The IC structure joins spin canting and phase separation as a mode for mixed-valent manganites to accommodate FM/AF competition. Transport in these materials is dominated by highly insulating behavior, although a region close to $x = 0.5$ exhibits metal-nonmetal transitions and an extreme sensitivity to oxygen content. We suggest two possible explanations for this transport behavior at doping just above $x=0.5$: localization by oxygen defects or charge ordering of Mn^{3+}/Mn^{4+} sites.

INTRODUCTION

Colossal magnetoresistive oxides have captured the interest of condensed matter scientists because of their strong coupling among spin, lattice, and charge degrees of freedom. The delicate balance of energy scales among these three aspects of manganite physics gives rise to the dramatic properties observed in the manganites, including insulator-metal (IM) transitions,

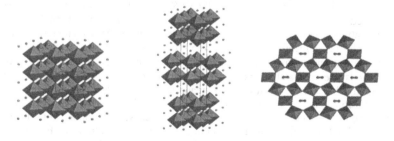

Figure 1.

Mat. Res. Soc. Symp. Proc. Vol. 602 © 2000 Materials Research Society

competing ferromagnetism (FM) and antiferromagnetism (AF), polaron dynamics, Jahn-Teller distortions, etc.

To date, there are three classes of oxide materials that exhibit colossal magnetoresistance; these three structure types are pictured in Fig. 1. On the left is the well-known perovskite phase, on the right is the pyrochlore structure of $Tl_2Mn_2O_7$, and in the middle is the layered n=2 Ruddlesden-Popper (R-P) phase. All three structure types have corner-linked MnO_6 octahedra as their building blocks, although the connectivity of these octahedra is somewhat different in the pyrochlore. It is generally accepted that the mechanism of CMR in the pyrochlore differs from that of the perovskite or layered compounds, as the pyrochlore is not a mixed-valent compound like the other two materials.

Although much of the focus in manganite research has centered around the perovskites, the discovery by Moritomo, et al.[1] of dramatically higher low-field magnetoresistance in the layered manganite $La_{1.2}Sr_{1.8}Mn_2O_7$ draws attention to the importance of these reduced dimensionality materials. In addition, the quasi-2D structure is expected to amplify the impact of charge and spin fluctuations in the region $T > T_C$. These stronger fluctuations can provide an inroad for probing the critical phenomena associated with the phase transition region, as has been shown recently by Doloc, et al.[2] and Rosenkrantz, et al.[3]

Unfortunately, the range of compositions available in the R-P phase $La_{2-2x}Sr_{1+2x}Mn_2O_7$ has been very limited. Single crystals and powders have only been successfully prepared in single-phase form in the region $0.3 < x < 0.5$,[4-6] and materials problems have been noted at each of these endpoints.[7,8] Extending the range of available phase space to $x > 0.5$ would provide the opportunity (a) to study the structure of the "baseline" compound $Sr_3Mn_2O_7$ (x=1.0) which has no Jahn-Teller active Mn^{3+} sites, and (b) to explore if and where antiferromagnetism and charge ordering impact the physics of the Mn^{4+} rich region of the phase diagram. We have fully described the crystal and magnetic structure of $Sr_3Mn_2O_7$ in an earlier work.[9,10] In this paper we describe the successful synthesis of the compounds with x between 0.5 and 1.0 *via* metastable oxygen deficient intermediates. We then present the structural and magnetic phase diagram, which shows an evolution of magnetic structure from layered A-type to "rocksalt" G-type *via* an incommensurate magnetic phase. As is the case in the ferromagnetic materials, lattice and magnetic structures are closely tied. Finally, we comment on the transport characteristics of these materials near $x = 0.5$, using the sensitivity to oxygen content to comment on the possible role of defect-induced localization or charge-ordering in this regime.

EXPERIMENTAL

Polycrystalline samples of $La_{2-2x}Sr_{1+2x}Mn_2O_7$ ($0.5 \leq x \leq 1.0$) were synthesized by a high-temperature, solid state route. La_2O_3 (Johnson-Matthey, 99.99%, dried at 1000 °C immediately before weighing), $SrCO_3$ (Johnson-Matthey, 99.999%), and MnO_2 (Johnson-Matthey, 99.999%) were combined in stoichiometric ratios and fired in air as powders first at 900 °C for 24 hours and subsequently at 1050 °C for an additional 24 hours. Samples were then pressed into 13 mm disks at 6000 lbs and ramped at 5 °C/min to 1650 °C. After 18 hours at synthesis temperature, each sample was quenched by dropping directly into dry ice. As is the case for $Sr_3Mn_2O_{7-\delta}$ [9], the compounds are metastable and must be rapidly cooled below 1000 °C to prevent decomposition. As described below, the as-made materials are oxygen deficient. To obtain stoichiometric compounds, each sample was annealed for 12 hours at 400 °C in flowing oxygen. The oxygen content of both as-made and annealed samples was determined both titrimetrically versus standardized sodium thiosulfate solution and thermogravimetrically. Battle, et al. have pointed out the difficulty in preparing single-phase samples of n=2 R-P phases near x=0.5.[8]

We have carefully examined our samples using high resolution synchrotron x-ray diffraction and find no evidence of the subtle phase separation these researchers discuss. Apparently the high temperature synthesis allows for more complete reaction and results in single phase samples.

Time-of-flight powder neutron diffraction data were collected on the Special Environment Powder Diffractometer (SEPD) and/or the High Intensity Powder Diffractometer (HIPD) at Argonne National Laboratory's Intense Pulsed Neutron Source (IPNS). The latter instrument is well suited for study of magnetic structures due to its high flux at long d-spacings. Measurements were taken as a function of temperature between 20 K and 300 K. All crystal structure analysis was performed by the Rietveld method using the GSAS program suite[11]

X-ray synchrotron measurements were made on the bending magnet beamline 12-BM at the Advanced Photon Source (APS). Samples were pulverized and ground to < 35 μm particle size. Measurements were taken at an incident energy of 9.2 keV in a θ-2θ mode.

Electrical resistance was measured as a function of temperature on sintered pellets using a standard four-lead technique. Measurements were typically repeated on cooling and heating to identify any thermal hysteresis.

RESULTS AND DISCUSSION

Synthesis of $La_{2-2x}Sr_{1+2x}Mn_2O_7$ ($0.5 < x \leq 1.0$) is complicated by the fact that this n=2 R-P compound is not the equilibrium phase in air below T ~ 1600 °C.[9,12]. Thus, preparation of these materials requires quenching from synthesis temperatures in excess of 1600 °C to prevent decomposition into perovskite or n=1 R-P phases. At these high temperatures, reduction of Mn^{4+} to Mn^{3+} becomes thermodynamically favorable, and the compounds accommodate this reduction by forming oxygen vacancies. Thus, the appropriate formulation for the as-prepared samples is $La_{2-2x}Sr_{1+2x}Mn_2O_{7-\delta}$. As shown in Fig. 2 for the x=1.0 compound, neutron diffraction unambiguously locates these vacancies in the MnO_2 conduction planes, forming a mixture of octahedra (Mn^{4+}) and square pyramids (Mn^{3+}) disordered throughout the structure[9]. It should be noted that the vast majority of n=2 R-P phases with oxygen vacancies order these vacancies between the planes, making the manganites described here unusual in their crystal chemistry.[13-14]

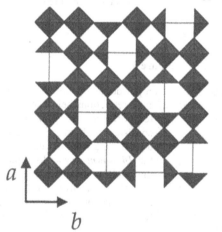

a

b

Figure 2.

Titrimetric analysis of the average Mn oxidation state versus standardized thiosulfate solution reveals a relationship between the oxygen nonstoichiometry, δ, and the dopant concentration x. This relationship suggests that the synthesis conditions (T = 1650 °C, p(O$_2$) =

0.21 atm) set the average Mn oxidation state at 3.5, i.e. $[Mn^{3+}] = [Mn^{4+}] = 50\%$. Fig. 3 shows the results of the titrimetic analysis, δ, as a function of x (error in δ is ± 0.02); TGA on selected compositions agrees within experimental error. Consideration of the chemical formula $La_{2-2x}Sr_{1+2x}Mn_2O_{7-\delta}$ shows that the average Mn oxidation state is given by $3 + x + \delta$. Thus, the linear relationship between x and δ with slope=1 implies a constant average Mn oxidation state, and the intercept at x=0.5 fixes this mean oxidation state at 3.5. It is conceivable that changing the synthesis conditions (e.g., changing $p(O_2)$) could shift this mean oxidation state.

Also shown on Fig. 3 are the results of titration on oxygen annealed samples. The outcome of this oxygen treatment has been to fill all of the oxygen vacancies, generating a stoichiometric n=2 R-P phase in the previously inaccessible region of composition space. It is important to note that these new compositions are metastable. Indeed, heating these stoichiometric compounds to 1400 °C in air slowly decomposes them into phase mixtures of n=1, 2 R-P phases and perovskite.

Magnetism

Rietveld refinement of neutron powder diffraction data measured on the stoichiometric compounds ($\delta = 0$) reveals an evolution of the magnetic structure from layered A-type to "rocksalt" G-type as x increases from 0.5 toward 1.0. However, this evolution is not smooth, as shown by the magnetic phase diagram of the $La_{2-2x}Sr_{1+2x}Mn_2O_{7.0}$ system in Fig. 4. In this diagram, the Néel temperatures are determined by the onset of intensity in AF peaks. Focusing only on the region x > 0.5, the diagram shows four regions: A-type, no long-range order, incommensurate (IC), and G-type. This IC structure is derived as a modulation of the G-type structure and represents a new response to the FM/AF competition in manganites.

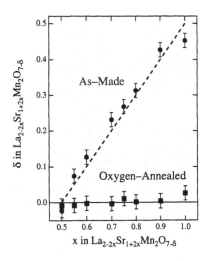

Figure 3.

Fig. 5 shows both the room temperature and 8 K neutron powder diffraction patterns of $Sr_3Mn_2O_{7.0}$, the x=1.0 endpoint composition. The presence of extra reflections below the Néel temperature of 160 K is consistent with an AF supercell $a_m = 2a_c$, $c_m = c_c$. (a_c and c_c are the *I4/mmm* crystallographic unit cell constants). The AF ordering temperature is approximately 100 K lower than that of the cubic perovskite $SrMnO_3$,[15] reflecting the quasi-2D nature of the layered compound. The refined magnetic structure in Fig. 5 reveals all AF Mn-Mn neighbor interactions—a "rocksalt" structure"—and a *c*-axis orientation of the moment. The magnetic

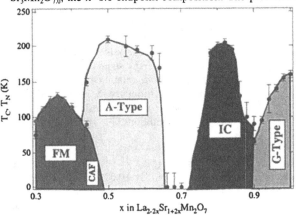

Figure 4.

structure results from t_{2g}^3–O_{2p}–t_{2g}^3 superexchange in this undoped parent compound.[19] However, in tetragonal symmetry this magnetic structure is frustrated. The spin labeled with an arrow in one bilayer sees in the neighboring bilayer a square of Mn ions with two "up" spins and two "down" spins across the diagonal of this square. This degeneracy suggests that the true symmetry of the magnetic phase may be lower (e.g., orthorhombic), but the resolution of the neutron data is inadequate to reveal any such symmetry-breaking at this composition.

However, consideration of the entire series from x=0.5 to x=1.0 clarifies the connection

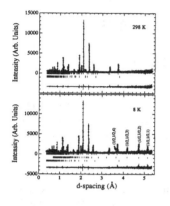 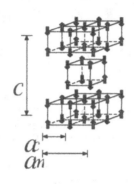 between crystal and magnetic structures by revealing the anticipated symmetry lowering. Fig. 6 shows the variation of the lattice constants with Sr content as determined by neutron powder diffraction at 300 K (circles) and at 20 K (triangles). At room temperature, all of the samples are tetragonal, but at low temperature samples with compositions between x=0.74 and 0.92 become orthorhombic. Previous to this

Figure 5.

work, n=2 R-P phases in the $La_{2-2x}Sr_{1+2x}Mn_2O_7$ series have only been found in tetragonal symmetry. As will be discussed shortly, this orthorhombic distortion is found in the same composition space where magnetic order based on the G-type magnetic structure is found (either true G-type or an incommensurate structure based on G-type). Furthermore, a high-resolution x-ray measurement at 300 K for an x=0.94 sample (previously cooled to 20 K) shows coexistence between orthorhombic and tetragonal phases. Additional temperature dependent, high resolution measurements will be required to properly map out the relationship between the structural and magnetic phase lines, in particular near x=1.0. Nonetheless, the current results support the hypothesis that the G-type magnetic structure is connected to the orthorhombic phase. At the other end of the region, near x=0.5, the A-type antiferromagnetic structure is stable. As pointed out by various researchers,[6,16,17] this structure consists of ferromagnetically coupled MnO_2 sheets coupled antiferromagnetically within a given bilayer with

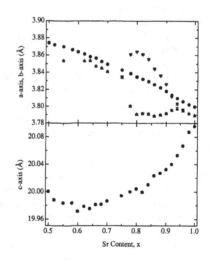

Figure 6.

319

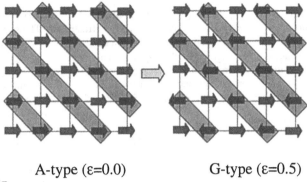

A-type (ε=0.0)   G-type (ε=0.5)

Figure 7.

the Mn spins lying in the a-b plane. This magnetic structure is consistent with an ordering of the x^2-y^2 orbitals on the Mn atoms in the a-b plane.[18] Our investigations have shown that this structure type is stable up to x ~ 0.65. In this region T_N is remarkably insensitive to Sr concentration, varying from 200 K at x=0.5 to 175 K at x=0.64 before plummeting dramatically to < 20 K at x=0.66.

A gap exists between x=0.66 and x=0.72 where no AF reflections are observed above 20 K, the lowest temperature measured by neutrons (5 K in the case of x=0.70). This gap is unprecedented in the manganite field and probably reflects the competing AF and FM interactions at the borders of the A- and G-type structures. At this time we have no detailed microscopic explanation for why such a gap appears. However, there are structural features that may shed light on this question. As shown in Fig. 6, the c-axis exhibits a shallow minimum between x=0.6 and 0.7. In addition, the Mn-O bond lengths (not shown) also have a cross-over in this region, with the MnO_6 octahedra becoming completely regular at x=0.62-0.64. Such structural effects are potential signatures of orbital ordering crossovers, which may be important for understanding the magnetic structure of these compositions. Clearly additional experimental and theoretical work are required to understand this particular region of composition space.

Finally, we consider the IC magnetic structure found between x=0.72 and x=0.90 and the relationship this structure bears to the A-type and G-type magnetic structures. In this region, Rietveld refinement shows that the moment lies in the a-b plane (the crossover is at x ~ 0.94). Fig. 7 shows schematically how the A- and G-type structures are related in a given MnO_2 plane; in both structures this plane will be AF coupled to the other member of the bilayer. Simply flipping the direction of the spins in alternate (110) planes changes the A-type to the G-type structure. Thus, the G-type structure may be considered as a q=(1/2,1/2,0) modulation of the A-type structure. This can be made more quantitative by introducing an incommensurability parameter, ε, which takes the value ε=0 for the A-type structure and ε=0.5 for the G-type structure. An intermediate value for ε gives rise to the IC structure. Fig. 8 shows a comparison of the selected portions of the neutron powder diffraction pattern for x=0.55 (A-type) and x=0.92 (G-type) samples below T_N with several magnetic peaks indexed. These patterns are clearly distinct, yet both can be related through the construction described above. Furthermore, note the three doubly degenerate peaks (degenerate because $-\varepsilon = \varepsilon = 1-\varepsilon$ for ε=1/2). As shown in Fig. 8, these peaks split as ε departs from 0.5 in the IC phase; here for x=0.82, ε ~ 0.4.

Fig. 9 shows a preliminary model of the IC structure as a commensurate approximation with ε=0.4. The key features of the model are the spin rotation angle that connects the A- and G-type structures and the pairwise interactions that are predominantly AF. With ε=0.4, the repeat unit is $5a_c$. Starting at any site, moving to the neighbor on the right or above is accompanied by a rotation of 144°; five such moves returns the spin to its original configuration. If this angle were 0°, the A-type structure would be produced; if it were 180°, the G-type structure would result. Presumably, any arbitrary value of this rotation angle should be allowed. However, in all cases we examined ($0.75 < x < 0.9$), a value of ε close to 0.4 is observed, despite T_N varying from 80 K at x=0.75 to 205 K at x=0.82. With $\varepsilon \sim 0.4$, the magnetic structure should closely approximate G-type ($\varepsilon=0.5$), and indeed the pairwise interactions between Mn spins are all close to AF as expected in the G-type phase. The weak FM component presumably arises because of the presence of doped electrons generating Mn^{3+}—O—Mn^{4+} FM double-exchange interactions.[19] This structure is quite unlike the canted state predicted by DeGennes[20] or a phase separation, both of which have $M \neq 0$. In contrast, compounds in this region adopt the complicated antiferromagnetic ($M=0$) IC structure to accommodate the competition between AF and FM exchange.

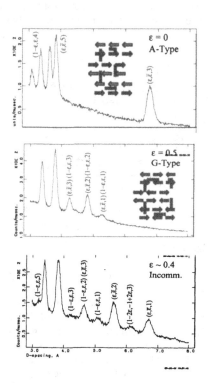

Figure 8.

Transport

In the 3-D perovskite $La_{1-x}Ca_xMnO_3$ (LCMO), the region $x > 0.5$ is characterized by insulating compounds undergoing charge order transitions whose existence displays little

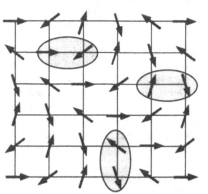

sensitivity to a Mn^{3+}/Mn^{4+} ratio differing from unity.[21] On the other hand, $La_{1-x}Sr_xMnO_3$ (LSMO) is metallic out to its solubility limit at x~0.6. The reduced dimensionality in the $La_{2-2x}Sr_{1+2x}Mn_2O_7$ system results in transport behavior unlike either of these cases. The majority of the compounds are highly insulating but do not show no the signature of charge-ordering seen in LCMO. On the other hand, a narrow composition region near x=0.5 shows metal-nonmetal transitions and an extreme sensitivity to oxygen content. Using electron diffraction, other researchers have shown the presence of charge-ordering in the x=0.5 compound $LaSr_2Mn_2O_7$.[22] Based on

Figure 9.

transport measurements, we suggest that similar behavior exists out to x~0.6, but that unlike the case of LCMO this property may be highly sensitive to the Mn^{3+}/Mn^{4+} ratio.

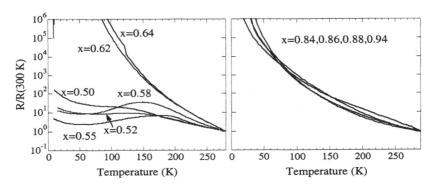

Figure 10.

The transport characteristics of $La_{2-2x}Sr_{1+2x}Mn_2O_7$ (x > 0.5) are divide cleanly into two classes: those that show metal-nonmetal transitions, and those that are extremely insulating. Furthermore, the insulating compounds do not show the typical signature of charge ordering seen in perovskites: an abrupt change in dR/dT.[23] Fig. 10 shows resistance measurements on several of the stoichiometric (δ=0.0) compounds. As shown in left panel, samples in the range $0.5 \leq x <$ 0.6 are far less resistive than their counterparts at x > 0.6. Furthermore, these compounds near x=0.5 typically show a maximum in resistance between 150-200 K, similar to the behavior seen by others in the x=0.5 compound.[22]. Samples with higher doping levels (shown in the right panel) are all highly resistive. As shown by other investigators, charge-ordering in perovskites (e.g., LCMO, $Nd_{1-x}Sr_xMnO_3$, $Pr_{1-x}Ca_xMnO_3$) causes a sudden change in the slope of $R(T)$ at the charge-ordering temperature. In the case of the $La_{2-2x}Sr_{1+2x}Mn_2O_7$ compounds, $R(T)$ is smoothly varying in the entire temperature region measured,[24] suggesting either that charge ordering is not playing a significant role in the physics of these layered compounds or that its signature is obscured in the transport measurement. In the former case, the localization presumably results from strong AF exchange producing a Mott-Hubbard insulator. Diffraction measurements showing the presence or absence of superlattice reflections will be required to distinguish between these two possibilities.

Charge-ordering of equal numbers of Mn^{3+} and Mn^{4+} in the $LaSr_2Mn_2O_7$ system is quite unlike that found in perovskites, as it is only stable in a narrow temperature window between 100 K and 200 K. It is currently believed that the A-type AF structure competes with charge ordering (which prefers a CE-type magnetic structure) and eventually wins out at low temperature.[25]. The resistive maximum seen in the samples 0.5 < x < 0.6 resembles the x=0.5 data and may imply that a similar effect is at play in this region as well. By exploiting the oxygen nonstoichiometry available in our materials we can comment on the impact of oxygen vacancies on transport and conjecture on the doping range where charge ordering may impact transport properties.

Fig. 11 shows $R(T)$ for two samples with x=0.55: as-made with δ=0.05(2) and annealed with δ=0.00(2). (A second annealed sample is also plotted to demonstrate reproducibility.) The dramatic difference between these curves demonstrates the high sensitivity of transport to oxygen content. In this case, an oxygen vacancy concentration of ~1% precipitates this change, which may arise because the vacancies are located in the MnO$_2$ conduction planes.

As shown in Fig. 3, the as-made samples contain oxygen vacancies whose concentration varies linearly with the Sr content. The structural implication of this variation is that the ratio of Mn^{4+}O$_6$ octahed-

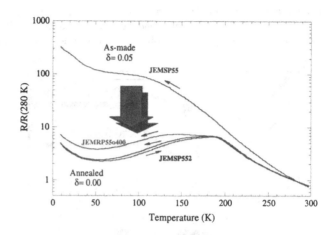

Figure 11.

ra to Mn^{3+}O$_5$ square pyramids varies systematically with x, as well. This variation is shown graphically in Fig. 12. The solid lines labeled "octa" and "pyr" are based on a model in which

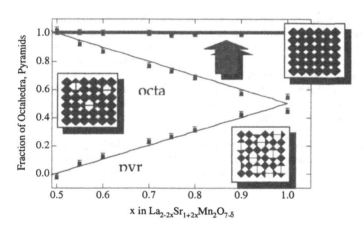

Figure 12.

each vacancy removed from the MnO$_2$ planes converts two octahedra into two square pyramids while keeping the mean Mn oxidation state fixed at 3.5. The points on these lines are based on the measured oxygen content of the compounds in the series and fit well to the model. Naturally, a direct determination of the fraction of octahed-

ra and pyramids would be desirable. Unfortunately the oxygen vacancies are disordered, precluding an actual measurement of these quantities by diffraction techniques. It is conceivable that the defects act as randomly distributed scattering sites, which in the reduced dimension of the layered compounds are likely to precipitate localization. For x~0.5 there are few such defects so resistance intermediate between x=0.50 and x > 0.60 results.

An alternative possibility for explaining the transport behavior rests on the conjecture that charge ordering in layered manganites is only possible when Mn^{3+}/Mn^{4+} = 1. The as-made x=0.55 compound contains 0.05 oxygen vacancies per formula unit. Recalling Fig. 3 above, this composition implies a mean Mn oxidation state of 3.5, or Mn^{3+}/Mn^{4+} = 1. Such a ratio is also found in the x=0.50 sample, which has no oxygen vacancies. As shown in Fig. 13, the $R(T)$ curves for the x=0.5 (δ=0.0) sample and the x=0.55 (δ=0.05) sample are qualitatively similar, yet both differ significantly from the x=0.55 (δ=0.0) sample, which has a clearly defined maximum

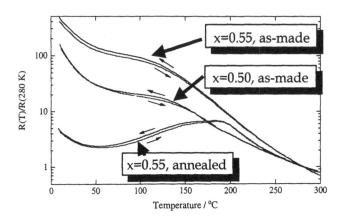

in R near 200 K. Thus, a picture develops in which charge-ordering, which has been established in the x=0.5 composition, is also stable in the oxygen deficient compounds x < 0.6. The charge ordering then disappears when the oxygen vacancies are filled. The vacancy-free materials with excess carriers ($Mn^{3+}/Mn^{4+} < 1$)

Figure 13.

then exhibit a pronounced nonmetal-metal transition in the absence of the charge-ordering.

There are several constraints implicit in this model beyond the requirement that $Mn^{3+}/Mn^{4+} = 1$ for charge ordering to be stable. First, it requires that charge ordering be robust to a small oxygen vacancy concentration (δ < 0.1). Second, it implies that small carrier concentrations beyond x=0.50 result in "metallic" like behavior, a feature not seen in LCMO perovskites. Finally, it requires that more highly doped compounds remain insulators even when oxygen stoichiometric, presumably because of a competing AF Mott-Hubbard ground state found at higher Mn^{4+} concentrations. Despite the speculative nature of this model, the experiments required to test it are clear. X-ray or electron diffraction patterns of x=0.5 – 0.6 as-made materials should all show charge-ordering superlattice reflections in the temperature range 100 – 200 K, as should the oxygen-annealed sample of x=0.5 (δ=0.0 for this composition both as-made or annealed). However, the patterns of the oxygen-annealed samples with x > 0.5 should show no such superlattice reflections.

SUMMARY

We have successfully synthesized n=2 R-P phases in the $La_{2-2x}Sr_{1+2x}Mn_2O_7$ series for x > 0.5 using a two-step process whose intermediate state is an oxygen deficient material. Efforts on understanding the structural, magnetic, and electronic behavior of the series are in their early stages. Nonetheless, despite the absence of ferromagnetic ground states, it is clear that the variety of magnetic and electronic phases in this Mn^{4+}-rich region compares favorably to that of the previously known composition space between x=0.3 and x=0.5.

As in the perovskites and the known layered manganites, structure and magnetism are tightly coupled. In particular, we have determined that the magnetic structure evolves from the A-type layered antiferromagnetic structure through an unprecedented "gap" in which no magnetism is observed, to an incommensurate phase, and finally to a G-type "rocksalt" antiferromagnet near the x=1.0 endpoint. The IC structure joins spin canting and phase separation as a mode for mixed-valent manganites to accommodate FM/AF competition. The existence of a tetragonal-orthorhombic phase line for x > 0.72 (i.e., in the IC and G-type regions) provides the mechanism necessary to relieve a magnetic degeneracy inherent to the G-type structure in the tetragonal symmetry of the paramagnetic phase. Likewise, the appearance of regular MnO_6 octahedra in the magnetic "gap" region may signify an orbital ordering crossover.

Transport measurements show that the majority of compositions in the Mn^{4+}-rich region are highly insulating (0.6 < x < 1.0). However, for x~0.5 transport is characterized by metal-nonmetal transitions and an extreme sensitivity to small O vacancy concentrations. Our preliminary results are consistent with a model in which charge-ordering is stable only at $Mn^{3+}/Mn^{4+} = 1$, but further measurements will be required to verify this thesis.

ACKNOWLEDGMENTS

The authors thank Mark Beno, Guy Jennings, and Jennifer Linton for assistance with the synchrotron measurements, and Simine Short for assistance at SEPD. This work was supported the the U.S. Department of Energy, Basic Energy Sciences—Materials Sciences under Contract W-31-109-ENG-38 and W-7405-ENG-36 (DNA)

REFERENCES

1. Y. Moritomo, A. Asamitsu, H. Kuwahara and Y. Tokura Nature **380**, 141 (1996).
2. L. Vasiliu-Doloc, S. Rosenkranz, R. Osborn, S. K. Sinha, J. W. Lynn, J. Mesot, O. H. Seeck, G. Preosti, A. J. Fedro and J. F. Mitchell Phys. Rev. Lett. **83**, 4393 (1999).
3   S. Rosenkranz, R. Osborn, J.F. Mitchell, L. Vasiliu-Doloc, J. Lynn, S.K. Sinha, D.N. Argyriou J. Appl. Phys. **83**, 7348 (1998).
4. R. Seshadri, c. Martin, M. Hervieu, B. Raveau, and C.N.R. Rao Chem. Mater. **73**, 1097 (1997).
5. M. Medarde, J. F. Mitchell, J. E. Millburn, S. Short and J. D. Jorgensen Phys. Rev. Lett. **83**, 1223 (1999).
6. K. Hirota, Y. Moritomo, H. Fujioka, M. Kubota, H. Yoshizawa and Y. Endoh J. Phys. Soc. Jpn. 67, 3380 (1998).
7. D.N. Argyriou, J.F. Mitchell, P.G. Radaelli, H. N. Bordallo, D.E. Cox, M. Medarde and J.D. Jorgensen Phys. Rev. B **59**, 8695 (1999).
8. P.D. Battle, D.E. Cox, M.A. Green, J.E. Millburn, L.E. Spring, P.G. Radaelli, M.J. Rosseinsky and J.F. Vente Chem. Mater. **9**, 1042 (1997).
9. J.F. Mitchell, J.E. Millburn, M. Medarde, S. Short, J.D. Jorgensen and M.T. Fernández-Diaz J. Solid State Chem. **141**, 599 (1998).
10. J.F. Mitchell, J.E. Millburn, M. Medarde, D.N. Argyriou and J.D. Jorgensen J. Appl. Phys. **85**, 4352 (1999).
11. A.C. Larson and R.B. Von Dreele, General Structural Analysis System, Los Alamos Internal Report No. 86-748, 1990.
12. N. Mitzutani, A. Kitazawa, O. Nobuyuki and M. Kato J. Chem. Soc. (Jpn.) Ind. Ed. **73**, 1097 (1970).
13. M. Itoh, M. Shikano, H. Kawaji and T. Nakamura Solid State Comm. **80**, 545 (1991).
14. S.E. Dann, M.T. Weller and D.B. Currie J. Solid State Chem. **97**, 179 (1992).
15. T. Takeda and S. Ohara J. Phys. Soc. Jpn. **37**, 275 (1974).
16. M. Kubota, H. Yoshizawa, Y. Moritomo, H. Fujioka, K. Hirota, Y. Endoh J. Phys. Soc. Jpn **68**, 2202 (1999).

17. M. Kubota, H. Fujioka, K. Ohoyama, K. Hirota, Y. Moritomo, H. Yoshizawa and Y. Endoh J. Phys. Chem. Solids 60, 11261 (1999).
18. R. Maezono and N. Nagaosa cond-matt/9904427 (1999).
19. J.B. Goodenough, *Magnetism and the Chemical Bond* (Interscience Publishers, New York, 1963).
20. P.G. DeGennes, Phys. Rev. **118**, 141 (1960)
21. A.P. Ramirez, P. Schiffer, S-W. Cheong, C.H. Chen, W. Bao, T.T.M. Palstra, P. L. Gammel, D.J. Bishop and B. Zegarski Phys. Rev. Lett. 76, 3188 (1996).
22. J.Q. Li, Y. Matsui, T. Kimura and Y. Tokura Phys. Rev. B **57**, R3205 (1998).
23. M. Roy, J.F. Mitchell, A. Ramirez, P. Schiffer J. Phys.: Cond. Matt. **11**, 4843 (1999).
24. The discontinuity in the x=0.64 curve is an experimental artifact.
25. D.N. Argyriou, private communication.

PHASE EQUILIBRIUM DIAGRAMS OF THE BINARY SYSTEMS LaMnO$_3$-SrMnO$_3$ AND LaMnO$_3$-CaMnO$_3$

PETER MAJEWSKI, LARS EPPLE, HEIKE SCHLUCKWERDER,
FRITZ ALDINGER
Max-Planck-Institut für Metallforschung, Pulvermetallurgisches Laboratorium,
Heisenbergstraße 5, 70569 Stuttgart, Germany

ABSTRACT

The quasi binary systems LaMnO$_3$ – SrMnO$_3$ and LaMnO$_3$ – CaMnO$_3$ have been studied. Both systems show a miscibility gap below about 1400°C in air. This phenomenon causes the decomposition of single phase (La,Sr)MnO$_{3-x}$ and (La,Ca)MnO$_{3-x}$ solid solution with intermediate La:Sr or La:Ca ratios into La rich SrMnO$_{3-x}$ or CaMnO$_{3-x}$, and Sr or Ca rich LaMnO$_{3-x}$ at lower temperatures. At 1400 °C a structure transformation of (La,Sr)MnO$_3$ from orthorhombic to rombohedral has not been observed and the structure of La$_{0.7}$Sr$_{0.3}$MnO$_3$ has been determined to be orthorhombic with a = 0.54927 ± 0.0009 nm, b = 0.54582 ± 0.0009 nm, and c = 0.76772 ± 0.0034 nm.

INTRODUCTION

The binary systems La$_2$O$_3$-Mn$_2$O$_3$ [1], SrO-Mn$_2$O$_3$ [2] and, CaO-Mn$_2$O$_3$ [3] contain the perovskite structured phase ABO$_3$ with A = Sr, Ca or La and B = Mn. For LM with high a Mn^{4+} content, e.g. Sr or Ca doped LM, a rombohedral structure has been reported for temperatures of 1000-1200 °C [1,4,5]. Whereas, at a lower Mn^{4+} content (low Sr or Ca content) LM crystallizes in an orthorhombic structure [1,5]. [4,5] observed the phase transition of Sr doped LM from the orthorhombic to the rhombohedral modification at 1200 °C in air at about x$_{Sr}$ = 0.175. [5] observed that at lower Mn^{4+} content (x$_{Sr}$ below 0.175) orthorhombic and rhombohedral LM coexist at 1000 °C in air.

SrMnO$_{3-x}$ (SM) exhibits a phase transformation from orthorhombic to orthorhombic distorted hexagonal at about 1400 °C and to hexagonal at 1035 °C due to uptake of oxygen combined with an increase of the Mn^{4+} content of the phase [6]. The crystal structure of CaMnO$_3$ (CM) is orthorhombic at high and low Mn^{4+} contents, but the axis parameter significantly varies with the Mn^{4+} content [7,8].

It is known that LM and SM as well as LM and CM form complete solid solution series at temperatures of about 1400 °C in air. The extensions of the solid solution series at lower temperatures have not been studied, yet.

EXPERIMENTAL

Samples with the compositions La$_{0.5}$Sr$_{0.5}$MnO$_{3-x}$ and La$_{0.5}$Ca$_{0.5}$MnO$_{3-x}$ were prepared using La$_2$O$_3$, SrCO$_3$, CaCO$_3$ and, MnO$_2$ (purity > 99 %), respectively. The powder mixtures were calcined at 1200 °C in air for 12 h, ground and cold isostatically pressed. The samples were than sintered at 800, 900, 1000, 1200, 1300, 1350 and 1400 °C in air for 48 h with intermediate regrinding as well as pressing and finally furnace cooled to room temperature

327

(batch 1). A second batch of samples with the composition $La_{0.5}Sr_{0.5}MnO_{3-x}$ and $La_{0.5}Ca_{0.5}MnO_{3-x}$, respectively, were sintered at 1400 °C in air and subsequently annealed at 1350, 1300, 1200, 1000, 900 and 800 °C in air for up to 500 h and furnace cooled (batch 2). In addition, samples with the composition $La_{1-x}Sr_xMnO_{3-y}$ and $La_{1-x}Ca_xMnO_{3-y}$ with $0 \leq x \leq 1$ were prepared in air at 1200 as well as 1400 °C and at 1300 as well as 1400 °C, respectively.

Phase identifications were performed using scanning electron microscopy (SEM) with EDX (Zeiss DSM 982 Gemini) and x-ray diffraction (Siemens D-5000, CuKα₁ radiation). The structure analysis were performed using the Rietveld analysis program DBWS-9807.

RESULTS

The samples prepared at temperatures above 1350 °C consist of single phase (La,Sr)MnO₃ and (La,Ca)MnO₃ solid solution. In the LM-SM system at 1400 °C the transition from the orthorhombic to the rombohedral modification of LM has not been observed. The Rietveld analysis of a sample with the composition $La_{0.7}Sr_{0.3}MnO_{3-x}$ sintered at 1400 °C and quenched in air shows that LM is orthorhombic (Figure 1, Table 1). However, in the system LM-CM at 1400 °C orthorhombic LM with intermediate Ca content has not been observed which is due to the fact that either at 1400 °C orthorhombic LM with intermediate Ca content does not exist or the transformation of Ca doped LM from the orthorhombic to rhombohedral structure is very fast and the orthorhombic structure can not

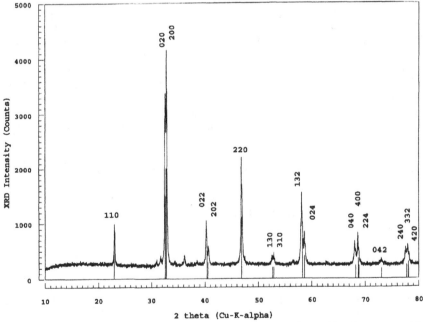

Figure 1: Rietfeld analysis of an orthorhombic samples with the composition $La_{0.7}Sr_{0.3}MnO_{3-x}$ sintered at 1400 °C in air.

be obtained by air quenching of the samples. At 1400 °C the increase of the Sr and Ca content of the solid solution, respectively, is monitored by a significant decrease of the a-axis parameter of LM (Table 2).

The decomposition of single phase $La_{0.5}Sr_{0.5}MnO_3$ (batch 1) prepared at 1400 °C in air into LM and SM at 1200 °C can be detected by the XRD diffraction pattern of the sample showing reflections of both phases, LM and SM (Figure 2). The XRD pattern of SM indicate that at 1200 °C in air La rich SM crystallizes in the orthorhombic distorted hexagonal modification determined by [6]. In Figures 3 and 4 the temperature concentration diagram for the quasi binary systems LM-SM and LM-CM show the miscibility gap between LM and SM as well as LM and CM at temperatures below 1400 °C in air.

Table 1:   Crystallographic data (see Fig. 1)

No	h k l	d value (obs)	d value (calc)	2 theta (obs)	2 theat (calc)
1	1 1 0	3.87000	3.87167	22.961	22.951
2	2 0 0	2.74600	2.74635	32.581	32.577
3	0 2 0	2.72900	2.72910	32.790	32.788
4	2 0 2	2.23400	2.23357	40.339	40.347
5	0 2 2	2.22400	2.22425	40.528	40.523
6	2 2 0	1.93700	1.93584	46.864	46.894
7	3 1 0	1.73600	1.73585	52.662	52.686
8	1 3 0	1.72750	1.72712	52.961	52.973
9	3 1 2	1.58150	1.58165	58.294	58.288
10	2 0 4	1.57300	1.57320	58.640	58.631
11	4 0 0	1.37350	1.37318	68.224	68.242
12	0 4 0	1.36540	1.36455	68.685	68.734
13	2 2 4	1.36300	1.36296	68.823	68.825
14	4 0 2	1.29320	1.29294	73.116	73.133
15	4 2 0	1.22700	1.22665	77.772	77.798
16	3 3 2	1.22370	1.22327	78.021	78.054
17	2 4 0	1.22260	1.22202	78.105	78.149

Table 2:   Determined a axis parameter

La:Sr ratio	a axis [nm]	La:Ca ratio	a axis [nm]
1:0	0.5526		
0.9:0.1	0.5506	0.9:0.1	0.5522
0.8:0.2	0.5500	0.8:0.2	0.5500
0.7:0.3	0.5493	0.7:0.3	0.5478
0.6:0.4	0.5488	0.6:0.4	0.5437
0.5:0.5	0.5482	0.5:0.5	0.5418
0.4:0.6	0.5478	0.4:0.6	0.5402
0.3:0.7	0.5470	0.3:0.7	0.5362
0.2:0.8	0.5464	0.2:0.8	0.5329
0.1:0.9	0.5458	0.1:0.9	0.5316
0:1	0.5454	0:1	0.5280

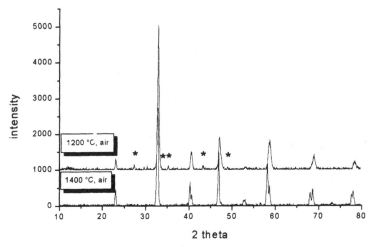

Figure 2: XRD pattern of a sample with the composition $La_{0.5}Sr_{0.5}MnO_3$ sintered at 1400 °C (below) and subsequently annealed at 1200 ° (top) in air (batch 1). The stars mark reflections of $SrMnO_3$ [7]. All other reflections can be attributed to $LaMnO_3$. The XRD pattern of $La_{0.5}Sr_{0.5}MnO_3$ sintered at 1400 °C also indicate that the phase is orthorhombic (see. Figure 1).

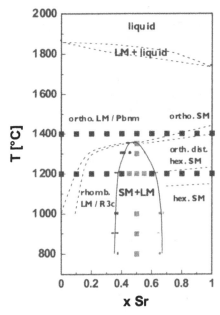

Figure 3: Temperature concentration diagram of the quasi binary system $SrMnO_3$-$LaMnO_3$. Black squares: single phase samples; grey squares: multi phase samples, results of EDX analysis of LM (black dots) and SM (grey dots) of

the samples of batch 1 and 2. Back rhombs: temperature of the onset of melting determined by DTA measurements. Phase transformations of La free SM after [5]. Melting temperature of LM after [1] and that of SM after [2].

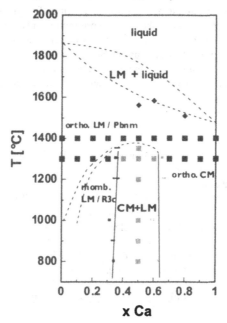

Figure 4:   Temperature concentration diagram of the quasi binary system CaMnO$_3$ - LaMnO$_3$. Marks: see Figure 3. Melting temperature of CM after [3].

DISCUSSION

The results of the experiments clearly show the occurrence of a miscibility gap between LM and SM as well as LM and CM at temperatures below about 1400 °C in air. The occurrence of the miscibility gap between LM and SM obviously correlates with the phase transformations of Sr doped LM from orthorhombic to rhombohedral with decreasing temperature. However, there is no evidence that Ca doped LM transforms from orthorhombic to rhombohedral with decreasing temperature. Therefore, the occurrence of the miscibility gap between LM and CM can not be explained, yet. The decomposition reaction below 1400 °C is slow and therefore, it does not affect the material during furnace cooling. Annealing times of above100 h are necessary to decompose the single phase samples into two phase samples at 1200 °C in air, respectively.

CONCLUSIONS

The existence of a miscibility gap influences the interpretation of magnetic phase diagrams of Sr or Ca doped LM. In samples with intermediate compositions prepared at temperatures below about 1400 °C ferromagnetic LM exists beside antiferromagnetic SM or CM. This aspect has to be taken into account for the drawing of magnetic phase diagrams of Sr or Ca doped LM. The observed rhombohedral to orthrhombic phase transition of Sr doped LM with increasing temperature allows to study the influence of the crystal structure of LM on its magneto resistive properties at constant Sr content. However, with respect to the system $SrMnO_3$-$SrMnO_{2.5}$ [6] a possible temperature dependence of the oxygen content of LM has to be taken into account which could influence the Mn^{3+}:Mn^{4+} ration of LM and its magneto resisitve properties.

REFERENCE

1] J.A.M. van Roosmalen, P. van Vlaanderen, E.H.P. Cordfunke, W.L. Ijdo, D.J.W. Ijdo, J. Solid State Chem., **114** (1995) 516.
2] T. Negas, J. Solid State Chem., **7** (1973) 85.
3] H.S. Horowitz, J.M. Longo, Mater. Res. Bull. **13**, 1359 (1978).
4] Proceedings of the Third International Symposium on Solid Oxide Fuel Cells, S.C. Singhal and H. Iwahara, Eds., The Electrochemical Society, Inc., Pennington, NJ, 1993.
5] J.F. Mitchell, D.N. Argyriou, C.D. Potter, D.G. Hinks, J.D. Jorgensen, and S.D. Bader, Physical Review **B54**, 6172 (1996).
6] T. Negas and R.S. Roth, J. Solid State Chem. **1**, 409 (1970).
7] J.B. MacChesney, H.J. Williams, J.F. Potter, and R.C. Sherwood, Phys. Rev. 164, 779 (1967).
8] K.R. Poeppelmeier, M.E. Leonowicz, and J.M. Longo, J. Solid State Chem. 44, 89 (1982).

ENHANCEMENT OF THE CMR EFFECT NEAR ROOM TEMPERATURE BY DEFECTS AND STRUCTURAL TRANSITIONS IN $La_{1-x-y}Ca_xSr_yMnO_3$

S. KOLESNIK, B. DABROWSKI, Z. BUKOWSKI, AND J. MAIS
Department of Physics, Northern Illinois University, DeKalb, IL 60115

ABSTRACT

We have studied magnetoresistance of a series of $La_{1-x}Sr_xMnO_3$ and $La_{1-x-y}Ca_xSr_yMnO_3$ samples, for which structural and ferromagnetic transformation temperatures are in close proximity. On cooling in zero magnetic field, we observe a rapid increase of resistivity just above T_C for $La_{1-x}Sr_xMnO_3$ samples with $x < 0.1425$ and $x \leq 0.1725$ due to the O*-O' and R-O* - structural phase transformations, respectively. This increase is followed by a rapid decrease due to the ferromagnetic transition. The applied magnetic field significantly shifts the ferromagnetic transition to higher temperatures and suppresses the structure-related resistivity increase. We show that a combination of structural and ferromagnetic transitions gives rise to an enhancement of the negative magnetoresistance due to strong spin-lattice coupling. By choosing a proper composition, the enhancement can be optimized to appear in relatively low magnetic fields. A proper selection of Sr and Ca contents in $La_{1-x-y}Ca_xSr_yMnO_3$ and preparation conditions leads to an enhancement of the magnetoresistance effect at room temperature.

INTRODUCTION

$La_{1-x}A_xMnO_3$ (A = Ca, Sr, Ba) compounds display strong coupling between magnetic, electronic, and structural properties. Magnetic interactions are classically described in terms of interplay between super- and double-exchange that leads to a competition between antiferromagnetic/insulating and ferromagnetic/metallic ground states.[1-3] The colossal magnetoresistive effect (CMR) observed near the Curie temperature (T_C) and frequently also near the Neel temperature (T_N), is believed to arise, in addition to magnetic interactions, from strong electron-phonon couplings of the Jahn-Teller (JT) and "breathing-mode" types; both producing charge localization. Usually magnetic fields of several Tesla are required to induce CMR. Competition between magnetic and electron-phonon interactions and structural transitions results in a very rich structure-property phase diagram for substituted $LaMnO_3$ materials.[4] Upon substitution of Sr, stoichiometric $La_{1-x}Sr_xMnO_3$ transforms from an antiferromagnetic insulator to a ferromagnetic insulator near $x = 0.1$ and to a ferromagnetic metal for $x = 0.145$. The Curie temperature increases from ~ 150 K for $x = 0.1$ to ~ 310 K for $x = 0.2$, and to ~ 380 K for $x \approx 0.3$. As the Curie temperature increases, the resistive drop at the transition from the insulating to metallic state at T_C becomes weaker, and as a consequence, the pure CMR effect gets too small for practical applications near room temperature, ~ 295 K. For $La_{1-x}Ca_xMnO_3$ the metallic phase appears for $x = 0.22$ ($T_C \sim 190$ K), and the highest $T_C \sim 270$ K is observed near $x = 0.375$. Three crystallographic phases have been identified at room temperature for $La_{1-x}Sr_xMnO_3$.[4] For pure and lightly substituted materials the crystallographic structure is orthorhombic Pbnm (O'), characterized by a large coherent orbital ordering of the JT- type.[5-10] For $x \geq 0.12$, a phase (O*) with the same orthorhombic Pbnm structure is present but with a considerably smaller coherent JT-orbital ordering.[5,6] At higher Sr substitution level, $x > 0.17$, the rhombohedral R3m structure (R), characterized by the absence of a coherent JT orbital ordering, is observed.[5] For $La_{1-x}Ca_xMnO_3$ the O'-O* transition at room temperature is present for $x = 0.145$, and

333

no R structure is observed below 400 K.[11] The substitution levels and temperatures of magnetic, resistive, and structural transitions may be altered by synthesis conditions that influence the effective oxygen content of the samples.[11] Crossing of magnetic and structural phase-transition lines results in six distinct phases with various interrelated structural and physical properties for $La_{1-x}Sr_xMnO_3$.[4] For $x < 0.145$ a strong magneto-lattice coupling below the Curie temperature suppresses the coherent JT- distortion of the orthorhombic O' phase but in turn increases the incoherent JT-distortion. For $x > 0.145$ the ferromagnetic transition is accompanied by a slight decrease of the incoherent JT distortion for the orthorhombic O* and rhombohedral R phases. As a result the metallic state occurs below the Curie temperature only when both coherent and incoherent JT distortions are suppressed and the insulating phase is formed above T_C as a result of polaronic, coherent or incoherent, JT-distortions that induce charge localization in addition to enhanced spin scattering in paramagnetic state.[4, 10, 12] We report here a study of the magnetoresistive properties for homogenous $La_{1-x}Sr_xMnO_3$ and $La_{1-x-y}Ca_xSr_yMnO_3$ samples prepared in air. Using knowledge of the relationships between magnetic, transport, and structural properties (see Fig. 1 in Ref. 4) we have designed compositions that display enhanced low-field CMR properties at room temperature. Enhancement of the CMR properties was accomplished by suppressing the structural O*-R transition present at temperatures just above T_C for zero magnetic field with an application of small fields that shift T_C above the structural transition. The CMR of 10% was achieved over a wide temperature range near 295 K in $\mu_0 H \sim 0.2$ T. Results indicate that additional increase of CMR effect and tuning of the transition may be achieved by controlled annealing in air.

EXPERIMENTAL

Polycrystalline samples were synthesized using a wet-chemistry method that leads to homogenous mixing of the metal ions. The final firing temperatures and oxygen atmosphere as well as holding times and cooling rates for obtaining stoichiometric samples were determined from TGA measurements.[11]

For this work, we selected $La_{1-x}Sr_xMnO_3$ samples within two narrow ranges of Sr content $x \sim 0.145$ and 0.1725, where the Curie temperature (T_C) coincides with the O'-O* and O*-R transformation temperatures, respectively. A series of $La_{1-x-y}Ca_xSr_yMnO_3$ samples with T_C close to room temperature has also been prepared. Powder x-ray diffraction at room temperature confirmed single-phase material in each case. Magnetoresistance measurements were performed using a Physical Properties Measurement System (Quantum Design) in the magnetic field up to 7 T.

MAGNETORESISTIVE PROPERTIES

In Fig. 1., we present resistivity of $La_{1-x}Sr_xMnO_3$ for $x = 0.14$ and 0.145, at several magnetic fields. For $x = 0.14$, in zero magnetic field, the O'-O* transformation temperature (seen on cooling the sample as an increase of resistivity at 230 K) is slightly higher than T_C. By increasing the magnetic field, the ferromagnetic transition is shifted to higher temperatures while the structural transformation temperature remains almost field-independent. For magnetic fields higher than 3 T, the structural transition takes place in the ferromagnetic state. For high fields, in the temperature range below T_C and above the O'-O* transformation temperature, resistivity shows metallic behavior consistent with the phase diagram of Ref. 4. The proximity of the O'-O* transformation temperature and T_C results in a slight enhancement of the magnetoresistance effect close to the ferromagnetic transition.

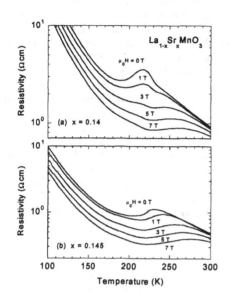

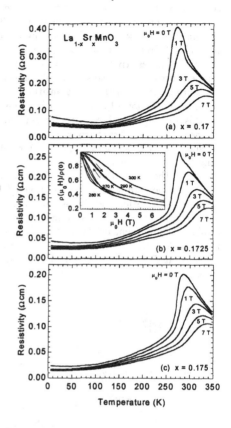

Fig. 1. Temperature dependence of resisitivty for $La_{0.86}Sr_{0.14}MnO_3$ and $La_{0.835}Sr_{0.145}MnO_3$.

Fig. 2. Resisitvity and magnetoresistance of $La_{1-x}Sr_xMnO_3$ for x = 0.17 - 0.175.

For $x = 0.145$, the O'-O* transformation temperature (≈ 215 K) is lower than T_C for all the values of the magnetic field, and hence, the magnetoresistance effect is less pronounced.

Resistivity of $La_{1-x}Sr_xMnO_3$, for $x = 0.17 - 0.175$, are presented in Fig. 2. For $x = 0.17$ and $x = 0.1725$ in zero magnetic field, the O*-R transformation temperatures, 295 and 280 K, respectively, are slightly above the ferromagnetic transition temperatures. As a result, a rapid increase of resistivity with decreasing temperature due to the structural transformation, is followed by a rapid decrease of resistivity due to the ferromagnetic transition. By increasing the magnetic field, the ferromagnetic transition temperature increases. At the same time, the resistivity increase due to the structural transformation becomes smaller. Above a certain value of the applied field, the structural transformation temperature is strongly shifted to lower temperatures, much below T_C. The resistivity increase due to the structural transformation becomes much less pronounced in the ferromagnetic state. The value of the magnetic field for which the structural transformation temperature shifts below T_C lies between 1 and 3 T for $x = 0.17$ and is lower than 1 T for $x = 0.1725$, for which the

335

zero-field structural transition temperature is closer to T_C. The combination of the O*-R transformation and the ferromagnetic transition leads to an enhancement of the magnetoresistance effect close to T_C. This is illustrated in the inset to Fig. 2(b). The resistivity ratio $\rho(\mu_0 H)/\rho(0)$ reaches a value as low as 0.6 at $T = 270$ K and $\mu_0 H = 1$ T, for $x = 0.1725$. A significant resistivity hysteresis with respect of the magnetic field can be observed at temperatures close to the structural transformation temperature, confirming that the O*-R transformation is a first-order transition. For $x = 0.175$, the structural transformation takes place below T_C for all the values of magnetic field, thus only pure CMR effect is observed for this and higher values of the Sr content. All the transitions observed for our polycrystalline samples are less sharp than those for single crystals.[13] However, the grain boundary resistance may be advantageous since for potential applications the operational temperature range of devices is not too narrow.

From the point of view of applications, it is extremely important to maximize the CMR effect at room temperature. The selected $La_{1-x-y}Ca_xSr_yMnO_3$ samples exhibit both the O*-R transformation and the ferromagnetic transition close to each other and close to room temperature. Resistivity vs. temperature for two of these samples is presented in Fig. 3. Similar enhancement of the CMR effect, as for $La_{1-x}Sr_xMnO_3$ samples can be observed, now near 295 K. The optimum magnetoresistance $\rho(\mu_0 H)/\rho(0) \approx 0.6$ in $\mu_0 H = 1$ T can be achieved at $T = 300$ and 295 K for $La_{0.7725}Ca_{0.08}Sr_{0.1475}MnO_3$ and $La_{0.78}Ca_{0.07}Sr_{0.15}MnO_3$, respectively.

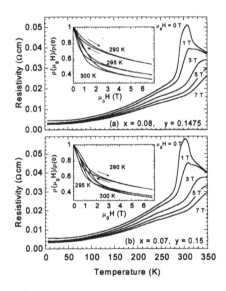

Fig. 3. Resisitvity and magnetoresistance of $La_{1-x-y}Ca_xSr_yMnO_3$.

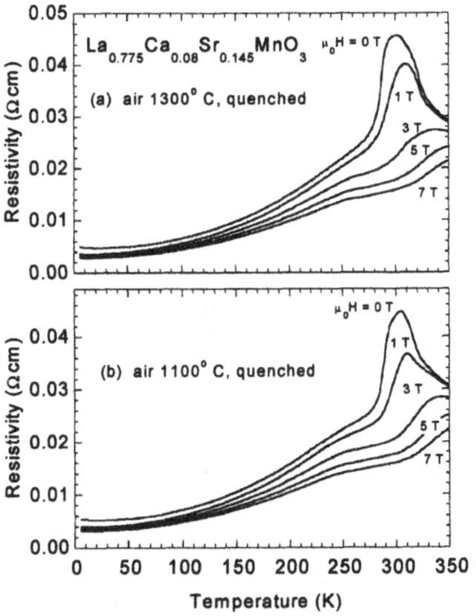

Fig. 4. Resistivity of $La_{1-x-y}Ca_xSr_yMnO_3$ samples after different heat treatments.

By varying the effective oxygen content through subsequent annealing steps at different temperatures, we were able to improve the magnetoresistive properties of the studied samples.[14] In Fig. 4., we show the resistivity vs. temperature dependencies for $La_{0.775}Ca_{0.08}Sr_{0.145}MnO_3$ sample after annealing in air at 1300°C and 1100°C. The sample, annealed at 1100°C, shows the structural transformation closer to the ferromagnetic transition and an improved CMR effect. Further optimization of the compositions and the preparation conditions is in progress.

In summary, we have investigated a series of $La_{1-x}Sr_xMnO_3$ and $La_{1-x-y}Ca_xSr_yMnO_3$ samples for which the structural transformation temperatures are close to the ferromagnetic transitions. The data clearly show that a combination of the structural and ferromagnetic transitions leads to a significant enhancement of the CMR effect near room temperature. When the O*-R transformation temperature is slightly higher than the ferromagnetic transition temperature, relatively low magnetic fields are sufficient to suppress the resistivity increase caused by the structural transformation, and enhance the CMR effect. A proper selection of Sr and Ca contents in $La_{1-x-y}Ca_xSr_yMnO_3$ and preparation conditions lead to an enhancement of the magnetoresistance effect at room temperature. Results indicate that much larger CMR effects can be achieved by these methods for single crystals and thin films that are free of the grain boundaries.

ACKNOWLEDGMENTS

Work supported by the ARPA/ONR and the State of Illinois under HECA.

REFERENCES

1. C. Zener, Phys. Rev. **82**, 403 (1951).

2. P. W. Anderson and H. Hasegawa, Phys. Rev. 100, **675** (1955).

3. J.B. Goodenough, *Magnetism and chemical bond*, Wiley, New York, 1963.

4. B. Dabrowski, X. Xiong, Z Bukowski, R. Dybzinski, P. W. Klamut, J. E. Siewenie, O. Chmaissem, J. Schaffer, and C. W. Kimball, Phys. Rev. B **60**, 7006 (1999).

5. A. Urushibara, Y. Moritomo, T. Arima, A. Asamitsu, G. Kido, Y. Tokura, Phys. Rev. B **51**, 14103 (1995).

6. H. Kawano, R. Kajimoto, and H. Yoshizawa, Phys. Rev. B **53**, R14709 (1996).

7. E. O. Wollan and W. C. Koehler, Phys. Rev. **100**, 545 (1955).

8. J. B. A. A. Elemans, B. Van Laar, K. R. Van Der Veen, and B. O. Loopstra, J. Solid State Chem. **3**, 238 (1971).

9. Q. Huang, A. Santoro, J. W. Lynn, R. W. Erwin, J. A. Borchers, J. L. Peng, and R. L. Greene, Phys. Rev. B **55**, 14987 (1997).

10. J. Rodriguez-Carvajal, M. Hennion, F. Moussa, A.H. Moudden, L. Pinsard, and A. Revcolevschi, Phys. Rev. B **57**, R3189 (1998).

11. B. Dabrowski, R. Dybzinski, Z. Bukowski, O. Chmaissem, and J.D. Jorgensen, J. Solid State Chem. **146**, 448 (1999).

12. J. F. Mitchell, D. N. Argyriou, C. D. Potter, D. G. Hinks, J. D. Jorgensen, and S. D. Bader, Phys.Rev. B **54**, 6172 (1996).

13. Y. Moritomo, A. Asamitsu, and Y. Tokura, Phys. Rev. B **56**, 12190 (1997).

14. Z. Bukowski et al., J. Appl. Phys., to be published.

ARTIFICIAL CONTROL OF MAGNETIC AND MAGNETORESISTIVE PROPERTIES IN THE PEROVSKITE MANGANITES SUPERLATTICES AND THEIR MULTILAYERS WITH ORGANICS

H.Tabata, K.Ueda, H.Matsui, H.Saeki and T.Kawai
The Institute of Scientific and Industrial Research, Osaka University,
8-1 Mihogaoka, Ibaraki, Osaka 567-0047, Japan.

ABSTRACT

Artificial superlattice of $LaFeO_3$-$LaMnO_3$ have been formed on $SrTiO_3(111)$, (110) and (100) substrates with various stacking periodicity using pulsed laser deposition. Their magnetic properties have been controlled by altering the ordering of magnetic ions (Fe or Mn). Charge disproportionate behaviors are also observed in these superlattices. For the superlattices on (111) plane, all the samples showed ferromagnetic (or ferrimagnetic) behaviors and the same Curie temperature of 230K. In the case of other superlattices formed on (110) and (100), oh the other hand, the increase of the spin frustration effect between LFO-LMO interface with decrease of the stacking periodicity causes reduction of Tc and magnetization. Specially, spin glass like behaviors observed in the superlattices of less than 3/3 stacking periodicity Furthermore, we have constructed heterostructures of Organic/Inorganic multilayers with a sequence of Copper phthalocyanine(CuPc), $BaTiO_3$ and $(La,Sr)MnO_3$. In this system, magnetoresistant properties have been controlled by the photo irradiation through the lattice strain and/or induced charges caused by the piezo effect and electric field effect. That is magnetoresistance in the $(La,Sr)MnO_3$ layer can be controlled by the shining the light.

INTRODUCTION

Perovskite-type transition metal oxides (ABO_3) exhibit various novel properties, such as ferroelectricity, ferromagnetism and high-T_C superconductivity. These materials can be grown layer-by-layer with atomic or molecular layer scale on substrates because they have similar lattice constants. Materials with unique properties are constructed by creating artificial superlattices through the combination of different perovskite-type transition metal oxides. Previously, we demonstrated the creation of new materials that have larger dielectric constants than (Ba, Sr)TiO_3 film by introducing lattice strain at the interface in artificial superlattice made of ferroelectric $BaTiO_3$ and dielectric $SrTiO_3$ layers [1].

It is possible to apply the method of creating artificial superlattices to form new magnetic materials with various magnetic structures by combining different magnetic layers. As is demonstrated in this study, materials with various spin structures can be constructed by the method of artificial superlattice with different orientations, making it possible to develop artificially controlled magnetism.

We have chosen $LaCrO_3$-$LaFeO_3$ and $LaMnO_3$-$LaFeO_3$ as starting materials. Both

LaCrO$_3$ and LaFeO$_3$ have G-type magnetic structures (inter- and intralayer spin coupling are antiparallel), and their Neel temperatures (T$_N$) are 280K and 750K, respectively (Fig. 1, left) [2-4]. If an artificial superlattice is synthesized by depositing one layer each of LaCrO$_3$ and LaFeO$_3$ alternately on (111) substrate, it is possible to form a three-dimensionally ordered structure of Fe^{3+} and Cr^{3+} ions at the B-site. Theoretically, the synthesis of ferromagnetic materials is predicted to be possible when Fe^{3+} and Cr^{3+} ions are introduced alternately in the B site of perovskite-type transition metal oxides [5-6]. Not only the conventional superexchange mechanism but also our ab-initio molecular orbital (MO) calculations support these predictions [9]. Ferromagnetism was observed in the artificial superlattice with one-layer by one-layer sequence (1/1) on the (111) plane for the first time [7].

On the other hand, when the artificial superlattice of LaCrO$_3$ - LaFeO$_3$ is synthesized on the (100) plane with a 1/1 sequence, the ferromagnetic spin-ordered structure of Fe^{3+}-O-Cr^{3+} is expected to form perpendicular to the substrate surface (along the direction of the c-axis). In contrast, in-plane (a-b plane) spins are ordered antiferromagnetically. The total magnetic structure of the superlattice should become C-type antiferromagnet (interlayer spin coupling--parallel, intralayer spin coupling--antiparallel). The larger the stacking periodicity, the higher the T$_N$ is expected due to the increment in the antiferromagnetic correlation. Therefore, the lowest Neel temperature is expected to be obtained in the 1/1 sequence superlattice on this (100) substrate.

In the case of (110) superlattice with 1/1 sequence, materials with A-type magnetic structure (interlayer spin coupling--antiparallel, intralayer spin coupling--parallel) can be obtained because the ferromagnetic spin-ordered structure of Fe^{3+}-O-Cr^{3+} is expected to form only in a-b plane and same ions are aligned (Fe^{3+}-O-Fe^{3+} or Cr^{3+}-O-Cr^{3+}) along the c-axis.

LaFeO$_3$ is antiferromagnetic and has a G-type magnetic structure as described before. On the other hand, LaMnO$_3$ films exhibit ferromagnetic behavior with a Curie temperature of 130 K for La deficiency (La$_{1-\delta}$MnO$_3$) [10]. In this paper, the ferromagnetic La$_{1-\delta}$MnO$_3$ is noted as LaMnO$_3$.

For artificial superlattices constructed on the (111) plane, ferromagnetic interactions should be introduced at the Mn-Fe interface because the Fe^{3+}-O-Mn^{3+} superexchange interaction is considered to be ferromagnetic according to the theory of Goodenough-Kanamori [5-6]. As a result, ferromagnetism should appear in the superlattice with one-layer by one-layer (1/1) stacking periodicity (see Fig. 1, right).

On the other hand, for superlattices constructed on (100) and (110) planes a spin frustration effect occurs at the LaMnO$_3$-LaFeO$_3$ interface because the LaMnO$_3$ film is ferromagnetic and LaFeO$_3$ is antiferromagnetic with a G-type spin structure (Fig. 1, right). The spin frustration effect increases as the stacking periodicity decreases. The spin frustration effect of (110) superlattices becomes larger than that of (100) superlattices in terms of their spin structure. This method even allows a spin frustration effect to be introduced artificially into the system. The LaMnO$_3$-LaFeO$_3$ artificial superlattices were formed according to these concepts. Based on this strategy, we control the spin order by arranging the atomic order in perovskite superlattices.

Furthermore, we have demonstrated highly harmonized multi layers with a combination of organic and inorganic (metal oxides) materials and created new memory effect.

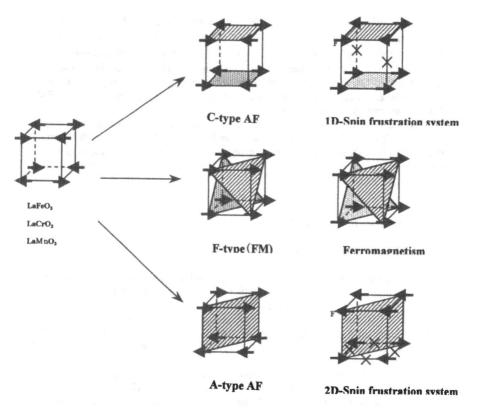

C-type AF

1D-Spin frustration system

F-type (FM)

Ferromagnetism

A-type AF

2D-Spin frustration system

LaFeO₃
LaCrO₃
LaMnO₃

Fig. 1 : Schematic models of spin structures in the LaCrO$_3$-LaFeO$_3$ and LaMnO$_3$-LaFeO$_3$ aartificial superlattices grown on (100), (110) and (111) surfaces. In the figure, AF and FM means antiferromagnetic and ferromagnetic, respectively. The $\times$ signs show the interactions where the spin frustration effect occurs.

EXPERIMENTAL

Magnetic artificial superlattices were constructed as follows. The LaCrO$_3$, LaMnO$_3$, and LaFeO$_3$ layers were stacked using multi target pulsed-laser deposition (PLD) technique. An ArF excimer laser pulse was focused on the targets to induce ablation, and the ablated atoms and ions were deposited on the substrates (LaAlO$_3$ or SrTiO$_3$ (100), (110) and (111)). The PLD technique is an excellent method that spatial and time conditions can be controlled. An atomic scale control of crystal growth is possible with the PLD method combined with reflection high energy electron diffraction (RHEED) observations [1, 8]. Targets of LaCrO$_3$, LaMnO$_3$, and LaFeO$_3$ were synthesized by usual ceramic techniques. Solid solution LaCr$_{0.5}$Fe$_{0.5}$O$_3$ films were also formed as reference samples. All the films were formed at 580 - 600°C and an oxygen/ozone (8%) pressure of 1×10^{-3} Torr. The deposition rate was 10 – 20 Å/min. The total thickness of the film is 700 - 1000 Å. Magnetic measurements were performed using a superconducting quantum interference devices (SQUID) magnetometer (Quantum design MPMS-5S) with the magnetic field applied parallel to the film plane. Surface morphology was observed by atomic force microscopy (AFM ; Digital Instruments – Nanoscope III).

Organic /inorganic materials cooperate at an atomic and molecular level, and a functional device with high performance have been formed. Copper phthalocyanine (CuPc) and $BaTiO_3$ (BTO) were used as organic and inorganic materials, respectively on Si(100) substrate. Each film were formed by thermal evaporate technique and laser ablation (ArF excimer laser) at 300 °C.

RESULTS AND DISCUSSION

<u>$LaCrO_3$-$LaFeO_3$ superlattices</u> [7, 9]

The crystal structures of the $LaCrO_3$-$LaFeO_3$ superlattices have been confirmed by X-ray diffraction (XRD). All the superlattices have a single phase c-axis orientation. Typical features of superlattices were observed in our superlattices [7, 9].

The RHEED patterns show streaks indicating that the superlattices are epitaxially formed on the substrates up to the topmost surface. These results of X-ray diffraction and RHEED patterns indicate that the $LaCrO_3$-$LaFeO_3$ superstructures are well constructed as desired.

The temperature dependence of magnetization for the $LaCrO_3$-$LaFeO_3$ artificial superlattice of 1/1 sequence and $LaCr_{0.5}Fe_{0.5}O_3$ films on $LaAlO_3$ (100) is shown in Fig. 2 (a). The cusp which corresponds to T_N appeared at 250 K for the 1/1 superlattice. In the case of superlattices constructed on (110) planes, the superlattice with 1/1 sequence show antiferromagnetic behaviors (Fig. 2(b)). The cusp which correspond to the Neel temperature appeared at 320 K. We think the different T_N in these superlattices reflects the different magnetic structures.

In contrast, the temperature dependence of the magnetization and hysteresis curve (M-H curve) of the $LaCrO_3$-$LaFeO_3$ artificial superlattice (1/1 sequence) grown on the (111) substrate exhibits characteristic of ferromagnetic (or ferrimagnetic) materials [Fig. 2(c)]. The saturation magnetization (M_S) is estimated to be 60 emu/g ($\fallingdotseq 2.5 \mu_B$/site) from the hysteresis curve measured at 6 K. It is considered that the material is ferromagnetic because the value of M_S is too large to attribute to ferrimagnetic order (theoretically estimation : 1 μ_B/site for Fe^{3+}-O-Cr^{3+} ferrimagnetic interaction). The decrease of our M_S from the theoretical estimation (4 μ_B/site for Fe^{3+}-O-Cr^{3+} ferromagnetic interaction) is considered to be due to the imperfection at the interface.

These results suggest that the control of the growth direction of the superlattices actually govern the magnetic properties as proposed in Fig. 1.

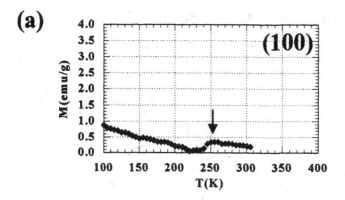

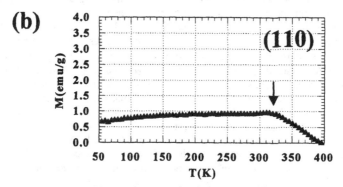

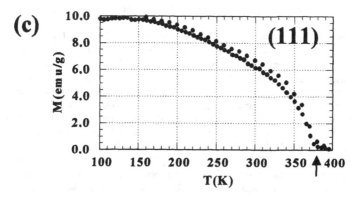

Fig. 2 : The magnetization vs. temperature curves of the LaCrO$_3$-LaFeO$_3$ superlattice formed on the (a) (100), (b) (110) and (c) (111) substrates with 1/1 sequence.

LaMnO₃-LaFeO₃ spin frustrated superlattices [11]

The crystal structures of the $LaMnO_3$-$LaFeO_3$ superlattices on $SrTiO_3$ (111), (100) and (110) were studied using X-ray diffraction. All the films showed a single phase and had a preferred orientation normal to the surface of the substrate. Typical features of superlattices were observed. In particular, for a superlattice formed on a (111) substrate with one-layer by one-layer (1/1) stacking periodicity, small peaks were observed at $2\theta = 19.8°$ and $61.6°$ due to the double perovskite features. The RHEED patterns show streaks which also indicate that the superlattices are epitaxially formed on the (111), (100) and (110) surfaces. The results of the X-ray diffraction and reflection high-energy electron diffraction (RHEED) measurements indicate that the $LaMnO_3$-$LaFeO_3$ superstructures were sufficiently well formed.

Magnetization versus temperature curves (M-T curves) of $LaMnO_3$-$LaFeO_3$ artificial superlattices with 1/1 sequence on $SrTiO_3$ (111) are shown in Fig. 3(a). A magnetic field of 0.1 T was applied parallel to the film surface. The Curie temperatures (T_C) of 230K were observed. The saturation magnetization (M_S) of the superlattice was measured to be about 30 emu/g ($\fallingdotseq 1.3 \mu_B$/site) from the hysteresis curve measured at 6 K. However, the value of M_S is estimated to be 103 emu/g (= 4.5 μ_B/site) for the $Mn^{3+}(d^4)$-O-$Fe^{3+}(d^5)$ state theoretically. The measured value, therefore, is relatively small compared with the above estimate. We suppose that the reduction of M_S is caused by the complex effect arising from the partial displacement between Fe and Mn ions, charge separation between Fe and Mn as seen in $LaMn_{0.5}Co_{0.5}O_3$ (or $LaMn_{0.5}Ni_{0.5}O_3$) ordered perovskites [12-13] and the deviation from stoichiometry due to La and/or oxygen deficiency. In the first case, a 10% displacement of Fe by Mn ions causes the value of M_S to be 80% of the theoretical value. When charge separation between Fe and Mn ions ($Fe^{3+}+Mn^{3+} \rightarrow Fe^{2+}+Mn^{4+}$) occurs, as described in the second case, M_S becomes 80 emu/g (=3.5 μ_B/site). Annealing of the 1/1 superlattice was performed at 500°C (less than deposition temperature) with O_2 flowing to remove the oxygen deficiency. The magnetic behavior after annealing did not change from that seen before.

The magnetization versus temperature curves for $LaMnO_3$-$LaFeO_3$ artificial superlattices with 2/2 sequence on $SrTiO_3$ (100) are shown in Fig. 3 (b). The inset shows the magnetization versus temperature curves of 2/2 superlattices on (100) substrate with different cooling (zero-field cooling (ZFC) and field cooling (FC)) processes and different magnetic fields. For superlattices formed on (100) (and (110)) substrates, the magnetic properties differed significantly from those of (111) due to the spin frustration effect. The magnetic behavior differs depending on whether the sample is cooled with (FC) or without (ZFC) an applied field and a sharp cusp at about 65K is observed in the ZFC sample when the applied field is 0.005 T, but this cusp loses its sharpness and becomes a broad maximum, and moves to a lower temperature when the applied field is increased to 0.1 T. This behavior is one of clear evidences for the spin glass state. The increased spin frustration effect caused by the reduced stacking periodicity leads to the formation of a spin-glass-like phase. This is caused by the competition between ferromagnetism in Fe-Mn and Mn-Mn, and antiferromagnetism in Fe-Fe.

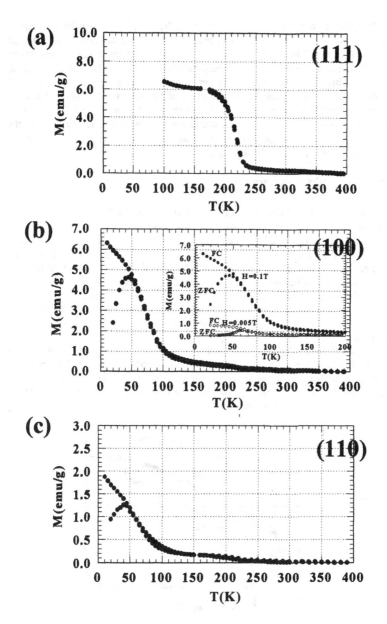

Fig. 3 : The magnetization vs. temperature curves of the $LaMnO_3$-$LaFeO_3$ superlattice formed on the (a) (111), (b) (100) and (c) (110) substrates with 1/1 sequence.

The superlattice formed on (110) substrates with 2/2 stacking periodicity also shows spin-glass-like behavior (Fig. 3(c)). The magnetic field was applied parallel to the [001] direction in the (110) and (100) superlattices to avoid the magnetic anisotropy effects. The magnetization at the glass temperature (T_g) is about 1/3 that of the (100) superlattice. This is caused by the larger frustration effect than that of (100) superlattices as shown in figure 1. The number of bonds where the spin frustration effect occurs in the (110) superlattice is twice as large as that in the (100) superlattice per eight-metal unit cell (Fig. 1), which correlates well with the suppression of the magnetization ($1/2 \sim 1/3$). In the case of (110) superlattices, the spin frustration effect is twice as great as that of the (100) superlattice.

Organic/inorganic photo memory

For the mimetic of eye function, CuPc/BTO heterostructured films were performed. In the dark condition, it shows the remanent polarization of 10 nC/cm^2 under the drive voltage below 2V.(Fig.4) In this voltage region, the BTO layer does not show saturate character. When the Hg lump irradiate, on the other hand, it increase to 20 nC/cm^2. It is twice as large as that of dark condition (without irradiation). This result indicate that the photo-signal can be memorized directly in the ferroelectric layer.[14-16]

So far, MO (magneto optical) disk is popular for the optical performed memory. But it is memorized by thermal phase transition. That is it is storaged with not optical process but thermal process. In our CuPc/BTO device, optical signal is written directly in the BTO layer.

Furthermore, we have constructed trilayers with combination of CiPc/BaTiO$_3$/(La,Sr)MnO$_3$. In this system, light information is changed to electron-hole carriers, dipole polarization, lattice strain and eventually to magnetoresistance.(Fig.5) These transforming character is quite similar to that of the information processing of humanbeeing.

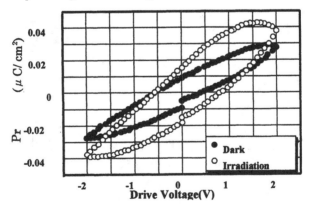

Fig.4 : Hysteresis loops of heterostructures of CuPc/BaTiO$_3$ multilayers with and without light irradiation.

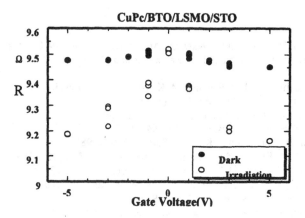

Fig.5 : Resistively of the source-drain vs. gate voltage of CuPc/BaTiO$_3$/(La,Sr)MnO$_3$ multilayers.

CONCLUSIONS

In summary, we demonstrated that artificial control of magnetic properties is possible by changing the stacking periodicity and direction in LaCrO$_3$-LaFeO$_3$ and LaMnO3-LaFeO$_3$ superlattices. Antiferromagnetic behaviors are observed in all the films deposited on (100) and (110) substrates. And, the T$_N$ is changed systematically as the stacking periodicity decrease. The film with 1/1 periodicity on a (111) substrate exhibits ferromagnetic character. On the other hand, in the case of LaMnO$_3$-LaFeO$_3$ superlattices, the magnetization of superlattices constructed on the (111) plane increases as the stacking periodicity decreases, and the superlattice with 1/1 stacking periodicity exhibited ferromagnetic (or ferrimagnetic) behavior. For (100) and (110) superlattices, on the other hand, the spin frustration effect increases with lower stacking periodicity and spin-glass-like behaviors was observed in superlattices with a stacking periodicity of less than 3/3. Though the total number of magnetic ions (Fe, Cr, Mn) are same in (100), (110) and (111) superlattices, quite different magnetic character can be created in the artificial superlattices.

We have also formed prototype of new type sensor and memory with a combination of ferroelectrics, ferromagnetics and photo-voltaic materials. These function harmonized materials have promising field of new type devices such as brain mimic memory and super five sense sensors.

RFERENCES

1. H. Tabata, H.Tanaka and T. Kawai, Appl. Phys. Lett. **65**, 1970 (1994)

2. R. Aleorard, R. Pauthenet, J. P. Rebouillat and C. Veyret, J. Appl. Phys. **39**, 379 (1968).

3. W. C. Koehler, E. O. Wollan and M. K. Wilkinson, Phys. Rev. **118**, 58 (1960).

4. D. Treves, J. Appl. Phys. **36**, 1033 (1965).

5. J. Kanamori, J. Phys. Chem. Solids **10**, 87 (1959).

6. J. B. Goodenough, Phys. Rev. **100**, 564 (1955).

7. K. Ueda, H. Tabata and T. Kawai, Science **280** (1998) 1064.

8. T. Kawai, M. Kanai and H. Tabata, Mater. Sci. Eng. **B41**, 123 (1996).

9. K. Ueda, H. Tabata and T. Kawai, Jpn. J. Appl. Phys. (1999) in press.

10. A. Gupta, T. R. McGuire, P. R. Duncombe, M. Rupp, J.Z.Sun, W. J. Gallagher, and Gang Xiao, Appl. Phys. Lett. **67**, 3494 (1995).

11. K. Ueda, H. Tabata and T. Kawai, Phys. Rev. B **60**, (1999) R12561.

12. G. Blasse, J. Phys. Chem. Solids **26**, 1969 (1965).

13. J. B. Goodenough, A. Wold, R. J. Arnott and N. Menyuk, Phys. Rev. **124**, 373 (1961).

14. H.Lee, H.Tabata and T.Kawai, Jpn.J.Appl. Phys. 36 (1997) 5156.

15. H.Tabata, H.Matsui and T.Kawai, ICIM 98, p.280 (Proc. 4th Int. Conf. Intelligent Materials)

16. H.Lee, Y.S. Kang, B.C. Choi, J.H.Jeong, H.Tabata and T.Kawai, J.Kor.Phys.Soc. 34 (1999) S64.

EPITAXIAL La$_{0.7}$(Pb$_{1-x}$Sr$_x$)$_{0.3}$MnO$_3$ THIN CMR FILM ROOM TEMPERATURE BOLOMETER

A. Lisauskas, S.I. Khartsev, and A.M. Grishin
Dept. of Condensed Matter Physics, Royal Institute of Technology, S-100 44 Stockholm, Sweden.

ABSTRACT

Using pulsed laser deposition technique we fabricated films of a continuous series of solid solutions La$_{0.7}$(Pb$_{1-x}$Sr$_x$)$_{0.3}$MnO$_3$, which undergo metal-to-insulator phase transition close and above room temperature. The optimal composition for uncooled bolometer applications was found at $x = 0.37$ with maximum of TCR = 7.4 %K^{-1} @ 295 K. Room temperature bolometer demonstrator has been built and tested. Relatively low excess noise (magnitude γ/n of $3 \cdot 10^{-21}$ cm^{-3}) and high TCR enabled achievement of high signal-to-noise ratio of $8 \cdot 10^6$ √Hz/K. The observed frequency dispersion of bolometer response has been ascribed to three relaxation mechanisms of the heat transfer: from film thermistor to substrate, from substrate to thermostat, and heat transfer via leads. For fabricated bolometer heat transfer from substrate to thermostat was found to be the slowest one with a time constant of 0.5 s (correspondent substrate-to-thermostat thermal conductance is $G = 3 \cdot 10^{-3}$ W/K). Bolometer performance test yields the responsivity $\Re = 0.6$ V/W, detectivity $D = 9 \cdot 10^6$ cm√Hz/W and noise equivalent power NEP of $3 \cdot 10^{-8}$ W/√Hz at 30 Hz frame frequency. Further improvement of bolometer thermal isolation using the micromachining technique is believed to achieve the responsivity about $4 \cdot 10^3$ V/W and detectivity higher than 10^9 cm√Hz/W@30 Hz.

INTRODUCTION

The great potential for IR and magnetic field sensor applications of the colossal magnetoresistive (CMR) manganite perovskites recently attracts great attention. [1-5] High temperature coefficient of resistance (TCR) in the vicinity of the phase transition and possibility to tailor material properties to achieve the maximum of TCR close to the room temperature make these materials suitable for uncooled bolometer applications. With the compositional and processing conditions control we have optimized CMR material to get high IR-detection performance. We report characteristics of the bolometer based on PLD-made thin CMR film operating at room temperature.

A bolometer is characterized as an absorber with the heat capacity (thermal mass) C connected to a thermostat with temperature T_{th} by a thermal conductance G. Bolometer temperature T relates to the thermostat temperature and an incident power W through the heat transfer equation:

$$C \cdot \frac{d(T - T_{th})}{dt} + G \cdot (T - T_{th}) = \eta \cdot W, \qquad (1)$$

where η is bolometer absorptivity. Bolometer performance is described by: optical responsivity $\Re$, noise equivalent power NEP and detectivity D. The responsivity, i.e., the output signal voltage ΔV per unit of incident infrared power, is given by:

$$\mathfrak{R} = \frac{\Delta V}{W} = I_b \frac{dR}{dT} \frac{\Delta T}{W} = \frac{\eta I_b \, dR/dT}{G\sqrt{1+\left(2\pi f \tau\right)^2}}, \tag{2}$$

where $\tau = C/G$ is the bolometer time constant, I_b is a bias current, R is a bolometer resistance, ΔT is the overheat temperature, and f is a frame frequency. For current biased bolometers with positive TCR, thermal feedback is positive and maximum bias current is limited by thermal runaway, which occurs when $G - I_b^2 dR/dT = 0$. Thus the optimum bias current is $I_{opt} \approx 0.3\sqrt{(G/dR/dT)}$. [6] Bolometer selfheating due to the bias current can be estimated from steady-state solution of Eq. 1. This gives $\Delta T \approx 0.1/\text{TCR}$. After those estimations one gets the following expression for the responsivity:

$$\mathfrak{R} \approx \frac{\eta \sqrt{0.1 \cdot R \cdot \text{TCR}}}{\sqrt{G}\sqrt{1+\left(2\pi f \tau\right)^2}}. \tag{3}$$

It states that to get higher responsivity material with higher resistance is needed. From Eq. 3 one can find the optimum thermal conductance $G = 2\pi f C$, which gives maximum responsivity at given chopping frequency f.

Bolometer sensitivity to fluctuations in incident power is often expressed as a noise equivalent power (NEP), given by

$$\text{NEP} = \sqrt{\frac{4kT^2 G}{\eta^2} + \frac{4kTR}{\mathfrak{R}^2} + \frac{S_V(f)}{\mathfrak{R}^2}}, \tag{4}$$

where k is Boltzmann constant and $S_V(f)$ is excess noise voltage fluctuations spectral density. The first two terms come from phonon and thermal fluctuations, while the last term in the most cases is predetermined by resistance fluctuations and is proportional to I_b^2 since $S_V = S_R I_b^2$.

The detectivity D is determined by NEP and detector area A:

$$D = \frac{\sqrt{A}}{\text{NEP}}. \tag{5}$$

Detectivity is limited mainly by Johnsson noise in the case of low G, by phonon fluctuations in the case of the high thermal conductance, while in the intermediate region the excess noise is starring. For this case:

$$D = \frac{\text{TCR}}{\sqrt{S_R/R^2}} \frac{\sqrt{A}}{G\sqrt{1+\left(2\pi f \tau\right)^2}}, \tag{6}$$

where S_R/R^2 is a normalized noise power and the first term at the right side of Eq. 6 is the signal-to-noise ratio (SNR).

EXPERIMENTAL

Films have been prepared by alternating pulsed laser ablation of two ceramic targets with the composition $La_{0.75}Sr_{0.25}MnO_3$ and $La_{0.7}Pb_{0.3}MnO_3$. Films of the basic chemical formula $(1-x)\cdot La_{0.7}Pb_{0.3}MnO_3 + x \cdot La_{0.75}Sr_{0.25}MnO_3$ with $x = 0$, 0.3, 0.37, 0.4, 0.5, 0.6, and 1 have been fabricated. Details of processing technique and crystalline film quality can be found elsewhere.

[8] Noise measurements have been performed using two probes soldered with indium to the silver contact pads. Contact noise was not observed. Responsivity has been measured by lock-in amplifier chopping the incident radiation flux of 640 nm laser. Voltage fluctuations across the sample were amplified by the nearly placed low noise, high impedance preamplifier on FET with a short junction noise of 2.6 nV/√Hz. Low frequency spectra in 1 to 20 kHz range have been recorded by the spectrum analyzer SR760.

RESULTS

We fabricated CMR films of various compositions using pulsed laser deposition technique (PLD) and magnetron sputtering. In spite of the fact that films have been made by two different techniques in all thin CMR films that have been made and characterized by ourselves we found universal behavior of the maximum value of TCR versus phase transition temperature T_c (see Fig. 1). It turns out, that material, which undergoes transition from the metal to the semiconducting state at higher transition temperature T_c, has smaller TCR. Fig. 1 includes also the results for PLD-made films from Venkatesan group. [2] We could not find TCR vs. T_c data in any other of numerous papers on CMR. Presented experimental data include results both for PLD and magnetron sputtered films of different compositions fabricated in different oxygen atmospheres and exposed to post-annealing at different conditions.

Also, we have succeeded to tailor transport properties by changing the film thickness: T_c has been found to decrease rapidly if film thickness reaches and becomes smaller than the critical one. For La$_{0.75}$Sr$_{0.25}$MnO$_3$(001) films grown onto SrTiO$_3$ and LaAlO$_3$ substrates the critical thickness has been found to be 100 and 300 Å, correspondingly. [9] It has been shown film-substrate lattice mismatch induces the nonuniform strain in ultra-thin films. As clearly seen from Fig. 1, thinning the film is the lost scenario, since one does reduce T_c meanwhile loosing TCR compared with the optimum composition. From our previous works we know, that films with

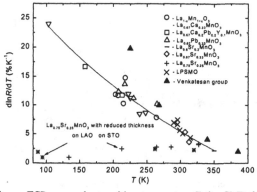

Fig. 1 The maximum TCR versus the transition temperature T_c for CMR thin films of different compositions and thickness.

oxygen deficit have extremely high level of the excess noise, which can kill performance of sensitive device. From the other hand, we tailored material properties by reducing film thickness down to 100 Å keeping TCR and signal to noise ratio almost constant. [3] This has been achieved because of the broadening of the phase transition caused by the stress in ultra-thin CMR films. The best solution to get film superior for bolometric applications is just to engineer the film having maximum of TCR at desired operation temperature. We found continuous series of solid solutions $La_{0.7}(Pb_{1-x}Sr_x)_{0.3}MnO_3$ ranges T_c close to room temperature thus can be explored as a material for uncooled bolometer.

The transition temperature of $La_{0.7}(Pb_{1-x}Sr_x)_{0.3}MnO_3$ solutions has been found to decrease continuously with increasing the Pb-to-Sr ratio (Fig. 2). Maximum of TCR of 10.2 %K^{-1}@266 K has been achieved for $La_{0.7}Pb_{0.3}MnO_3$ film while $La_{0.75}Sr_{0.25}MnO_3$ film showed TCR of 3.2 %K^{-1} @327 K. The composition with $x = 0.37$ was found to have maximum of TCR of 7.4 %K^{-1}@295 K. We have chosen this composition to make room temperature bolometer demonstrator.

The main transport characteristics of 0.45 μm thick LPSMO film with $x = 0.37$ are shown in Fig. 3. Both electrical resistance and excess noise voltage exhibit similar temperature behavior increasing with temperature increase. Due to this the signal-to-noise ratio (SNR) demonstrates peak, which is wider that TCR's. The SNR and TCR maxima appear to be close to each other. The maximum SNR value of 8·10^6 √Hz/K@30 Hz indicates, that with 1 Hz frequency bandwidth at 30 Hz frame frequency amplitude of electrical fluctuations is equal to signal produced by 120 nK temperature change (Noise Equivalent Temperature Difference, NETD).

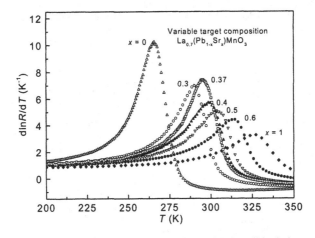

Fig. 2. Temperature dependence of TCR for $La_{0.7}(Pb_{1-x}Sr_x)MnO_3$ solid solutions.

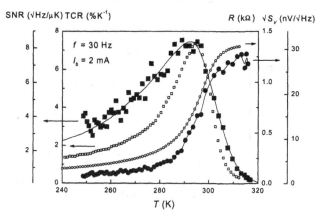

Fig. 3 Temperature dependences of basic properties of $La_{0.7}(Pb_{0.63}Sr_{0.37})MnO_3$ 0.45 μm thick film, used for bolometer demonstrator.

The frequency dependence of the excess noise was measured in 60 K temperature range close to room temperature. Insert in Fig. 4 shows voltage fluctuations spectral density. It appears to be proportional to I^2 indicating the excess fluctuations arise from the resistance fluctuations. From excess noise frequency dispersion shown in Fig. 4 it becomes evident that these fluctuations has $1/f^\alpha$ origin, where the exponent α is close to 1.

Normalized noise power $\gamma/n = f \cdot V \cdot S_R/R^2$ (n is carriers concentration, V is a film volume) has been found to be of $3 \cdot 10^{-21}$ cm^3 at 295 K that is about 2-4 orders of magnitude lower than reported

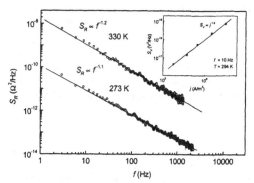

Fig. 4 Excess noise spectra for LSPMO film at different temperatures. In the insert dependence of the excess noise on bias current is shown at room temperature.

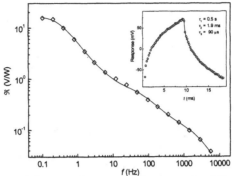

Fig. 5 Bolometer responsivity versus frequency at bias current I_b = 1.2 mA. Solid curve is the fitting function from the Eq. 7. In the insert time dependence of voltage response on single radiation pulse is shown.

in Refs. [4,5]. We suggest that so significant improvement of noise performance was achieved due to the optimization of deposition temperature and *in situ* annealing made in oxygen atmosphere. To make reasonable comparison of SNR for different bolometric materials we have scaled noise measurement results to the volume of 50×50×0.5 μm^3. For this volume we expect SNR of $6 \cdot 10^4$ $\sqrt{Hz}/K$@30 Hz which is comparable and even better than calculated SNR values for the same volume of amorphous Si ($2.7 \cdot 10^3$ $\sqrt{Hz}/K$, [10]), VO_x ($1.8 \cdot 10^4$ $\sqrt{Hz}/K$, [11]), and YBCO in semiconducting state ($4.2 \cdot 10^4$ $\sqrt{Hz}/K$, [12]).

Frequency dependence of the responsivity for our bolometer is shown in Fig. 5. We found this dependence can be fitted to three-relaxation process. Oscilloscope record of voltage response on single radiation pulse also shows multi-parametric relaxation process. This fact we explain suggesting there are three main heat transfer processes: from film to substrate, from substrate to thermostat, and additional heat exchange through the wire leads. [13] Thus three time constants appear in the fitting function:

$$\Re = \frac{A_1}{\sqrt{1 + (2\pi f \tau_1)^2}} + \frac{A_2}{\sqrt{1 + (2\pi f \tau_2)^2}} + \frac{A_3}{\sqrt{1 + (2\pi f \tau_3)^2}}. \tag{7}$$

They have been found to be 0.5 s, 2.1 ms and 90 μs to get almost perfect fit to experimental time response. Thermal conductance G at zero frequency was found to be $3 \cdot 10^{-3}$ W/K.

CONCLUSIONS

In summary, universal dependence of maximum TCR on transition temperature suggests the optimum CMR material for bolometric application has to have maximum of TCR at the operating temperature. Continuous series of solid solutions with variable composition $La_{0.7}(Pb_{1-x}Sr_x)_{0.3}MnO_3$ has been grown and film with composition $x = 0.37$ has been found to possess the maximum TCR = 7.4 %K^{-1} @295 K. Normalized noise magnitude γ/n was found to be $3 \cdot 10^{-21}$ cm^3, which is comparable and even better than same noise parameter of VO$_x$ or semiconducting YBCO thin films, believed to be the best materials for uncooled bolometric applications. Room temperature thin film bolometer exhibits the responsivity $\Re = 0.6$ V/W, detectivity $D = 9 \cdot 10^6$ cm$\sqrt{\text{Hz}}$/W, noise equivalent power NEP = $3 \cdot 10^{-8}$ W/$\sqrt{\text{Hz}}$ W at 30 Hz frame frequency. Using micromachining technique and La$_{0.7}$(Pb$_{0.63}$Sr$_{0.37}$)$_{0.3}$MnO$_3$ film one can expect to get responsivity as high as $4 \cdot 10^3$ V/W and detectivity higher than 10^9 cm$\sqrt{\text{Hz}}$/W@30 Hz.

ACKNOWLEDGEMENT

This work was supported by the Göran Gustafssons Foundation and the Swedish Superconductivity Consortium. One of us (A.L.) is most obliged to Svenska Institutet for scholarship he got under Visby program.

REFERENCES

1. M. Rajeswari, C.H. Chen, A. Goyal, C. Kwon, M.C. Robson, R. Ramesh, T. Venkatesan, and S. Lakeou, Appl. Phys. Lett. **68**, 3555 (1996).
2. A. Goyal, M. Rajeswari, R. Shreekala, S.E. Lofland, S.M. Bhagat, T. Boettcher, C. Kwon, R. Ramesh, and T. Venkatesan, Appl. Phys. Lett. **71**, 2535 (1997).
3. A. Lisauskas, S.I. Khartsev, A.M. Grishin, and V. Palenskis, Mat. Res. Soc. Symp. Proc. **574**, 365 (1999).
4. G.B. Alers, A.P. Ramirez, and S. Jin, J. Appl. Phys. **68**, 3644 (1996).
5. S.K. Arora, Ravi Kumar, D. Kanjilal, Ravi Bathe, S.I. Patil, S.B. Ogale, and G.K. Metha, Solid State Comm. **108**, 959 (1998).
6. P.L. Richards, J. Appl. Phys. **76**, 1 (1994).
7. P. Dutta, P.M. Horn, Rev. Mod. Phys. **53**, 497 (1981).
8. A.M. Grishin, S.I. Khartsev, P. Johnsson, Appl. Phys. Lett. **74**, 1115 (1999).
9. S.I. Khartsev, P. Johnsson, A.M. Grishin, J. Appl. Phys. **87**, 2394 (2000).
10. C.E. Parman, N.E. Israeloff, and J. Kakalios, Phys. Rev. B **47**, 12578 (1993).
11. A.P. Gruzdeva, V.Yu. Zerov, O.P. Konovalova, Yu.V. Kulikov, V.G. Malyarov, I.A. Khrebtov, and I.I. Shaganov, J. Opt. Technol. **64**, 1110, (1997).
12. C.M. Travers, A. Jahanzeb, D.P. Butler, and Z. Çelik-Butler, J. Microelectromech. Syst. **6**, 271 (1997).
13. E.M. Wormser, J. Opt. Soc. Am. **43**, 15 (1953).

PROSPECTS OF ROOM-TEMPERATURE MICROBOLOMETER BASED ON CMR THIN FILMS

A. Verevkin
Applied Physics Department,
Yale University, New Haven, CT06520

N. Noginova and E.S. Gillman
Center for Materials Research
Norfolk State University, Norfolk,
VA 23504

ABSTRACT

We discuss performance of CMR-based bolometers for pulse and modulated electromagnetic radiation detection. Ultimate characteristics of CMR thin film-based bolometers are considered based of current experimental studies of photoresponse.

INTRODUCTION

During past few years, much attention has been paid to investigation of interrelated processes of magnetic and charge ordering in the manganites of doped rare earth elements with perovskite crystal structure. The manganites demonstrate "colossal" magnetoresistance (CMR) at temperatures near the Curie point T_c [1]. Most of the application research works are devoted to magnetic properties as determining the usage of those compounds in magnetic sensing applications. Other possible directions are study and design of CMR thin film-based microbolometers ([2],[3]). In particular, a sharp temperature dependence of the resistance of CMR materials can be utilized for possible microbolometer application. Temperature variations of fast photoresponse to intensive pulsed laser radiation [4-5] as well as slow bolometric response [6-7] in CMR films have been studied recently.

Why may CMR films be attractive for microbolometer applications? In this paper we discuss two types of microbolometers, slow response and fast response CMR-based detectors.

Slow CMR-based detectors.

Here we discuss possible advantages of the slow quasi equilibrium heating in CMR-based thin film bolometers, when the electron subsystem temperature T_e is equal to the phonon subsystem temperature T_p, or the time dependence of $T_e(t)$ is close to the time dependence $T_{ph}(t)$. Note, that in the CMR materials, at the temperatures below T_c, the electron temperature T_e is related to the electron spin system temperature for more correct description (see below).

The effective times of the "slow" bolometric response observed in CMR films are typically in the range of milliseconds. These times are determined mainly by slow phonon cooling of both of the film and the substrate. Temperature dependence of the response observed in this regime is strictly proportional to dR/dT at 50<T<300K ([6]). At low temperatures 4K<T<100K, the additional features in bolometric response appear due to glass-like state transition ([6]).

In bolometers, the absorbed pulsed or modulated radiation leads to heating of the electron subsystem. After electron-phonon interaction the energy dissipates in the substrate through nonequilibrium phonons. Then, the substrate is cooled to the bath temperature or temperature of cold metal contacts if an air-gap membrane structure is used.

The most sensitive among existing devices are detectors based on superconducting films. However, unfortunately they require cooling far below room temperature for operation. Semiconductor bolometers have good efficiency only at the optical range close to the band gap energy and can be used for narrowband devices only.

Third group is represented by microbolometers based on thin metal films. This type of microbolometers demonstrates nonselective photoresponse from microwaves up to UV spectrum range. Usually this type of microbolometers is based on high-resistance material films (bismuth, vanadium oxide and others).

As the result, the sensitivity of substrate-supported devices is determined by conduction of heat out of the detector into both the substrate and the metal planar contacts.

In simple substrate supported devices the most obvious path of heat conduction is the conduction directly into the substrate material. If the thermal conductivity, K, of the bolometer material itself is small, this may contribute significantly to heat retention in the bolometer and air-based structure design can be more preferable. Because in a real broadband device the impedance of antenna is a fixing value $R \sim 10^2$ Ω, the device dimensions must satisfy to relation

$$\frac{L}{dw} = \sigma R \ , \tag{1}$$

where d is the thickness, w is the width, L is the length of the bolometer, and σ is a material electrical conductivity.

To maximize the thermal resistance to increase the detector response, it is necessary to choose materials with the largest value of σ/K. The ratio of the electrical conductivity to the thermal conductivity K, is approximately the same constant for the most metals; these two properties are fundamentally related. The reason for this relation is in the electron-phonon interaction in a metal film and cooling of the electron subsystem through phonons. This relation is embodied in the Wiedemann-Franz law, which reads

$$\frac{\sigma}{K} = \frac{3}{\pi^2} \left(\frac{e}{k_B} \right)^2 \frac{1}{T} \ , \tag{2}$$

where k_B is the Boltzmann constant, e is the electron charge, and T is the temperature. Due to similar values of the electron-phonon interaction constant and phonon propagation speed in metals, for a fixed device resistance, almost all metals would give approximately the same bolometer thermal resistance.

The other material constant that enters the responsivity (voltage or current response to fixing radiation power) of the detector is the temperature coefficient of resistance determined as

$$\alpha = \frac{1}{T} \frac{dR}{dT} \ . \tag{3}$$

The larger α leads to the higher responsivity. Once again, this value is very the same for all metals because the resistivity ρ near room temperature is proportional to temperature due to effective electron-phonon interaction in dirty metals. For $T \sim 300$ K, Eq. (3) yields the temperature coefficient of about .003 K^{-1}; almost every metal has the coefficient within a factor of two of this value. It should be noticed that semiconductors demonstrate much higher value of α in comparison to metals. However, low electrical conductivity and thermal conductivity of semiconductors yield a figure of merit actually smaller than a typical metal. If the semiconductor is doped to increase its conductivity, the resistivity is no longer proportional to exponential dependence with inverse temperature due to increase of ionized impurity scattering. Due to this reason high-doped semiconductors are not attractive for bolometer application.

Note, that for estimation of the bolometer performance in some applications it is preferable to use the detectivity rather then the responsivity. The detectivity, D^*, is defined as $D^*=S^{1/2}(NEP)^{-1}$, where S is the operating area of the detector and NEP is the noise equivalent power. This parameter is more useful for multi-element detectors, but for a first estimation of merit it would be enough to compare responsivities of traditional and CMR based bolometers.

In our opinion, CMR materials represent a new class of materials advantageous for bolometers application due to dominating of non-phonon mechanism of carriers scattering at room temperature or close to room temperature range. In CMR compounds, below the temperature of insulator-metal transition, the resistivity is determined by the single-magnon scattering mechanism. The most reliable experimental evidence of this hypothesis was suggested recently in [8]. This mechanism increases α in one-two orders of magnitudes with comparison to conventional metals.

The temperature independent term in the resistivity can be ascribed to scattering on impurities and different defects as grains, domain walls, and so on. In thin films the main origin of this type of scattering is grain-boundary formation during film deposition. The temperature dependent term in the resistance below T_c can be described by the $R{\sim}T^2$ term related to the single-magnon scattering, with additional much smaller terms of T^5 due to the contribution of electron-phonon processes. According to the model suggested in [8], a single-magnon scattering contribution to the resistivity is modified by spin-split bands. Spin-subband splitting in CMR materials is under strong examination supported by different experimental and theoretical studies recent years (see, for example, [8]).

In CMR materials, the temperature dependent term $R{\sim}T^2$ is well pronounced in single-crystals and high quality films. It is suppressed by scattering on defects in non-uniform films and polycrystals. Nevertheless we can summarize, that resistance dependence in perfect CMR thin films is determined mainly by magnon dynamics and dissipation of spin waves.

Neutron inelastic measurements of magnon spectrum [9] demonstrate, that the characteristic magnon energy is at the range of $\sim$kT at room temperature at the presence of the temperature independent modes with energy close to 5 meV. Up to now, many optical measurements have been done in CMR materials. However, the spin gap structure in CMR materials is not explained well enough in details to present reliable picture. Spin-gap splitting in CMR materials seems to be small enough for bolometers applications in middle far-infrared range $\sim$ 100 micron wavelengths as well as for higher energies. Note that the spectral dependence of bolometer photoresponse in infrared range would reflect spin wave spectrum details of CMR films.

The photoresponse in CMR based bolometers has maximum at the temperature just below T_c, where the T^2 term starts to exceed near-transition fluctuations [6].

According to well-known heat capacity and resistivity experimental data, we estimated the detectivity of the CMR-based bolometers as $D^*{\approx}(1{\div}5)*10^8$ cmHz$^{1/2}$ at T=300K (we used a film with R$\sim$10^2 Ohm, d$\sim$100 nm, and w=L=10μm, input amplifier spectrum density noise level 1nV*Hz$^{1/2}$). This value is close to well-known data for high-temperature superconducting thin film-based detectors and it can be sufficient for different applications. However, the open question remains concerning the noise level in CMR-based bolometers. This problem needs an additional study for more accurate estimation of the best NEP level.

Fast CMR thin film-based bolometers.

We discuss here scale of times shorter than the time of steady state attainment between electron (spin) subsystem and phonon subsystem in the film. Fortunately, recent experimental studies of CMR-based photoresponse allow one to discuss possible limitations for characteristic response time τ in this situation.

First measurements of the photoresponse times in CMR thin films in strictly nonequilibrium state in film were presented in [4]. According to [4], the response had nonbolometric origin close to T_c. Authors of Ref. [4] suggested a possible explanation for this observation associating the effect with excitation of spin-flip processes. Detail study of the photoresponse times in CMR films below and above T_c was presented also in more recent work [5]. Authors of [5] discussed the photoresponse in terms of photoinduced thermoelectric effect and observed the time of the transient thermoelectric effect about 10 microsecond close to T_c. It was found that this time was practically independent of temperature. The sign of the voltage photoresponse (or so called transient thermoelectric effect) observed after pulse laser excitation corresponded to the bolometric model in whole temperature range from helium liquid temperature up to $T > T_c$.

We can conclude that all the experimental results point to the fast excitation processes of picosecond range in the spin system and relatively slow cooling of the electron subsystem after laser pulse. According to [10], the fast rise time corresponds to spin-wave excitations in the Mn^{3+}- Mn^{4+} ion system. After some intermediate steps of interactions in the spin system, microsecond range processes of collapsing to ferromagnetic ground state are developed. According to [5] the microsecond range times are related to the process of spin-wave propagation with modifications due to the spin gap developed in ferromagnetic phase. Such long decay times may be the main problem for future fast CMR-based bolometer design and applications.

CONCLUSION

Slow response substrate-based CMR room temperature microbolometers may be attractive due to higher responsivity value in comparison to common metals. High values of responsivity in CMR based bolometers are related to single-magnon scattering nature of resistivity.

Fast CMR-based bolometers would demonstrate picosecond range rise times but long microsecond range fall times. The origin of the relatively long fall times can be associated with slow dynamics of ferromagnetic phase reassembling after the excitation pulse. Correct description of fast heating processes in CMR films should be done in terms of spin system temperature variations in contrast to superconducting or normal conventional metal thin film.

ACKNOWLEDGMENTS

N.N. and E.S.G. wish to acknowledge support from NASA through grant NRA-99-OEOP-4.

REFERENCES
1. R.M.Kusters, J.Singleton, D.A.Keen, R.Mcgreevy, and W.Hayes, Physica **B 155**, 362 (1989)
2. J.H.Hao, X.T.Zeng, and H.K.Wong, J.of Appl.Phys. **79**, 1810(1996).
3. M. Rajeswari, C.H.Chen, A.Goyal, C. Kwon, M.C.Rohsor, R. Ramesh, T.Venkatesan, and S.Lakeou, Appl.Phys.Letters **68**, 3555 (1996).

4.Y.G.Zhao, JJ.Li, R.Shreekaa, H.D.Drew, C.L.Chen, W.L.Cao, C.H.Lee, M. Reajeswari, S. B. Ogale, R. Ramesh, G. Baskaran and T. Venkatesan, Phys.Rev.Letters **81**, 1310 (1998).

5. M. Sasaki, G.R. Wu, W.X. Gao, H. Negishi, M. Inoue, and G.C. Xiong, Phys. Rev. B, **59**, 12425 (1999).

6. Yu.P.Gousev, A.A.Verevkin, H.K. Olsson, N.Noginova, E.S.Gillman, and K.-H. Dahmen, Journal of Low-Temp. Phys., in press.

7. A.M. Grishin, Yu.V. Medvedev, Yu.M. Nikolaenko, Physics of the Solid State **41**, 1260 (1999).

8.M.Jaime, P.Lin, M.B.Salamon, and P.D.Han, Phys.Rev.B **58**, R5901(1998).

9.T.G.Perring, G.Aeppli, S.M.Hayden, S.A.Carter, J.P.Remeika, and S.W.Cheong, Phys.Rev.Lett. **77**, 711(1996).

10. K. Matsuda, A. Machida, Y. Moritomo, and A. Nakamura, Phys. Rev. B **58**, R4203 (1998).

DEPOSITION AND ELECTRICAL CHARACTERIZATION OF DIELECTRIC/FERROMAGNETIC HETEROSTRUCTURE

T. WU, S.B. OGALE, J.E. GARRISON, B. NAGARAJ, Z. CHEN, R. RAMESH, T. VENKATESAN
Center for Superconductivity Research, Department of Physics, University of Maryland, College Park, MD20740-4111

ABSTRACT

We grow $Pb(Zr,Ti)O_3$ (001)/$La_{1-x}Ca_xMnO_3$ (001) hetereostructure epitaxially on Nb doped STO substrate using pulsed laser ablation. A field effect device configuration is formed with the manganite as the channel and the Nb:STO substrate as the gate. Channel resistance modulation by the gate pulsing is studied both with and without magnetic field. We not only find a remarkably large electroresistance effect of 76% at 4×10^5V/ cm, but also the complimentarity of this ER effect with the widely studied CMR effect. The large size of this effect and the complimentarity of ER and MR effects strongly suggest a percolative phase separated picture of manganites.

INTRODUCTION

While rare earth manganites were studied almost fifty years ago [1], recently there has been a rejuvenation of interest in these perovskite oxide materials which may be potential candidates for magnetic sensors and detectors because of the colossal magnetoresistance effect (CMR) [2]. Although the theoretical studies have shown that double exchange (DE), electron-phonon coupling and orbital ordering effect play important roles in this system, there are still a lot of open questions regarding the the mechanism of this interesting effect. Recently both computational and experimental studies have revealed a rich phase diagram, showing phase separation tendencies both above and below T_p [3]. Direct evidences supporting this scenario have been given by TEM [4] and STM [5] observations which show intrinsic electronic inhomogeniety in these materials. The electrical transport should be able to be tuned by external magnetic and electrical field, presumably.

We study this idea by using a field-effect device structure using LCMO as channel and find a remarkably large resistance modulation (~76%) at T_p. It is noteworthy that this large electroresistance effect is complimentary with the well-studied CMR effect. We propose that this electrical effect is brought about by the modulation of the percolative phases in the manganites, which is different from and complimentary with the spin alignment mechanism of CMR effect.

EXPERIMENT

In Fig. 1(a), we show our device configuration. This inverted structure helps to grow better dielectric film and is different from structures used in earlier studies [6] [7]

(a)

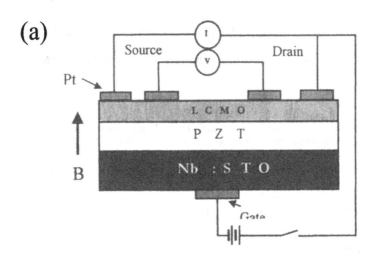

(b)

1. (a) The device configuration; (b) Cross section TEM picture of the device.

[8]. Both the 150 nm layer of PZT and 50 nm layer of LCMO are grown by pulsed laser deposition technique using a KrF excimer laser ($\lambda=248$ nm). A (100)-orientated single crystal of STO doped with 1 at. % of Nb is treated with diluted HF solution to achieve atomic flat surface [9] and served both as substrate for film growth and as gate for the device. Growth conditions including the heater temperature and the oxygen pressure are optimized to achieve good film quality for both layer and epitaxial interfaces between them. XRD patterns show (001) orientation for both layers. After the film deposition, a device structure is then defined using standard photolithography and chemical etching. Fig. 1(b) is the TEM picture of our device showing the smooth interfaces, which are crucial for reproducible device performance. Resistivity of the LCMO channel is obtained by standard four probe technique with temperature from 4K to 300K after pulsing the gate with positive and negative 6 volts. Experiments are performed both with and without a magnetic field of 6T.

RESULTS

Fig. 2 shows the temperature dependence of resistivity of the LCMO channel measured before and after pulsing the gate with positive and negative 6 Volts, which, basically, induce additional electrons and holes in the channel, respectively. The ER, defined as $(R(E)-R(0))/R(0)$ is about 76% at T_p after negative 6V pulsing. The negative pulse produces much larger effect than the positive one, reflecting the hole conducting nature of the CMR channel. In Fig. 3 we present data on the combined consequences of the electrical and magnetic fields for the transport in the CMR channel. We can reduce the channel resistance by using either -6V volts pulsing (from A to B) or 6T magnetic field (from A to C). Most interestingly, both B and C can be further changed into D if both electrical pulse and magnetic fields are used. In Fig. 4, we show the ER and MR data separately. It is noteworthy that not only their sizes are comparable, but also they are complimentary with each other. There is an expected temperature shift in the MR data, caused by the spin aligning effect of magnetic field. But the ER peak temperature remains the same.

Our device shows reasonable large ER effect even at room temperature. In Fig.5 we show a series of I-V curves of the CMR channel after gate pulsing of -2, -4, -6 Volts respectively.

Obviously, this large modulation of channel resistance cannot be cause by carrier density modulation only, which is less than 0.25%. Our experimental data strongly suggests that the induced holes are not doped homogeneously, indicating the electronic inhomogeity of the manganite channel. According to the phase reparation picture, in the CMR manganites the FM (metallic) and the charge ordered (CO) states coexist in a broad range of phase space [3]. The size and the shape of the FM clusters are determined by the competition between the attractive double exchange tendencies among carriers and the Coulomb force. Microscopic droplets, stripes or polarons may arise as the likely configurations. The electrical current is transported by spin-dependent tunneling between the ferromagnetic clusters, whose spin orientation can be controlled by applied magnetic field. Our data clearly indicates that electrical and magnetic fields affect different phases in the phase-separated channel separately and have little interference. Presumably, the induced holes are not doped into the channel homogeniously, but affect the FM phase

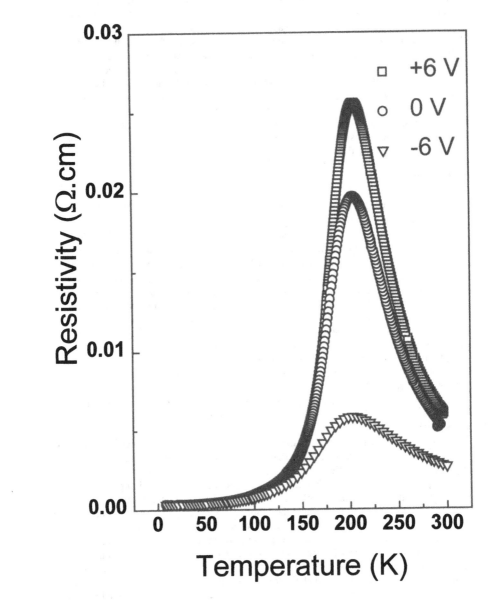

2. Temperature dependence of resistivity of the CMR channel both before and after applied electrical pulses on the gate.

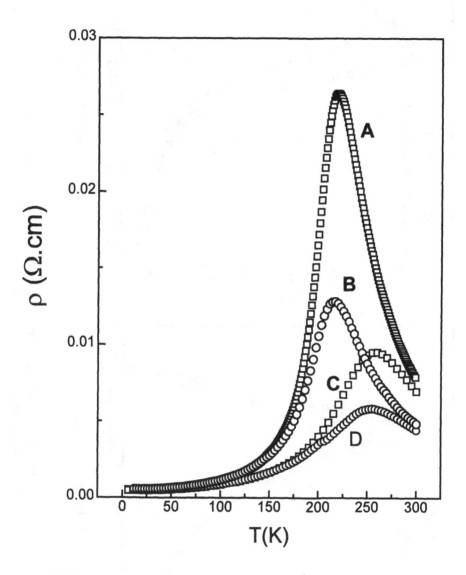

3. Dependence of channel resistivity on temperature for the unbiased (A) and biased channel (B). Also for the unbiased channel under magnetic field (C) and biased channel under magnetic field (D).

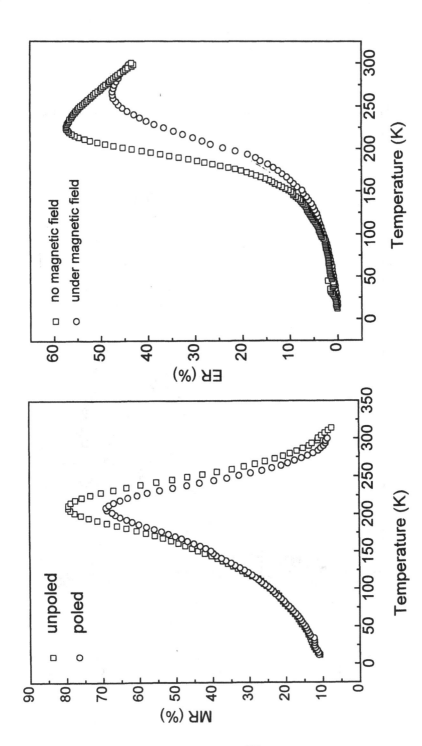

4. (a) MR vs T in 6T magnetic field for unbiased and biased CMR channel; (b) ER vs T for the channel with and without a magnetic field of 6T.

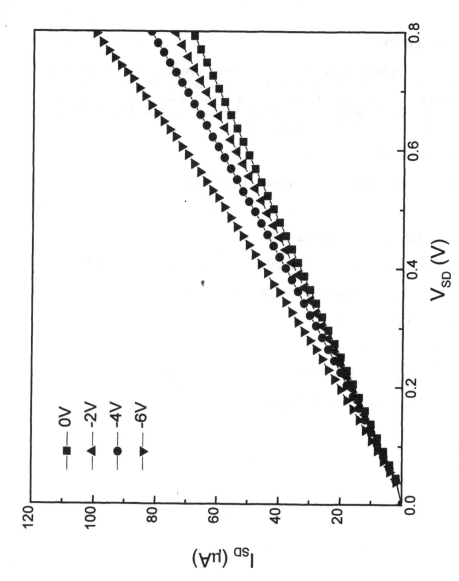

5. I-V curves of CMR channel after gate pulsing of -2, -4, -6 volts, respectivelty.

only and change the domain boundary, which will bring the largest effect around T_p where FM phase starts to connect with each other. On the other hand, the magnetic field reduces the channel resistance by aligning spin of the FM phase and increasing tunneling probability.

CONCLUSIONS

In conclusion, we have observed a large electroresistance effect in LCMO, which is complimentary with magnetoresistance effect. All our observations indicate the electronic inhomogeneity of CMR thin film and support the phase separation scenario. Our study not only helps to understand the basic mechanism of the CMR effect, but also may open a new avenue for oxide applications.

REFERENCES

1. G. H. Jonker and J. H. Santen, Physica **16**, 337 (1950).

2. J. M. D. Coey, et al., Adv. Phys. **48**, 167 (1999).

3. A. Mereo and S. Yunoki, Science **283**, 2034 (1999).

4. M. Uehara, S. Mori, C. H. Chen and S.-W. Cheong, Nature **399**, 560 (1999).

5. M. Fäth, et al., Science **285**, 1540 (1999).

6. S. B. Ogale et el., Phys. Rev. Lett. **77**, 1159 (1996).

7. S. Mathews et el., Science **276**, 238 (1997).

8. H. Tabata and T. Kawai, IEICE Tran. Electron. **E88-C**, 918 (1997).

9. M. Kawasaki, Science **107**, 102 (1996).

EPITAXIAL PbZr$_{0.52}$Ti$_{0.48}$O$_3$/La$_{0.7}$(Pb,Sr)$_{0.3}$MnO$_3$ FERROELECTRIC/CMR MEMORY OPTIMIZED FOR ROOM TEMPERATURE OPERATION

Daniel Lundström [a], Jan Yilbar [a], S.I. Khartsev [a], Alex Grishin [a], and Masanori Okuyama [b]
[a] Dept. of Condensed Matter Physics, Royal Institute of Technology, Stockholm, Sweden;
[b] Dept. of Physical Science, Graduate School of Engineering Science, Osaka University, Japan.

ABSTRACT

Epitaxial ferroelectric/colossal magnetoresistive PbZr$_{0.52}$Ti$_{0.48}$O$_3$/La$_{0.7}$(Pb,Sr)$_{0.3}$MnO$_3$ (PZT/LPSMO) thin film heterostructures have been grown onto SrTiO$_3$ single crystals by KrF pulsed laser deposition technique to fabricate nonvolatile magnetosensitive memory. Colossal magnetoresisitivity (CMR) in LPSMO film has been tailored to room temperature by compositional control to get the maximum temperature coefficient of resistivity of 7.3 %K^{-1} @ 295 K and maximim magnetoresistivity of 27% @ 7 kOe and 300 K. The main processing parameters have been optimized to preserve CMR performance in LPSMO film after deposition of the top ferroelectric layer. Vertical Au/PZT/LPSMO/STO capacitor cell possesses very high dielectric permittivity about 1500 and rather low loss of 5% at 1 kHz, saturation polarization of 40.4 µC/cm^2, remnant polarization of 20.6 µC/cm^2, coercive field of 22.8 kV/cm, and no visible fatigue after $1.33 \cdot 10^8$ reversals. Three top contact metals: Au, Pt, and Ta, deposited at room temperature, have been examined. As compared with Ta, Pt and Au top contacts show superior performance regarding to combined properties: high remnant and saturation polarization, low loss and no fatigue while top Ta contacts have been found to be more efficient to reduce leakage in ferroelectric film.

INTRODUCTION

Perovskite ferroelectric thin films have a tremendous potential for data storage, sensor, and microelectromechanical system technologies. Recently, doped rare-earth manganites exhibiting colossal magnetoresistivity have been proposed to be used as semiconductor channel material for Ferroelectric Field Effect Transistors (FET). [1] The operation of the FET is based on the depletion and accumulation of hole carriers in compounds of mixed valence, such as manganites with low Mn^{3+} content, by an electric stray field of reversed polarization in adjacent ferroelectric. Furthermore, colossal magnetoresistivity (CMR) of manganite can endow such FET with high magnetosensitivity. We have already demonstrated *magnetosensitive nonvolatile memory* (MSM) based on epitaxial ferroelectric/colossal magnetoresistive heterostructures made by pulsed laser deposition. [2] In this paper we report new PbZr$_{0.52}$Ti$_{0.48}$O$_3$/La$_{0.7}$(Pb,Sr)$_{0.3}$MnO$_3$ film structure optimized for room temperature MSM operation.

EXPERIMENTAL

A description of the processing technique is as follows: at first a 248 nm KrF excimer laser (Lambda Physik-300) was used to ablate two ceramic targets with the composition La$_{0.75}$Sr$_{0.25}$MnO$_3$

and La$_{0.7}$Pb$_{0.3}$MnO$_3$. Films of continuous series of solid solutions with the basic chemical formula (1-x)·La$_{0.7}$Pb$_{0.3}$MnO$_3$ + x·La$_{0.75}$Sr$_{0.25}$MnO$_3$ with x = 0, 0.3, 0.37, 0.4, 0.5, 0.6, and 1 have been fabricated and LPSMO film with x = 0.37 has been chosen for further processing. Deposition of LPSMO onto SrTiO$_3$ (001) single crystal was carried out in an oxygen pressure of about 300 mTorr at substrate temperature of 730 °C, was followed by annealing at the same temperature in 600 Torr for 20 min, and cooled to room temperature with the rate of 20 °C/min. Then PbZr$_{0.52}$Ti$_{0.48}$O$_3$ film has been grown on the top of LPSMO at 560 °C in 250

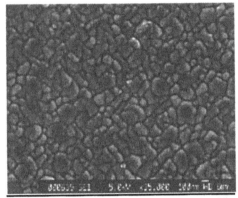

Fig. 1 Au electrode, top view.

mTorr of O$_2$, cooled from 560 °C to 450 °C with the rate of 10 °C/min, kept at 450 °C in 600 Torr of oxygen for 30 min, and finally cooled down to room temperature with the rate 10 °C/min. LPSMO film has been made at a laser radiation energy density of 3 J/cm^2 while the energy has been increased up to 3.5 J/cm^2 to fabricate PZT layer.

Since PLD process was interrupted to change manganite targets for PZT, already fabricated LPSMO/SrTiO$_3$ structure has been heated up for PZT deposition in two different ways: 1) keeping LPSMO/STO structure in vacuum at 750 °C for 2 hours to increase chamber outgassing rate, and 2) with a quick rise to 560 °C in 200 mTorr of oxygen. Thickness of deposited oxide films controlled using AFM and α-step meter has been found to be 350 nm and 500 nm for LPSMO and PZT layers correspondingly. To fabricate top electrodes, contact pads (∅ = 0.55 mm) have been deposited through a contact mask on the top of PZT layer at room temperature: Au by thermal evaporation, while Ta and Pt have been deposited by a magnetron sputtering.

RESULTS AND DISCUSSION

Comprehensive X-ray diffraction analyses (θ-2θ scan, ω-scans, and φ-scans for off-normal planes of substrate, LPSMO and PZT layers) reveal highly crystalline performance of fabricated heterostructures with the following cube-on-cube epitaxial

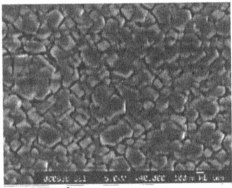

Fig. 2 PZT surface.

relationship of PZT/LPSMO bilayer on SrTiO₃ substrate:

$$(001)\ PZT \parallel (001)\ LPSMO \parallel (001)\ SrTiO_3,$$
$$[100]\ PZT \parallel [100]\ LPSMO \parallel [100]\ SrTiO_3.$$

The detail description of crystalline quality of prepared heterostructures can be found elsewhere. [2]

Preheating of LPSMO layer in vacuum at 750 °C for 2 hours leads to significant Pb lost and results in CMR film conductance and phase transition temperature increase. On the contrary to this, the tour 2) with a quick rise of substrate temperature was found to be very effective to prevent Pb-dopant loosing. Pb:Zr:Ti ratio in PZT film measured by XPS/ESCA technique has been found to be 1:0.73:0.67 before and 1:0.60:0.93 after surface etching in UHV conditions.

SEM was used to study surface morphology and grain size in LPSMO and PZT layers as well as in the top metallic electrodes. Figs. 1-3 present SEM images of Au top electrode, PZT surface, and Au/PZT/LPSMO/STO sample cross-section. Although manganite LPSMO film in the cross-section (Fig. 3) looks as continuous layer, PZT film exhibits the columnar growth. Characteristic cross-sectional area of these needle-like crystallites was estimated from the Fig. 2 to be 200×200 nm² in average.

The capacitance of vertical capacitive cell (ferroelectric film thickness of 0.5 μm, top electrode ∅ = 0.55 mm) was found to be 6.3 nF at 100 Hz, which yields surprisingly high dielectric permittivity of 1500. Loss $tan\ \delta$ was less than 0.05.

RT66A ferroelectric measurements of capacitors with the gold top electrode showed resistivity as high as $2 \cdot 10^{11}$ Ω·cm, leakage current of $1.5 \cdot 10^{-3}$ A/cm² @ 160 kV/cm, and nicely shaped $P\text{-}E$ hysteresis loops (see Fig. 4) with saturation polarization of 40.4 μC/cm², remnant polarization of 20.6 μC/cm², and coercive field of 22.8 kV/cm. Fig. 5 shows there no visible fatigue has been observed

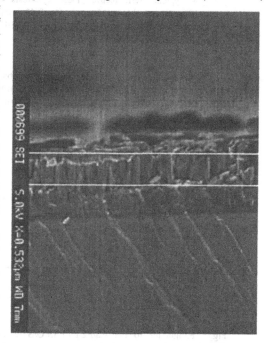

Fig. 3 Au/PZT/LPSMO/STO sample cross-section. The distance between the two horizontal white marking lines (PZT thickness) is 0.532 μm.

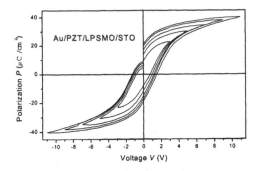

Fig. 4 Typical ferroelectric *P-E* hysteresis loops in Au/PZT(500nm)/ $La_{0.7}(Pb_{0.63}Sr_{0.37})_{0.3}MnO_3$(350nm), $\varnothing = 0.55$ mm capacitive cell. Saturation polarization is 40.4 $\mu C/cm^2$ at 8 V, remnant polarization is 20.6 $\mu C/cm^2$, coercive field is 22.8 kV/cm.

after $1.33 \cdot 10^8$ polarization reversals. Fatigue has not been observed for both structures: 1) and 2), fabricated with the different LPSMO preheating tours, although capacitors with the Pb-rich LPCMO bottom electrode exhibit higher remnant polarization.

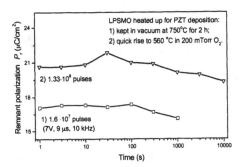

Fig. 5 Fatigue test for two Au/PZT/LPSMO/STO capacitive cells. RT66A supplied 7 V, 9 μs pulses with 10 kHz repetition rate. 1) LPSMO/STO has been kept in vacuum at 750 °C for 2 hours, 2) a quick temperature rise to 560 °C in 200 mTorr of oxygen.

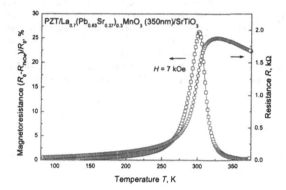

Fig. 6 Temperature dependences of the resistance $R(T)$ and magnetoresistance 1 - R_{7kOe}/R_0 of 350 nm thick $La_{0.7}(Pb_{0.63}Sr_{0.37})_{0.3}MnO_3$ bottom electrode. The temperature coefficient of resistivity is 7.3 %K^{-1} @ 295 K and maximum magnetoresistivity of 27% @ 7 kOe was reached at 300 K.

Fig. 6 presents the transport characteristics of the bottom CMR LPSMO layer. Temperature dependence of the resistance R vs. T shows very sharp transition from the semiconducting to metallic behavior. Temperature coefficient of resistivity (TCR) reaches the maximum value of 7.3 %K^{-1} at 295 K. Magnetoresistance measured in the magnetic field of 7 kOe has a maximum of 27% at 300 K. Very high TCR in LPSMO film has been used to fabricate sensitive IR-bolometer operating at room temperature. [3]

Properties of ferroelectric/CMR capacitors have been examined for three different metals Au, Ta and Pt used as top electrodes. Pt capacitor possesses superior saturation and remnant (28 $\mu C/cm^2$) polarizations, leakage current of $1.25 \cdot 10^{-4}$ A/cm^2@160 kV/cm, though higher coercive field of 43 kV/cm and small offset voltage of 0.8 V. Although Ta capacitor showed extremely low leakage current density of $4.2 \cdot 10^{-9}$ A/cm^2@160 kV/cm, however it was not able to open the polarization loop completely even under applied 15-20 V. The voltage drop across the oxide TaO_x, which forms onto Ta-PZT interface, could be the reason of worse ferroelectric performance at low voltages.

CONCLUSION

Magnetosensitive nonvolatile memory based on PLD-made epitaxial thin film ferroelectric/CMR heterostructures has been optimized for room temperature operation. Phase transition temperature in CMR film has been tailored by alloying $La_{0.75}Sr_{0.25}MnO_3$ and $La_{0.7}Pb_{0.3}MnO_3$ solid solutions. In the continuous series of solid solutions $La_{0.7}(Pb_{1-x}Sr_x)_{0.3}MnO_3$, the composition with $x = 0.37$ has been found to possess superior performance as a bottom magnetosensitive electrode with the temperature coefficient of resistivity as high as 7.3 %K^{-1} @ 295 K and maximum of magnetoresistivity of 27%@7kOe and 300 K. Dielectric characterization of the $Au/PZT/LPSMO/SrTiO_3$ vertical capacitor cell yields very high dielectric permittivity of 1500 and rather low loss of 5% at 1 kHz, saturation polarization of 40.4 $\mu C/cm^2$, remnant polarization of 20.6 $\mu C/cm^2$, coercive field of 22.8 kV/cm, and no visible fatigue after $1.33 \cdot 10^8$ reversals.

ACKNOWLEDGEMENT

This work was supported by the Göran Gustafssons Foundation and the Swedish Superconductivity Consortium.

REFERENCES

[1] S. Mathews, R. Ramesh, T. Venkatesan, J. Benedetto, Science **276**, 238 (1997).
[2] A.M. Grishin, S.I. Khartsev, P. Johnsson, Appl. Phys. Lett. **74**, 1015 (1999).
[3] Alvydas Lisauskas, S.I. Khartsev, Alex Grishin, MRS Fall99 Meeting Proceedings (this issue JJ10.6), Boston, Nov.29-Dec.3, 1999.

THE OXYGEN MOBILITY AND CATALYTIC ACTIVITY OF LaMO$_{3\pm\delta}$ (M=Cr, Mn, Co) PHASES.

I.A. KOUDRIASHOV*, L.V. BOROVSKIKH*, G.N. MAZO*, S. SCHEURELL**,
E. KEMNITZ**
*Moscow State University, Chemistry Department, Moscow, Russia
**Humboldt University, Institute of Inorganic Chemistry, Berlin, Germany

ABSTRACT

The activity of LaMO$_3$, where M=Cr, Mn, Co, perovskite-type complex oxides in oxygen diffusion and catalytic processes was investigated. For sample preparation freeze-drying technique was used and dynamic thermal isotope exchange method was used to study exchange reaction between ^{18}O from the gas phase and ^{16}O from the samples synthesized, and to investigate the methane catalytic oxidation reaction. The results obtained allowed to indicate temperature intervals of different types of reaction taking place and that LaCrO$_3$ is far less active in mentioned reactions than LaCoO$_3$ and LaMnO$_3$.

INTRODUCTION

Highly defective nonstoichiometric perovskite-like complex oxides of 3d-transition elements are promising objects for elaboration of multi-functional materials. The compounds of the general formula La$_{1-x}$Sr$_x$MO$_3$, where M=Mn, Co, Cr, are interesting because of their unusual physical properties (e.g. CMR-effect for manganites [1,2], metal-insulator transition for cobaltates [3-6], etc.). These compounds are of interest as catalysts of different redox reactions [7-9], as the activation energies of oxygen diffusion are rather low, but the question about the role of solid phase oxygen in catalytic processes is not completely solved yet. The results of studies of catalytic properties are not in good agreement with each other, as in a number of works concerning the problem amorphous [10] or containing impurities [9] samples were investigated.

The reaction of isotope exchange between oxygen molecules ^{32}O$_2$ and ^{36}O$_2$ occurring in the presence of a solid oxide phase can be considered the simplest catalytic reaction with oxygen participation [11]. From the course of the oxygen exchange reaction, conclusions can be drawn about the way in which oxygen takes part in the catalytic process [12]. In this work, the mobility of oxygen in multicomponent oxides was studied by the dynamic-thermal isotope exchange method, which made it possible to determine the temperature intervals of the occurrence and the characteristics of various processes in the oxygen–solid system.

The purpose of this work was to investigate oxygen mobility in the LaCrO$_3$, LaMnO$_3$, and LaCoO$_3$, phases and its correlation with the catalytic activity of these phases in the oxidation of CH$_4$.

EXPERIMENT

In order to obtain samples with a high surface area, the freeze-drying method was used. The method allows preparing substances with a high level of homogeneity in cation distribution at rather low temperatures [8,13]. The freeze-drying precursors obtained from solutions of nitrates of the appropriate cations were heated up to 773K at 0,5 K/min and annealed for 20h at that temperature. Then the powders synthesized were fired at 873K, 973K, 1073K with intermediate grindings. The whole firing time was 80h. When preparing LaCrO$_3$, we also used additional sintering at 1173K and 1273K, for this sample, the time of synthesis was 120h.

377

The phase composition of the samples prepared was controlled by the XRD method (Guinier camera, $CuK_{\alpha 1}$-radiation, $\lambda=1,54056$, Ge as internal reference). The determination of the oxidation state of Mn and Co ions was performed by iodometric titration [14].

Powder specific surface areas were measured by the BET method on an ASAP 2000 (Micrometric, USA) instrument.

The average size of particles was determined by the SEM technique on a microscope JEM-2000FXII (U=200kV).

The dynamic-thermal isotopic exchange between the prepared samples and the ^{18}O-enriched gas phase was investigated in a special glass vessel with separate control valves for the isotope gases $^{32}O_2$ (99,95%, Messer-Griessheim) and $^{36}O_2$ (95,97%, IC Chemikalien). A schematic of the setup for isotope exchange experiments is described in detail in [15]. The methane oxidation was studied under polythermal conditions in static experiments using the instrument mentioned. The samples (100-500mg depending on the surface area) were placed into a plate holder. Before measurements, it is necessary to remove adsorbed molecules such as H_2O and CO_2 from the solid surface, which disturb the oxygen isotope exchange, so the samples have to be heated up to 673K for 4h in the $^{32}O_2$-flow at a pressure of 200Pa. The initial pressure in the experiments on isotope exchange was 150 Pa (the mixture of 20Pa Ar, 65Pa $^{32}O_2$, 65Pa $^{36}O_2$). In methane oxidation experiments, the working gas was the mixture of 21Pa Ar, 43Pa CH_4, and 86Pa $^{36}O_2$ (the initial whole pressure was 150Pa). The concentration of all oxygen-containing molecules in the system were detected by a quadrupole mass spectrometer (Balzers, QMG 421 I), these molecules are $H_2^{16}O$, $H_2^{18}O$, $^{32}O_2$ ($^{16}O^{16}O$), $^{34}O_2$ ($^{16}O^{18}O$), $^{36}O_2$ ($^{18}O^{18}O$), CO_2-44 ($C^{16}O^{16}O$), CO_2-46 ($C^{16}O^{18}O$), CO_2-48 ($C^{18}O^{18}O$). The change in CH_4 concentration was determined according to the signal of the CH_3^+ particle (m/z=15), as the signal m/z=16 is the sum of the signals of CH_4 and $(^{16}O^{16}O)^{2+}$ particles.

RESULTS AND DISCUSSION

In studies of the complex oxides, the measurement results are functions of a number of parameters, thus comparison of the results is only possible when the experimental conditions are standardized and all the factors influencing the mechanism and kinetics are considered. Such factors as surface condition and phase composition of the samples are possibly of the primary importance. For investigations of catalytic reactions, the materials with a high surface area are preferable. As a rule, such samples are obtained at low temperatures and may contain amorphous impurities. The well-crystallized samples, whose phase composition may be characterized by XRD unambiguously, are usually prepared at high temperatures, e.g. >1173K for manganites. The materials obtained at high temperatures usually have low specific surface areas ($<1m^2/g$). In order to prepare samples with a high surface area, we used firings at gradually increased temperatures. The firing temperature was increased until samples after the next sintering were single-phase (and did not contain amorphous impurities also). The X-ray diffraction patterns of all the samples synthesized were indexed on the basis of a hexagonal unit cell. The results of calculating unit cell parameters are shown in Table I.

Table I.

Unit cell parameters, value of the formal oxidation state of Mn or Co and specific surface areas of the prepared samples.

Composition	Unit cell parameters		$Mn^{n+}(Co^{n+})$	Specific surface area, m^2/g
	a, A	c, A		
$LaCrO_3$	5,5332(8)	6,6847(9)		1,30(3)
$LaMnO_{3+\delta}$	5,5192(6)	13,330(2)	3,20(2)	2,57(1)
$LaCoO_{3-\delta}$	5,4427(8)	13,091(3)	2,88(2)	2,51(2)

Cobaltites and manganites are characterized by mixed valences of Mn and Co ions ($Mn^{3+}- Mn^{4+}$, $Co^{2+}- Co^{3+}- Co^{4+}$). The oxidation state of the transition metal is an important characteristic of complex oxides of 3d-elements, determining the sample composition. The observed the formal oxidation states are presented in Table I. The presence of the Mn ion in the oxidation state higher than +3 is a characteristic feature for lanthanum manganite. The formula of an undoped manganite $LaMnO_{3+\delta}$ is a formal one. In fact, the formation of vacancies in of La and Mn sites is observed for this structure; this is accompanied by an increase of the formal oxidation state of Mn [16-19]. Since, there is no position for the extra oxygen in this structure, the chemical formula for the compound should be written $La_{(3/3+\delta)}Mn_{(3/3+\delta)}O_3$. The oxidation state of Mn depends on the conditions of the synthesis of manganites (e.g. temperature and oxygen partial pressure). In case of cobaltite, changes in the Co oxidation state are accompanied by oxygen vacancy formation [5].

According to the SEM data, the average particle size is 300 nm for $LaMnO_3$, and $LaCoO_3$, $LaCrO_3$, whose average particle size is the highest (450нм), is characterized by the lowest value of surface area ($1,3 m^2/g$). this may be explained by the highest sintering temperatures required for this sample. Thus, the distinction in particle size is in good agreement with the specific surface area values (Table I).

In polythermal experiments carried under unequilibrium conditions with the high heating rate (10K/min), the main parameter influencing the rate of the reaction is the sample surface area [15], that is why the sample weight was varied in order to make the whole surface area constant ($\sim 0,8 m^2$). From the variation of isotopic composition of the gas phase, it is possible to draw conclusions on the mechanism of oxygen exchange and diffusion on the basis of the model of Musikantov et al. [20-22]. The data corresponding to the temperature intervals of various processes of oxygen exchange are presented in Table II.

Table II.

Temperature intervals of predominant occurrence of surface reaction and out-diffusion.

Composition	Completely heterogeneous exchange	Partially heterogeneous exchange	Completely heterogeneous and homogeneous	Out-diffusion of oxygen
$LaCrO_3$	675-755K	755-845K	>845K	
$LaMnO_{3+\delta}$	535-575K	575-725K	725-830K	>830K
$LaCoO_{3-\delta}$	515-585K	>585K		>585K

The exchange processes start at different temperatures: at 675K for $LaCrO_3$ (Fig. 1), at 535K in the case of $LaMnO_3$ (Fig. 2) and at 515K in the case of $LaCoO_3$ (Fig. 3). In the interval I the ionic current of $^{36}O_2$ decreased and that of the oxygen $^{32}O_2$ increased, while the ionic current of $^{34}O_2$ remains constant. This character of ionic current changing is caused by the completely heterogeneous mechanism:

$$^{36}O_2(g) + 2\,^{16}O(s) = {}^{32}O_2(g) + 2\,^{18}O(s) \qquad (1)$$

As it can be seen in Table II, the process is the first stage of the exchange for all studied samples.

If the temperature is increased (interval II), increasing of $^{34}O_2$ ionic current and reduction of $^{36}O_2$ ionic current is observed (Figs. 1-3). This changing of ionic current may be caused by the fact that, in interval II, besides process (1), there is also partially heterogeneous exchange for all the compounds investigated:

$$^{36}O_2(g) + {}^{16}O(s) = {}^{34}O_2(g) + {}^{18}O(s) \qquad (2)$$

For $LaCrO_3$ and $LaMnO_3$ in the interval III, oxygen $^{34}O_2$ is formed as a result of homogeneous exchange reaction (the solid oxide surface reaction between molecules $^{32}O_2$ and $^{36}O_2$ without participating of oxygen of solid oxide):

$$^{32}O_2(g) + {}^{36}O_2(g) = 2\,^{34}O_2(g) \qquad (3)$$

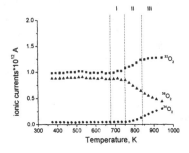

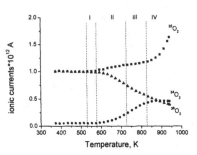

Fig.1 Change of the composition of the gas phase during the exchange reaction with LaCrO₃.

Fig. 2.Change of the composition of the gas phase during the exchange reaction with LaMnO₃₊δ.

Process (3) is accompanied by a decrease in the amount of $^{32}O_2$ in the gaseous phase.

Thus if there is an almost constant ionic current of $^{32}O_2$ (Figs. 1-2), one can make the conclusion that, for LaCrO₃ and LaMnO₃, the exchange in the interval III is the sum of mechanisms (1) and (3).

The absence of the interval of homogeneous exchange for LaCoO₃ is probably consequence of processes beginning at far lower temperatures, and it excludes the existence of the interval where the ionic current of $^{32}O_2$ is constant or decreases (Fig.3).

For studied cobaltite (T>585K) and manganite (T>830K), we also observed (Fig. 4) the process of oxygen release from the samples, i.e. out-diffusion (without substance decomposition):

$$2^{16}O(s) = {}^{32}O_2(g) \qquad (4)$$

An increase in the amount of all forms of oxygen corresponds to this process. The change of the pressure of all oxygen forms (p_i/p_0) was determined according to the total oxygen ionic current during the experiment:

$$\frac{\left\{\left[^{32}O_2\right]+\left[^{34}O_2\right]+\left[^{36}O_2\right]\right\}_{t=i}}{\left\{\left[^{32}O_2\right]+\left[^{34}O_2\right]+\left[^{36}O_2\right]\right\}_{t=0}} = \frac{p_i}{p_0} \qquad (5)$$

The out-diffusion of oxygen was not observed in the case of LaCrO₃ along the whole studied temperature range (373-943K).

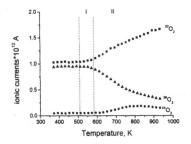

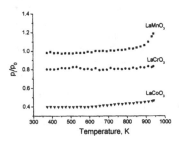

Fig. 3. Change the composition of the gas phase during the exchange reaction with LaCoO₃₋δ.

Fig.4. Temperature-dependent of the relative oxygen partial pressure (p_i/p_0) for investigated compounds. (for LaCrO3 0,8* p_i/p_0 and for LaCoO3 0,4* p_i/p_0)

As it can be seen in Table II, the exchange processes on the phase boundary "oxygen-oxide" and out-diffusion for $LaMnO_3$ and $LaCoO_3$ is observed at far lower temperatures than in the case of $LaCrO_3$.

There are no intervals, where the pressure decreases, on the temperature dependence of the oxygen partial pressure (Fig. 4). This proves that there are no in-diffusion processes for the substances under investigation

$$^{32}O_2(g) = 2\ ^{16}O(s) \tag{6}$$
$$^{34}O_2(g) = \ ^{16}O(s) + \ ^{18}O(s) \tag{7}$$
$$^{36}O_2(g) = 2\ ^{18}O(s) \tag{8}$$

From methane catalytic oxidation by $^{36}O_2$ experiments, the same temperature intervals of the isotope exchange are obtained. Decreasing of methane concentration is observed at temperatures of 675K for manganite (Fig. 5), 695K in the case of cobaltite (Fig. 6) and 845K for $LaCrO_3$ (Fig. 7), while CO_2 formation is observed only at T=735K for manganite (Fig. 5) and cobaltite (Fig. 6), and is not observed in the case of $LaCrO_3$ (Fig. 7). The break of C-H bonds is likely to be the slowest step of the oxidation; this is proved in [9].

At the onset of methane oxidation, there are three forms of molecular oxygen $^{32}O_2$, $^{34}O_2$, $^{36}O_2$ in the gas phase due to the isotope exchange processes. Despite this fact, formation of CO_2-48 ($C^{18}O^{18}O$) is not observed. The oxidation process leads to the formation of CO_2-44 ($C^{16}O^{16}O$) and a little of CO_2-46 ($C^{18}O^{16}O$). This allows us to make a conclusion that oxygen from the gaseous phase does not react directly with methane (otherwise, the formation of CO_2-48 must be

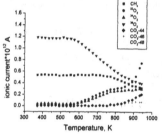

Fig. 5. Change of the composition of gas phase during the $CH_4 + ^{36}O_2$ reaction on $LaMnO_{3-\delta}$.

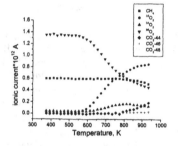

Fig. 6. Change of the composition of the gas phase during the $CH_4 + ^{36}O_2$ reaction on $LaCoO_{3-\delta}$.

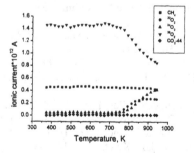

Fig. 7. Change of the composition of the gas phase during the $CH_4 + ^{36}O_2$ reaction on $LaCrO_3$.
(CO_2-46, CO_2-48 are on CO_2-44 curve)

381

observed), and in oxidation process, the solid phase oxygen takes place. The observed at higher temperatures (775), than for CO_2-44 (740 K) (Figs. 5,6), formation of a little of CO_2-46, is probably the result of isotope exchange between oxygen containing isotope ^{18}O and CO_2-44.

The rate of methane conversion calculated from the respective decrease of ionic current of CH_3^+ particles was highest for manganite (the reducing of CH_3^+ signal in the temperature interval of 675-943K ~45%) and lowest for chromite (decreasing of the signal of CH_3^+ ~5%). In the case of cobaltite, a decrease in the CH_3^+ signal was ~15%.

CONCLUSIONS

Thus the correspondence between the activity of the studied compounds in isotope exchange reactions and CH_4 oxidation is observed. The lower temperatures of exchange and diffusion processes for $LaMnO_3$ and $LaCoO_3$ evidence that oxygen mobility in $LaMnO_3$ and $LaCoO_3$ is higher than in $LaCrO_3$.

ACKNOWLEDGMENTS

The authors are thankful to the Ministry of Science and Technological Politics of RF (project No139/97 RUS), to RFFI and INTAS (97-04-02) for the financial support of this work.

REFERENCES

1. R.von Helmot, J.Wecker, B.Holzapfel, L.Schultz, K.Samwer, Phys. Rev. Lett. **71**, 2331 (1993).
2. S.Jin, T.H.Tiefel, M.McCormack, R.A.Fastnacht, R.L.Ramesh, H.Chen, Science, **64**, 413 (1994).
3. J.B.Torrance, P.Lacorre, C. Asavaroengchai, R.M.Metzger, Physica C **182**, 351 (1991).
4. T.Arima, Y.Tokura, J.B.Torrance, Phys.Rev. B **48**, 17006 (1993).
5. A.Mineshige, M.Inaba, T.Yao, Z.Ogumi, J.Solid State Chem.**121**, 423 (1996).
6. S.Yamaguchi, Y Okimoto, Y.Tokura, Phys. Rev. B **54**, R11 022 (1996).
7. L.G.Tejuca, J.L.G.Fierro, J.M.D.Tascon, Adv. Catal. **36**, 237 (1989).
8. D.V.Johnson, P.K.Gallacher, F.Schrey, W.W.Rhodes, Ceram. Bull. **55,** 520 (1976).
9. L.Marchetti, L.Forni, Appl. Catal. B **15**, 179 (1998).
10. N. Guilhaume, M.Primet, J.Catal. **165**, 197 (1997).
11. L.A.Kasatkina, G.V.Antoshin, Kinet.Catal. **4**, 252 (1963).
12. C.Doornkamp, M.Clement, V.Ponec, J.Catal, **182**, 390 (1999).
13. L.Wachowski, Z. phys. Chemie, Leipzig, **269**, 743, (1989).
14. L.V.Borovskikh, G.N.Mazo, V.M.Ivanov, Vestnik MGU, Khimiya, **40**, 402 (1999).
15. A.A.Galkin, G.N.Mazo, V.V.Lounin, S.Scheurell, E.Kemnitz, Rus. J. Phys. Chem, **72**, 1459 (1998).
16. C.Ritter, M.R.Ibarra, J.M.De Teresa, P.A.Algarabel, C.Marquina, J.Blasco, J.Garcia, S.Oseroff, S.-W.Cheong, Phys. Rev. B **56**, 8902 (1997).
17. J.Toepfer, J.B.Goodenough, J.Solid. State Chem. **130**, 117 (1997).
18. J.F.Mitchell, D.N.Argyriou, C.D. Potter, D.G.Hinks, J.D.Jorgensen, S.D.Bader, Phys. Rev. B **54**, 6172 (1996).
19. M.Hervieu, R.Manesh, N.Rangavittal, C.N.R.Rao, Eur. J. Solid State Chem. **32**, 79 (1995).
20. V.S.Musikantov, V.V.Popovski, G.K.Boreskov, Kinet. Catal. **5**, 624 (1964).
21. V.S.Musikantov, G.I.Panov, G.K.Boreskov, Kinet. Catal. **10**, 1047 (1969).
22. V.S.Musikantov, G.I.Panov, G.K.Boreskov, Kinet. Catal. **14**, 948 (1973).

AUTHOR INDEX

Adams, P.M., 263
Aldinger, Fritz, 327
Andrus, A.E., 9
Argyriou, D.N., 315
Arthur, E., 239
Ausloos, M., 269

Babalola, K., 107, 239
Bailleul, M., 137
Bendersky, L.A., 29
Billinge, S.J.L., 177
Biswas, Amlan, 23, 81, 195, 231
Blamire, M.G., 3
Boerio-Goates, J., 245
Borca, C.N., 75, 301
Bordallo, H.N., 315
Bordet, P., 41
Borovskikh, L.V., 377
Bougrine, H., 269
Browning, V.M., 283
Bukowski, Z., 333

Carpenter, E., 283
Cheikh-Rouhou, W., 41
Chen, K., 131
Chen, Z., 363
Cheng, R.H., 301
Cheong, S-W., 35, 113, 167
Chippaux, D., 277
Cho, N-H., 93
Choi, Jaewu, 301
Chuprakov, I.S., 47
Cloots, R., 269
Cohn, J.L., 29
Colla, E., 9
Cooper, S.L., 167

Dabrowski, B., 333
Dahmen, K-H., 47
Davidson, B.A., 9
Dorbolo, S., 269
Dowben, P.A., 75, 301
Dulli, H., 75

Eckstein, J.N., 9
Enoki, T., 17
Eom, C.B., 55, 69, 99, 201, 207
Epple, Lars, 327
Evetts, J.E., 3

Fisk, Z., 167
Fontcuberta, J., 41

Galley, C., 23

Garrison, J.E., 363
Gershenson, M.E., 113
Gillman, E.S., 357
Gopalakrishnan, J., 23
Greene, R.L., 23, 231
Grishin, Alex M., 349, 371
Gross, G.M., 225

Habermeier, H-U., 125, 225
Haghiri-Gosnet, A.M., 277
Hakim, N., 35
Hanfland, M., 41
Heo, H., 93
Hervieu, M., 145
Hong, C.S., 159
Hu, S.F., 257
Hu, Y.F., 63, 87, 185
Huang, C.Y., 257
Huang, S., 131
Huang, X., 131
Hur, Nam Hwi, 159

Idzerda, Y.U., 301
Isobe, M., 17
Izumi, M., 289

Jeong, Yoon Hee, 159
Jo, Moon-Ho, 3
Joh, Keon Woo, 159

Kang, D.J., 81, 195
Kawai, T., 307, 339
Kawasaki, M., 289
Kemnitz, E., 377
Kent, A.D., 207
Khan, H.R., 251
Khartsev, S.I., 349, 371
Kiyama, T., 17
Kolesnik, S., 333
Kolody, M.R., 263
Koo, T.Y., 113
Koudriashov, I.A., 377
Krajewski, J.J., 55
Kravets, A.F., 251
Kusko, C., 35
Kwei, G.H., 177
Kycia, S., 177

Lecoeur, Ph., 277
Lee, Chang Hoon, 159
Lee, Cheol Eui, 159
Lee, M.K., 55, 69, 201
Li, Qi, 63, 87, 185, 219
Li, Xiaohang, 125

383

SUBJECT INDEX

Printed in the United States
By Bookmasters